MAGIC POLYGONS

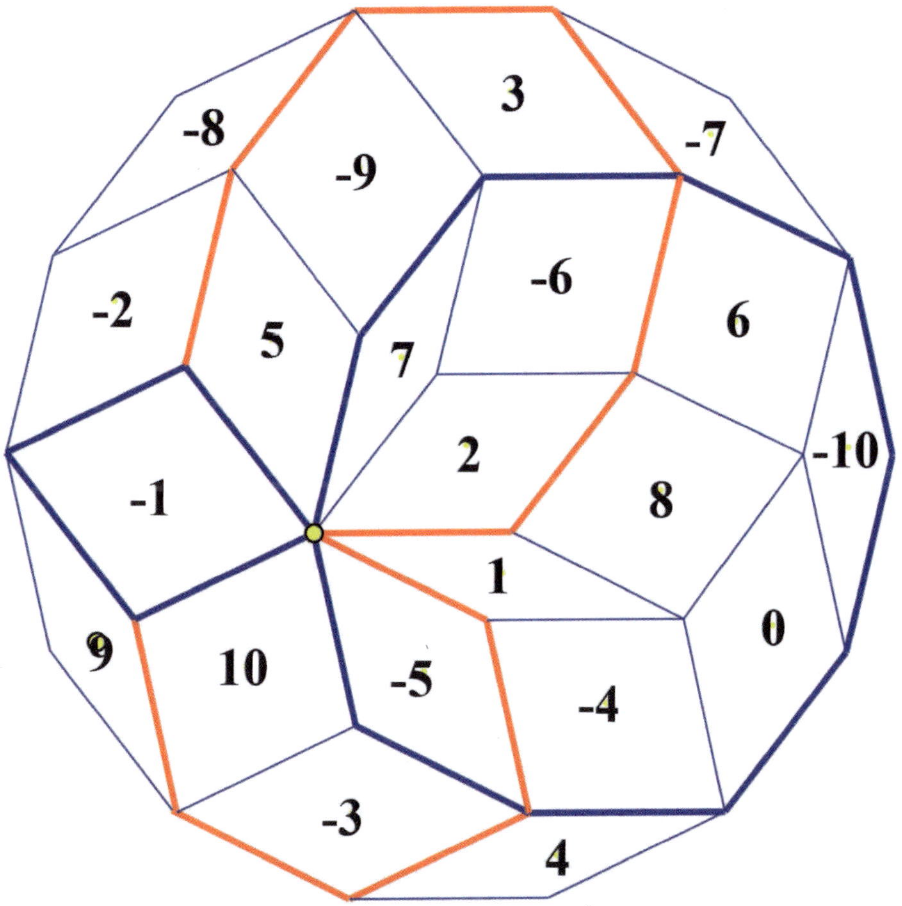

RAYMOND R. FLETCHER III

MAGIC POLYGONS

RAYMOND R. FLETCHER III

Printed in the United States of America.

Library of Congress Control Number: 2022930675

ISBN	Paperback	978-1-68536-267-6
	Hardback	978-1-68536-268-3
	eBook	978-1-68536-269-0

Westwood Books Publishing LLC
Atlanta Financial Center
3343 Peachtree Rd NE Ste 145-725
Atlanta, GA 30326

www.westwoodbookspublishing.com

Contents

Preface

For thousands of years the concept of a *magic square* has intrigued mankind. In the following pages this concept is generalized in two ways. Traditionally, a magic square of order n is a square array of positive integers {1. 2. …, n^2} with the property that all rows, columns and corner diagonals have the same sum. By subtracting 1 from each entry the elements {0, 1, 2, …, n^2-1} are obtained and the common sums of rows, columns and diagonals is maintained. The advantage of this small change is that now the magic square is populated with the elements of the abelian group Z(n^2). This leads to a natural question: can magic squares be constructed over other finite abelian groups? This is our first generalization.

The second generalization arises from a study of parallelogram tilings of polygons. A square is, of course, a regular polygon with an even number (4) of sides and a magic square can be regarded as a tiling of a square with labeled parallelograms. Any regular polygon P with an even number of sides can be tiled with parallelograms. These tiles form sequences which stretch from one side of P to an opposite side and which consist of parallelograms with mutually parallel sides. We call such a sequence a *stream*. The streams of P form the analog of rows and columns in a magic square. To make P into a *magic polygon*, elements from an abelian group G are assigned in one to one fashion to the tiles of P so that all streams have the same sum. Except for our treatment of magic squares, corner diagonals, are not considered.

Although a regular polygon forms the closest geometric analog to a square, the theory is greatly enriched by the consideration of *semiregular* polygons. A convex polygon with an even number of sides and whose opposite sides are equal and parallel is called *semiregular*. These also can be tiled with parallelogams which form streams which extend between opposite sides. If a semiregular polygon P is not regular, then the streams will not all have the same length but it still may be possible to turn P into a magic polygon.

If P is a semiregular polygon tiled with parallelograms, then the *order n of P* is the number of tiles used to tile P. Thus n is the order of the abelian group G used to label the tiles of P. We call G the *base* of P. If G = Z(n), the integers (mod n), then we distinguish three ways that G can be used: (1) addition of stream values occurs exactly as in Z(n); (2) Z(n) just denotes the set of integers {0, 1, …, n-1}. In computing stream sums these values are just added as integers, (no consideration of modulus). (3) If n is odd, then Z(n) is represented using inverses: {0,1,-1, 2,-2, …, (1/2)(n-1)}, and if n is even then the representation: {0, 1,-1,2,-2, …, n/2 -1, -(n/2 – 1), n/2} is used. We use the notation Z*(n) to refer to these representations. In either case these elements are added as integers without regard to the modulus. Method (1) is called a *basic solution*; method (2) is a *natural solution*, and method (3) is called an *integral solution.* In order to obtain a closer analog to magic squares natural and integral solutions will be sought. Basic solutions sometimes will appear as an intermediate stage in the construction of a natural solution. Integral solutions are especially attractive since they allow constructions to be obtained by extending integral solutions for smaller polygons.

Chapter 1 covers the theory of parallelogram tilings of semiregular polygons. The theory covers tilability, composition sequence and transformations. If P is a semiregular polygon with two parallelogram tilings T_1, T_2 then it is possible by a series of simple hexagon inversions to transform T_1 into T_2. The significance for the construction of magic polygons is that these transformations preserve stream sums. Thus if P has a magic assignment with tiling T_1 then it also has a magic assignment with tiling T_2. This allows us to select a suitable generic tiling, (a *polar tiling),* to use in the construction of magic solutions.

Chapter 2 resulted from an observation of Dr. Fabio Guerinoni regarding the connection between parallelogram tilings of semiregular polygons and line configurations. We show that a certain type of line configuration can always be represented by a parallelogram tiling, and Conjecture that every parallelogram tiling of a semiregular polygon can be represented by a line configuration.

Chapter 3 concerns magic squares which are interesting in their own right, but also serve as building blocks for magic polygons with more than 4 sides. In this Chapter the focus is on constructing magic squares over various square order abelian groups. A *universally magic square* of order n is a square which is magic over every abelian group of order n^2. Special attention is given to the construction of magic squares with base $Z(n)\times Z(n)$. In these constructions the concepts of orthogonal Latin and semilatin squares are useful. The Chapter closes with a construction of nested integral magic squares.

Chapter 4 concerns magic (non-square) rectangles. It is impossible for such a rectangle to have all natural column and row sums equal, so instead the concept of a *2-sum rectangle* is introduced. In a 2-sum rectangle all rows have the same natural sum and all columns have the same natural sum, the two sums being necessarily different. It is Conjectured that if p, q have the same parity and $q>p\geq 2$, then a 2-sum solution over Z(pq) exists for any p by q rectangular polysquare.

Let S be a subset of an abelian group G. A *near magic square over S* is a square whose cells are populated with elements from the set S, and whose rows and columns, (but not necessarily corner diagonals), have the same sum in G. The construction of a partition of $Z(mp^2)$ into m near magic squares of order p is given also in Chapter 4. It is useful not only in the construction of 2-sum rectangles, but also in later Chapters in the construction of magic polygons with more than 4 sides.

The notion of a *magic box* is also introduced in this Chapter. It is a generalization of the well studied *magic cube.* It is a 3-dimensional polycube whose cells are occupied by elements from an abelian group G so that all linear polycubes in the x-direction have the same sum; all linear polycubes in the y-direction have the same sum, and all linear polycubes in the z-direction have the same sum. Thus a sort of 3-sum structure is obtained.

Perhaps the most satisfying result in this Chapter concerns the construction of 0-sum integral rectangles. Unlike natural rectangles, it is possible to construct integral rectangles with the same sum, (0), in all columns *and* rows. The general procedure for creating these is not clear but it is Conjectured that if p, q are odd integers both larger than or equal to 3, then a p by q 0-sum integral rectangle can be constructed. One important construction method involves extending a given integral rectangle usually by structures whose rows and/or columns consist if an element and its inverse. This type of extension is used extensively in later Chapters for the construction of integral hexagons, octagons, decagons, etc.

Chapter 5 deals entirely with the construction of magic hexagons. The first part concerns the construction of regular magic hexagons. The concepts of orthogonal hexagons, near magic partitions and dilations are used to create examples. The remainder of this Chapter is devoted to the construction of integral hexagons. A progression is made from the simple *chevron sequence* to the *platform* to more complex constructions involving a series of extensions. It is shown that each member of a certain class of integral hexagons can be obtained by a series of extensions from three small, (*primary),* members. It is also shown that regular integral hexagons can be constructed in nested fashion. Along the way, an interesting procedure we call a *slide transformation* is introduced as a means for moving from one hexagon tiling to another. This transformation assumes more significance in Section 6.1.

Chapter 6 is devoted entirely to the construction of magic octagons. One interesting phenomenon occurs for the first time. This has to do with the order of the sides of a semiregular octagon. We use the notation O(m,p,q,r) to denote a semiregular octagon whose semiperimeter consists of 4 sides with lengths m,p,q,r. These sides occur in the order indicated by the notation. The octagon O(m,q,p,r) has an essentially different shape than O(m,p,q,r). These are *dual* semiregular octagons. It is shown that there is a simple way to pass from a magic solution of an octagon to a magic solution for its dual. The procedure may not in general preserve an extension sequence, but we show that the same extension sequence can be recovered in the dual by means of slide transformations.

Natural and integral regular octagons are then constructed. Any semiregular polygon with more than 4 sides has a minimal dissection into parallelograms. We call these *major parallelograms.* A regular octagon O has a dissection into 6 major parallelograms consisting of 3 pairs which have only a single point in common. If O is given a magic solution then it is easily shown that the members of each of these pairs have the same sum. In case all 6 major parallelograms have the same sum, then we call the solution *perfect.* Many of the solutions in this Chapter and in Chapter 7 are perfect and usually easier to find. Two 2 by 2 squares we call *α and β-squares* are used in this Chapter and the next to construct perfect natural and integral regular polygons. Chapter 6 concludes with a foray into the construction of odd order integral octagons.

In Chapter 7 the α and β-squares are used to construct a large class of natural and integral magic polygons. A brief study of integral decagons, dodecagons and tetradecagons is included in this Chapter. The constructions concentrate on the smallest examples most of which are composite. The integral tetradecagon constructed in this Chapter is the largest magic simple,(i.e, one basic edge per side) regular polygon so far discovered. (The simple regular dodecagon is the smallest, also illustrated in this Chapter.)

The concept of a *basis* is introduced in this Chapter. Given a tiling T of an integral regular polygon P, the existence of a *basis pair* in T implies the existence of subregular polygons in T which must have the same sum. The *capacity of T* is the number of such basis pairs which can be found in T. It is conjectured that a certain tiling, a *flower tiling,* yields the maximum capacity among all tilings of a given regular polygon.

In Chapter 8 the concept of a *Latin polygon* is introduced. If a regular polygon P is tiled with parallelograms and then labeled with the elements of an abelian group G so that each stream contains exactly one occurrence of each element of G, then P is a *Latin polygon over G.* P must have an even number s of tile streams and any member x of G must occur in exactly s/2 positions. Each tile stream contains the full set of elements of G and thus all tile streams have the same sum, (ΣG). We can thus say that P is magic over the *multiple group* $G^{s/2}$. Various methods for constructing Latin polygons are given.

Two Latin squares over G are *orthogonal* in case their superimposition produces the full set of elements in the direct product G×G. The construction of orthogonal Latin squares and sets of mutually orthogonal Latin squares has been studied extensively, e.g., in [5]. *Orthogonal Latin polygons* are defined similarly and are used in Chapter 8 to construct several magic polygons with direct product bases.

Chapter 9 represents a digression into a sort of three dimensional version of a magic polygon which, owing to its specific structure, we call a magic cylinder. Two planar representations of a magic cylinder P are provided: (1) a graph, G(P), whose points represent tiles of P and whose edges are determined by tiles which share a common side. This graph is regular of degree 4 and has a factorization into equal length cycles; (2) a *magic circle system* in which each cycle of G(P) is represented by a circle in which the order of elements in any given cycle is preserved. Several examples of magic cylinders and their planar representations are illustrated.

The subject of magic circle systems is then taken up in its own right. The matrix representation of such a system leads, in turn, to the concept of a *circular magic square.* Instead of all rows and columns having the same sum as in a standard magic square, the sum of the combined elements in the row and column through any given cell is constant.

In Chapter 10 another interesting offshoot is introduced: a *Latin circle system.* The contact points of a circle system Ω are labeled with the full set of elements from the group Z(n) so that no circle in Ω contains two equal values. Latin circle systems are represented by a square matrix we call a *circular Latin square.* Several Latin circle systems are represented by their corresponding circular magic squares. Certain of these circular magic squares are dissected into upper and lower triangular sections which we call *Latin triangles.* Various constructions of Latin triangles constitute the remainder of the Chapter.

This work is intended as an introduction to a new field of mathematical research. Whereas the subject of magic squares may be construed as fodder for the mathematical hobbyist, the present work indicates many research possibilities for the serious mathematician. There are almost as many Conjectures as there are Theorems and thus many avenues for future research. I am thankful to all my teachers both mathematical and spiritual, who have, in one way or another, equipped me with what was needed to complete this work. I am thankful also to my parents and wife Sylvie for their support in this endeavor.

Raymond R. Fletcher III

CHAPTER 1

PARALLELOGRAM TILINGS OF POLYGONS

1.1 Tilability of Semiregular Polygons

In this Chapter we will define a certain kind of polygon which will serve as the setting for our magic polygons. Instead of the rows and columns of squares used to define magic squares, we will use sequences of parallelograms which occur in a parallelogram tiling of a polygon. To describe these tilings it is useful to use the following vector notation. A set V of vectors is *basic* if (i) each vector in V has a different direction, and (ii) if $\mathbf{v} \in V$ then $-\mathbf{v} \in V$. A *V-polygon* or *V-gon* is a polygon P each of whose sides has the form e**v** where e is a positive integer and $\mathbf{v} \in V$. The integer e is the *edgecount* of the side. If the edgecount is 1 for each side then P is called *simple.* A V-gon which is convex, has an even number of sides and whose opposite sides are parallel and equal is called *semiregular over V.* It is precisely this type of polygon which will serve to generalize the concept of a magic square.

Let $P_0(V)$ denote the collection of simple parallelograms whose edges consist of vectors from a basic vector set V. A V-polygon is *tilable over* $P_0(V)$ in case it can be tiled completely with parallelograms in $P_0(V)$. Unless otherwise indicated these parallelograms will simply be called *tiles.* Our first Theorem concerns the tilability of semiregular polygons using parallelograms.

Theorem 1.1: *Let V be a basic vector set. Then every semiregular V-polygon is* $P_0(V)$ *tilable.*

Proof Let $P = (x, \mathbf{s}_1, \mathbf{s}_2, \ldots, \mathbf{s}_n, -\mathbf{s}_1, -\mathbf{s}_2, \ldots, -\mathbf{s}_n, x)$ be a V-polygon with sides $\mathbf{s}_i = e_i\mathbf{v}_i$ for each $i \in \{1, \ldots, n\}$ and where $\{\mathbf{v}_i : i \in \{1, \ldots, n\}\}$ is a set of distinct basic vectors. The point x can be chosen anywhere on the boundary of P. The *crescent* C(x,y) with *endpoints* x, y on the boundary of P, is defined by

$$C(x,y) = (x, \mathbf{s}_1, \mathbf{s}_2, \ldots, \mathbf{s}_n, -\mathbf{s}_1, y, -\mathbf{s}_n, -\mathbf{s}_{n-1}, \ldots, -\mathbf{s}_2, x)$$

as outlined in Figure 1.1(I). The crescent C(x,y) has a dissection into n-1 *major parallelograms*

$$\{P_t = (\mathbf{s}_1, \mathbf{s}_t, -\mathbf{s}_1, -\mathbf{s}_t) : t = 1, \ldots, n\},$$

and P itself has a dissection into two regions, namely the crescent C(x,y) and the semiregular polygon $Q = (x, \mathbf{s}_2, \mathbf{s}_3, \ldots, \mathbf{s}_n, y, -\mathbf{s}_2, -\mathbf{s}_3, \ldots, -\mathbf{s}_n)$ as indicated in Figure 1.1(II), which has two fewer sides than P. So we can employ an inductive argument: If n = 2 then $P = (x, \mathbf{s}_1, \mathbf{s}_2, -\mathbf{s}_1, -\mathbf{s}_2, x)$ is itself a major parallelogram and has a unique $P_0(V)$ tiling consisting entirely of simple parallelograms ($\mathbf{v}_1$, $\mathbf{v}_2$, $-\mathbf{v}_1$, $-\mathbf{v}_2$). This is the basis. We then assume inductively that every semiregular V-polygon with fewer than 2n sides has a $P_0(V)$ tiling. Thus Q has a $P_0(V)$ tiling. Since each major parallelogram of C(x,y) also has a $P_0(V)$ tiling, we can conclude that P has a $P_0(V)$ tiling. □

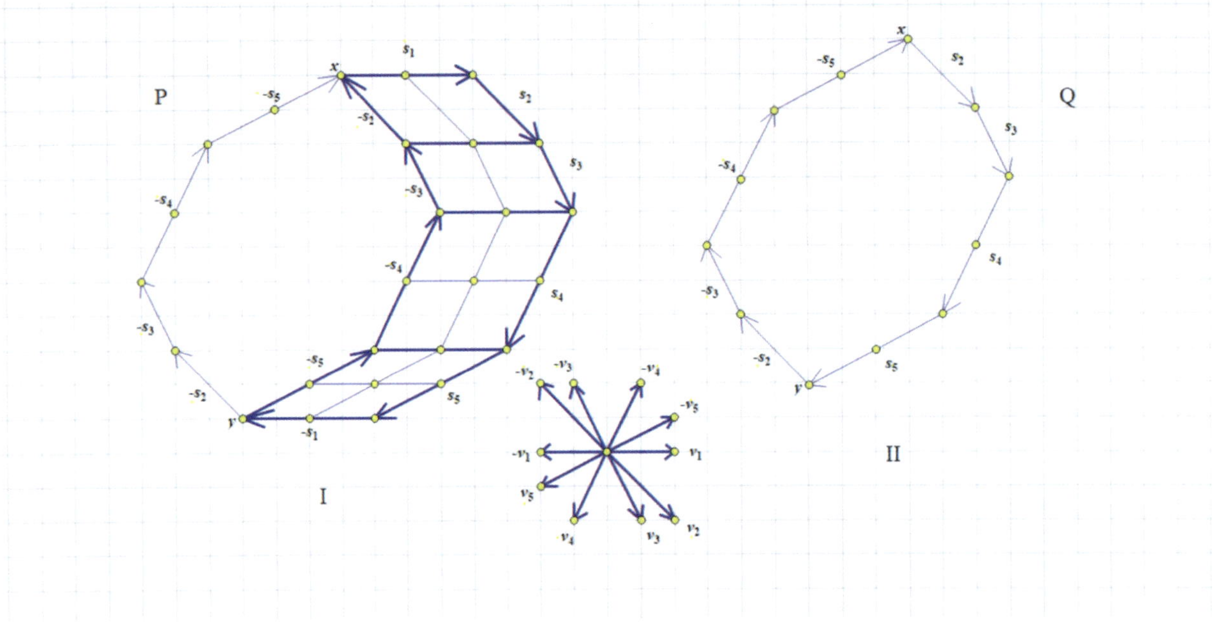

Figure 1.1

1.2 The Composition Sequence

A V-gon along with a parallelogram tiling T will be regarded as a plane graph G(T). The sides of the parallelograms constitute the edge set of G(T) and the four corners of each parallelogram form the vertex set of G(T). Let P be a V-gon with a $P_0(V)$ tiling T. A *stream* is a sequence of mutually parallel edges in G(T) which extends between two boundary edges of P. The streams form a partition of the edge set of G(T). The tiles which lie between the edges of a stream form a *tile stream.* Each tile belonging to T lies at the intersection of exactly two tile streams. The *composition sequence* of T has the form (p_1^{m1}, p_2^{m2}, ..., p_t^{mt}) where p_1, ..., p_t are mutually noncongruent members of $P_0(V)$ and the superscripts indicate the number of occurrences in T of the associated parallelograms. In case a distinction is made between the different orientations of each constituent parallelogram, the *refined composition sequence* is obtained.

In Figure 1.3 a semiregular V-gon P with a $P_0(V)$ tiling T is used to illustrate these concepts. The basic vector set V consists of the full set of lattice vectors with length 2 or √5. This set will be denoted by V = L(2, √5). In Figure 1.2 the full set of parallelograms belonging to $P_0(V)$ is named. In Figure 1.3 these names are used to express the composition sequence $A^5B^9C^4E^4F^9S^2$ as well as the refined composition sequence $A_1{}^1A_2{}^4B_1{}^1B_2{}^1B_3{}^3B_4{}^4C_1{}^2C_2{}^2E_1{}^2E_2{}^2F_1{}^4F_2{}^2F_3{}^1F_4{}^2S^2$ of T.

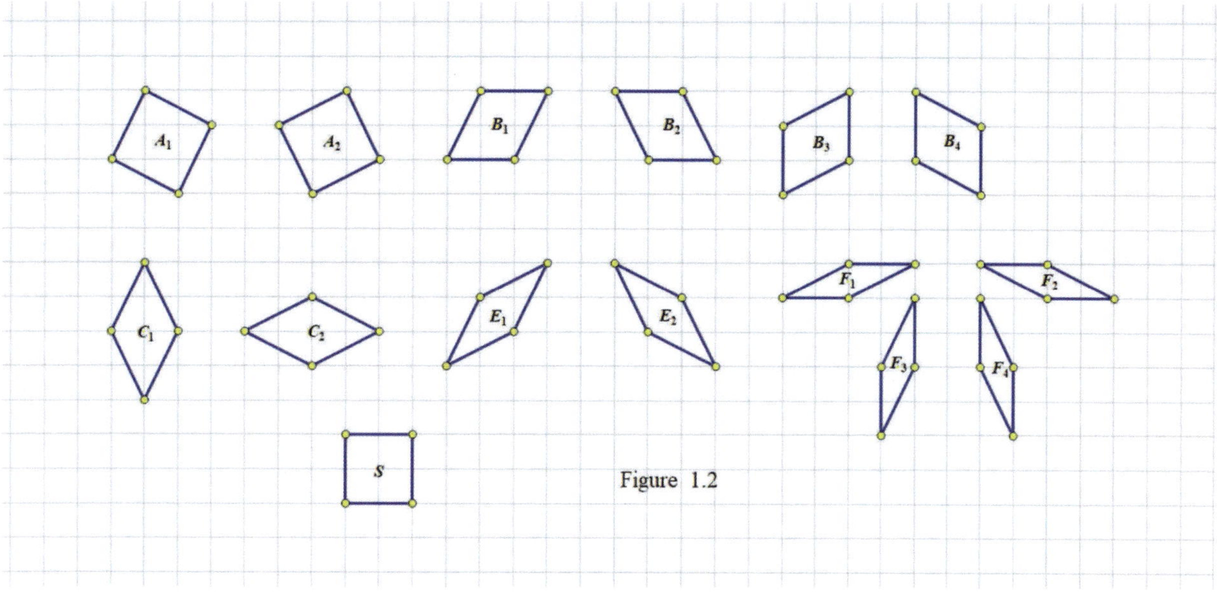

Figure 1.2

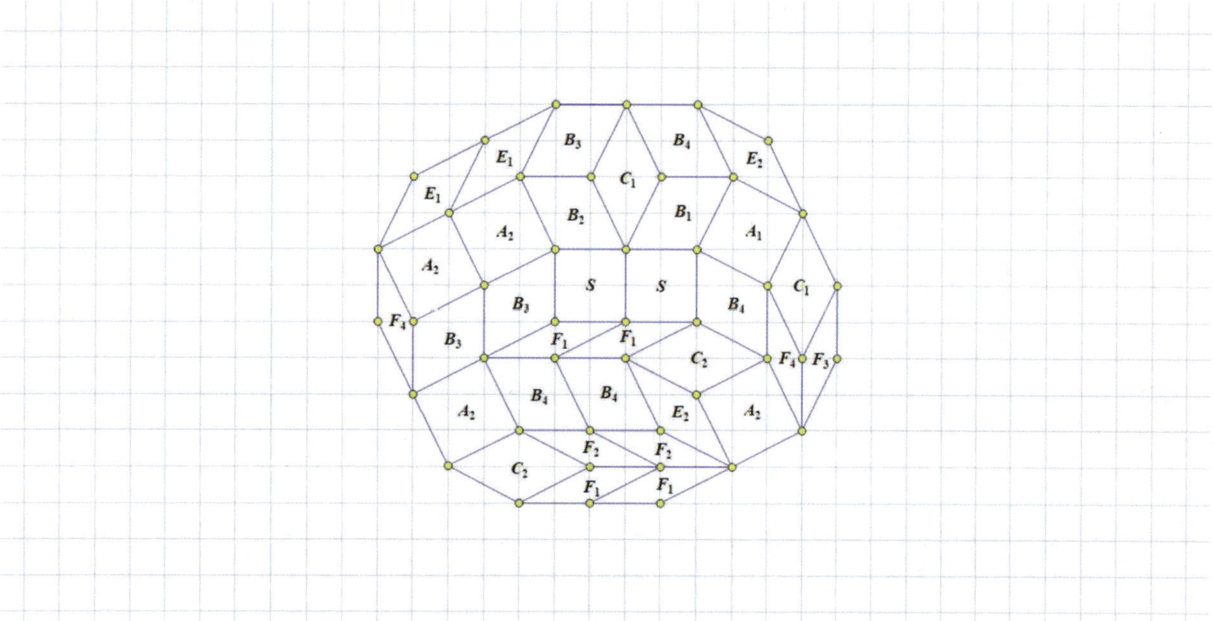

Figure 1.3

There certainly exist a large number of ways to obtain a parallelogram tiling of a given semiregular polygon, but surprisingly all such tilings have the same composition sequence.

Theorem 1.2: *Let P be a semiregular V-gon with two tilings T_1, T_2 over $P_0(V)$. Then T_1, T_2 have the same refined composition sequence, (and consequently the same composition sequence).*

Proof. Let P = ($\mathbf{s}_1$, $\mathbf{s}_2$, …, $\mathbf{s}_n$, $-\mathbf{s}_1$, $-\mathbf{s}_2$, …, $-\mathbf{s}_n$) where $\mathbf{s}_i = e_i\mathbf{v}_i$, $\mathbf{v}_i \in V$ and e_i is a positive integer for i = 1, …, n. If $\mathbf{v} \in V$ and T is any parallelogram tiling of P, let ***v***- *stream* denote the stream in T consisting of (mutually parallel) v-edges, and let ***v****-tile stream* denote the corresponding sequence of tiles. At the unique intersection of a $\mathbf{v}_i$ –tile stream and a $\mathbf{v}_j$ –tile stream, (i≠j), the parallelogram ($\mathbf{v}_i$, $\mathbf{v}_j$, -$\mathbf{v}_i$, -$\mathbf{v}_j$) occurs whether these tile streams occur in T_1 or T_2; in fact each of T_1, T_2 contain exactly e_ie_j such parallelograms since there are exactly e_i *parallel*,(i.e., nonintersecting) $\mathbf{v}_i$ - tile streams, and exactly e_j parallel $\mathbf{v}_j$ - tile streams. For each i,j with i≠j, both T_1 and T_2 contain exactly e_ie_j ($\mathbf{v}_i$, $\mathbf{v}_j$, -$\mathbf{v}_i$, -$\mathbf{v}_j$) parallelograms, and thus have the same refined composition sequence. □

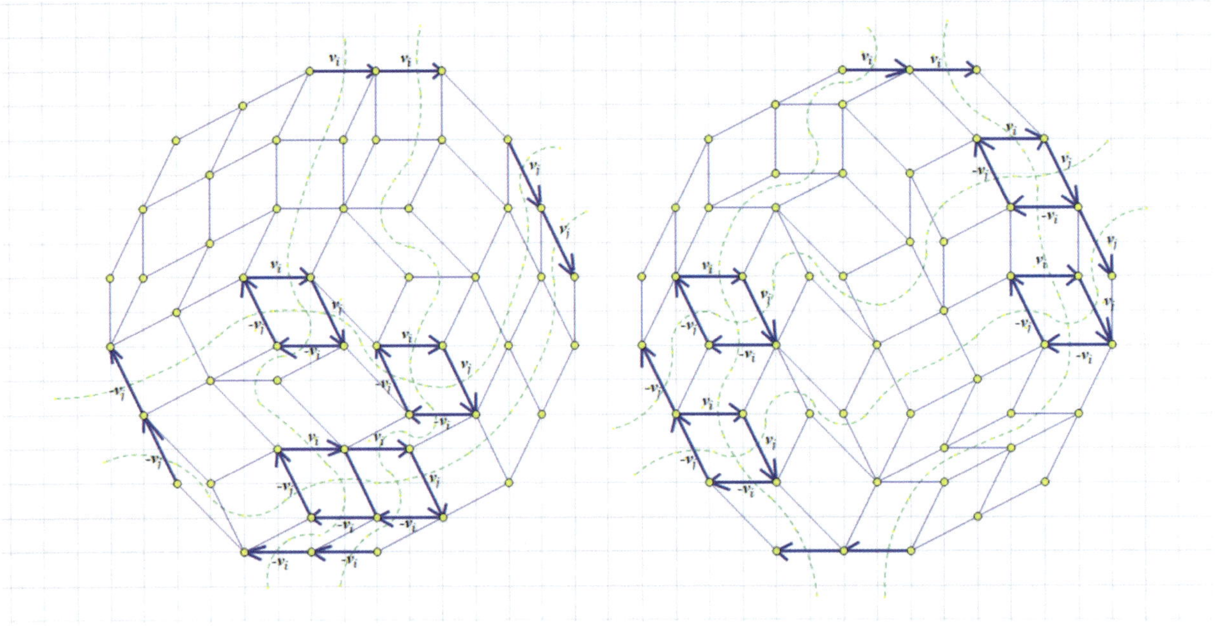

Figure 1.4

In Figure 1.4 two copies of a semiregular polygon are illustrated each with a different parallelogram tiling. The $\mathbf{v}_i$ and $\mathbf{v}_j$ tile streams are indicated in each tiling by dashed lcurves, and the tiles which occur at the intersections of these streams are marked with bold vectors.

From Theorem 1.2 the following formulas are easily obtained. Let r(P) denote the number of parallelograms which occur in any $P(V_0)$ tiling of P; let $e(\mathbf{v}_i)$ denote the number of edges in a $\mathbf{v}_i$ - stream; let e(P) denote the number of edges in any $P(V_0)$ tiling of P; let o(P) denote the number of vertices which occur in any $P(V_0)$ tiling of P, and let $M = \Sigma\{e_i : i = 1, \ldots, n\}$ denote the *semiperimeter* of P . Then

$$r(P) = \Sigma\{e_ie_j : i,j \in \{1, \ldots, n\}, i\neq j\}$$

$$e(\mathbf{v}_i) = 1 + M - e_i$$

$$e(P) = \Sigma\{e(\mathbf{v}_i) : i = 1, \ldots, n\} = (n-1)M + n$$

$$o(P) = e(P) - r(P) +1$$

The formula for o(P) is Euler's Theorem.

1.3 Isomorphic Tilings

Two parallelograms of historical interest are rhombs, one with interior angles 36°, 144°, and the other with interior angles 72°, 108° . These were used by Roger Penrose [1] to construct aperiodic tilings of the plane. In [2] it is queried as to whether our $L(2,\sqrt{5})$ lattice parallelograms can be used to create a tiling T of a polygon such that G(T) is isomorphic to a corresponding tiling which uses only the two Penrose rhombs. This query is answered in the affirmative by our next result.

Theorem 1.3: Let $V = \{\mp\mathbf{v}_1, \mp\mathbf{v}_2, \ldots, \mp\mathbf{v}_n\}$ and $W = \{\mp\mathbf{w}_1, \mp\mathbf{w}_2, \ldots, \mp\mathbf{w}_n\}$ be basic vector sets, and let $\phi{:}V\rightarrow W$ defined by $\phi(\mathbf{v}_i) = \mathbf{w}_i$ be a one to one correspondence. Suppose

$$P = (X, e_1\mathbf{v}_1, e_2\mathbf{v}_2, \ldots, e_n\mathbf{v}_n, -e_1\mathbf{v}_1, -e_2\mathbf{v}_2, \ldots, -e_n\mathbf{v}_n, X)$$

is a semiregular V-gon with X a vertex of P, and

$$Q = (Y, e_1\mathbf{w}_1, e_2\mathbf{w}_2, \ldots, e_n\mathbf{w}_n, -e_1\mathbf{w}_1, -e_2\mathbf{w}_2, \ldots, -e_n\mathbf{w}_n, Y)$$

is a semiregular W-gon with Y a vertex of Q. Let T be a $P_0(V)$ tiling of P. Then there exists a $P_0(W)$ tiling T′ of Q such that G(T) is isomorphic to G(T′).

Proof: Let A be a vertex of G(T) as in Figure 1.5. Then $A = (X, k_1\mathbf{v}_1 +, \ldots, +k_n\mathbf{v}_n)$ for some integers k_i. Let Π denote the Euclidean plane and define a mapping ψ from the vertex set of G(T) into Π by $\psi(A) = (Y, k_1\mathbf{w}_1 +, \ldots, +k_n\mathbf{w}_n)$. Suppose B is a vertex of G(T) which is adjacent to A. If $B = (Y, j_1\mathbf{v}_1 +, \ldots, +j_n\mathbf{v}_n)$ then $(j_1 - k_1)\mathbf{v}_1 +, \ldots, + (j_n - k_n)\mathbf{v}_n$ must be an element of V, say $\mathbf{v}_i$ or $-\mathbf{v}_i$,i.e., $j_t - k_t = 0$ if $t \neq i$ and $j_i - k_i = \mp 1$. Then

$$(k_1\mathbf{w}_1 +, \ldots, +k_n\mathbf{w}_n) - (j_1\mathbf{w}_1 +, \ldots, +j_n\mathbf{w}_n) = \mp\mathbf{w}_i,$$

and the edge corresponding to $\mathbf{w}_i$ is inserted between the points ψ(A) and ψ(B). (These are labeled A′, B′ in Figure 1.5.) These edges along with the points {ψ(U): U is a vertex of G(T)} form a plane graph H which is isomorphic to G(T).

Let $p = (A, \mathbf{v}_s, B, \mathbf{v}_t, C, -\mathbf{v}_s, D, -\mathbf{v}_t, A)$ denote a particular parallelogram tile of T. Then

$$p' = (\psi(A), \phi(\mathbf{v}_s), \psi(B), \phi(\mathbf{v}_t), \psi(C), -\phi(\mathbf{v}_s), \psi(D), -\phi(\mathbf{v}_t), \psi(A)) =$$
$$(\psi(A), \mathbf{w}_s, \psi(B), \mathbf{w}_t, \psi(C), -\mathbf{w}_s), \psi(D), -\mathbf{w}_t), \psi(A))$$

is the corresponding parallelogram region of H, and the collection of tiles {p′ : p is a tile of T} form a tiling T′ of Q with G(T) isomorphic to G(T′). □

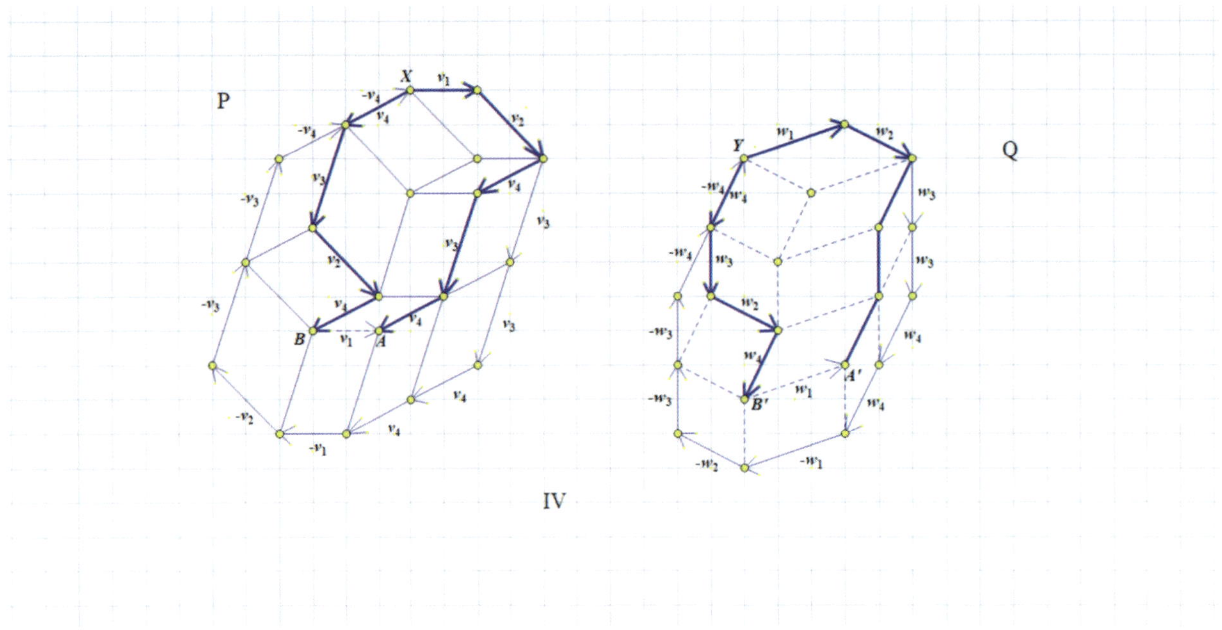

Figure 1.5

In Figure 1.6(I) a Penrose decagon is illustrated with a tiling T which uses only the two Penrose rhombs. In Figure 1.6(II) an L(2,√5) decagon is illustrated which uses the A,B,C,E,F tiles to obtain a tiling T' with G(T') clearly isomorphic to G(T). Note that each tiling uses exactly 5 basic vectors and their negatives.

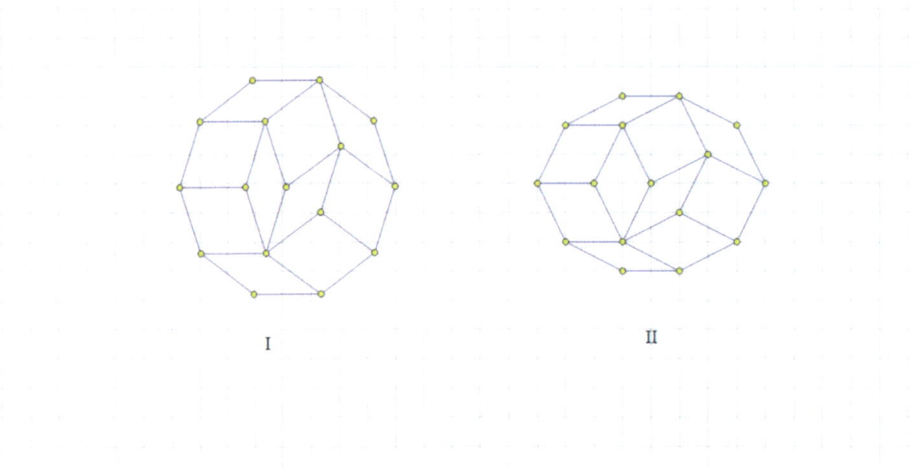

Figure 1.6

1.4 Hexagon Inversions

The final result for this Chapter has the greatest significance for our main topic of magic polygons. By means of a series of local transformations we move between any two tilings of a given semiregular polygon. Let P be a semiregular V-polygon with a $P_0(V)$ tiling T and let $\theta = \angle pqr$ be an angle in P which is covered by more than one tile of T as in Figure 1.7. Let the pq tile stream A and the qr tile stream B intersect in parallelogram E = agbc with bc || ag || pq and ab || gc || qr as in Figure 1.7. Let A' denote the set of tiles belonging to the A tile stream between edge pq and edge bc, and let B' denote the section of the B tile stream between edges qr and ab . The collection of tiles in A'∪B' along with the tiles of T which lie between A', B' is called a *(θ,E)-interval.* The angle θ is the *generator* and tile E is the *endtile* of the interval. Let I = (θ,E) be an interval in a semiregular V- polygon P with a $P_0(V)$ tiling T . A retiling of I in which the single tile congruent to E covers θ is called an *inversion* of *I.* The smallest interval consists of a simple semiregular hexagon H which has exactly two distinct tilings. The retiling of H which moves between these two tilings is called a *hexagon inversion.*

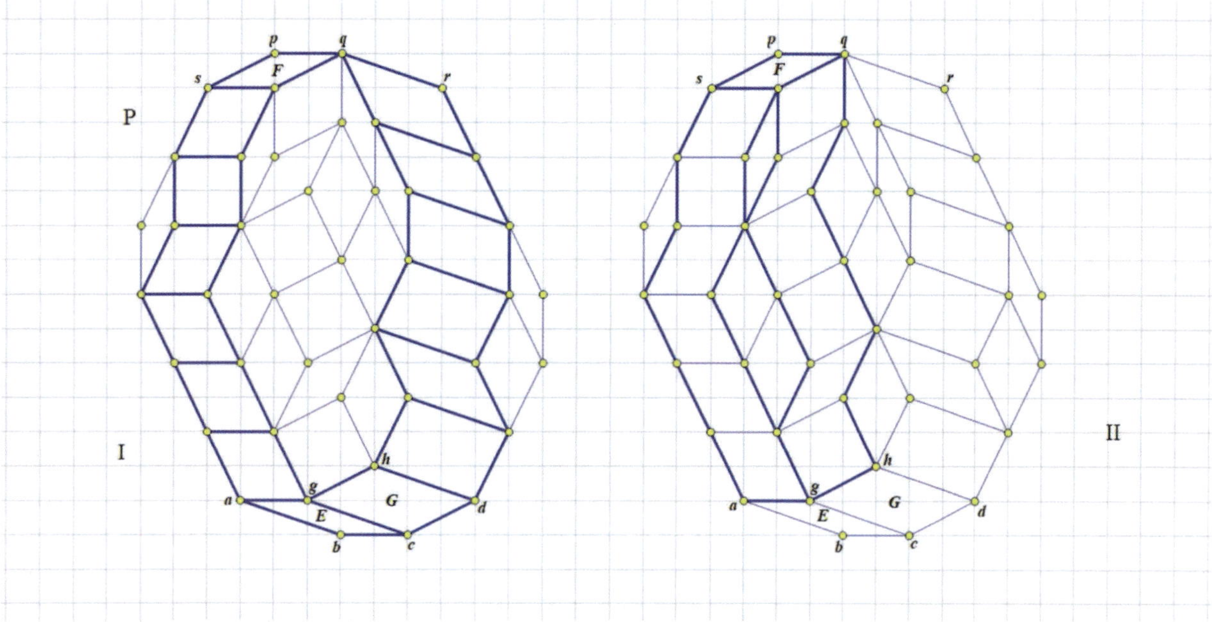

Figure 1.7

Lemma 1.4: Let P be a semiregular V-polygon with a $P_0(V)$ tiling T . Then every interval in T can be inverted.

Proof: Let I = (θ,E) be an interval in T with θ = ∠pqr <180°, endtile E = agbc and bounding streams A, B; let $θ_1$ = ∠bcd and $θ_2$ = ∠agh as in Figure 1.7(I). Clearly $θ_1 \leq 180°$. Let C denote the cd tile stream. If $θ_1 = 180°$ then tile streams A and C are parallel in I making it impossible for the A, B tile streams to touch at point q. So we must have $θ_1 < 180°$. As already mentioned, the smallest interval has the shape of a simple hexagon and so is easily inverted. Assume inductively that every interval in P which has fewer tiles than I is invertible. Note that $θ_2 = θ_1$, and so is less than 180°. In case $θ_2$ is covered by a single tile in T then the hexagon aehdcb is created as in Figure 1.7 (III). If $θ_2$ is covered by more than one tile, then consider: the ag tile stream A and the gh tile stream C must intersect at a tile F in the interior of I since the A, B streams form the boundary of I. Thus $I_1 = (θ_2,F)$ is a subinterval of I consisting of fewer tiles than I, (see Figure 1.7(II)). By the inductive hypothesis I_1 is invertible and so can be retiled so that a tile F' congruent to F covers angle $θ_2$ as in Figure 1.7(III). In this case we also obtain the hexagon aehdcb.

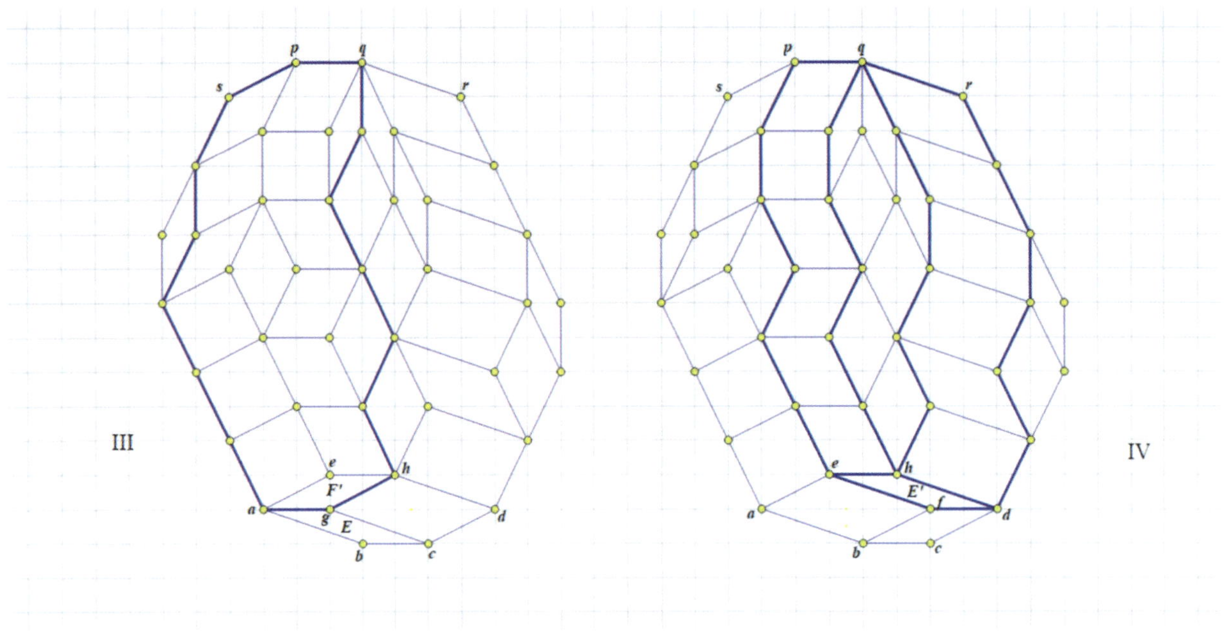

Figure 1.7

When hexagon aehdcb is inverted, the tile E'=ehdf is created as in Figure 1.7(IV). Let D denote the new stream that runs through edges bc and eh. Then this stream and the original stream B intersect at E' and exit I at the adjacent boundary edges pq, qr. We thus obtain an interval I_2 = (θ,E') which is bounded by streams B, D, and which is smaller than I. By the inductive hypothesis, I_2 can be inverted with the result that a tile E" congruent to E becomes the sole cover for θ. □

In the following Theorem it will be shown that an arbitrary parallelogram tiling T of a semiregular polygon P can be transformed by a series of hexagon inversions into a special tiling T' of P which consists of a series of crescents with a common endpoint. An example of such a tiling, which we call a *polar tiling,* is illustrated in Figure 1.8. It consists of three crescents: $C(X_0,Y)$, $C(X_1,Y)$, $C(X_2,Y)$ consisting of major parallelograms and a residual parallelogram with opposite points X_2,Y. The point Y which serves as the common endpoint of the crescents is the *pole* of the tiling.

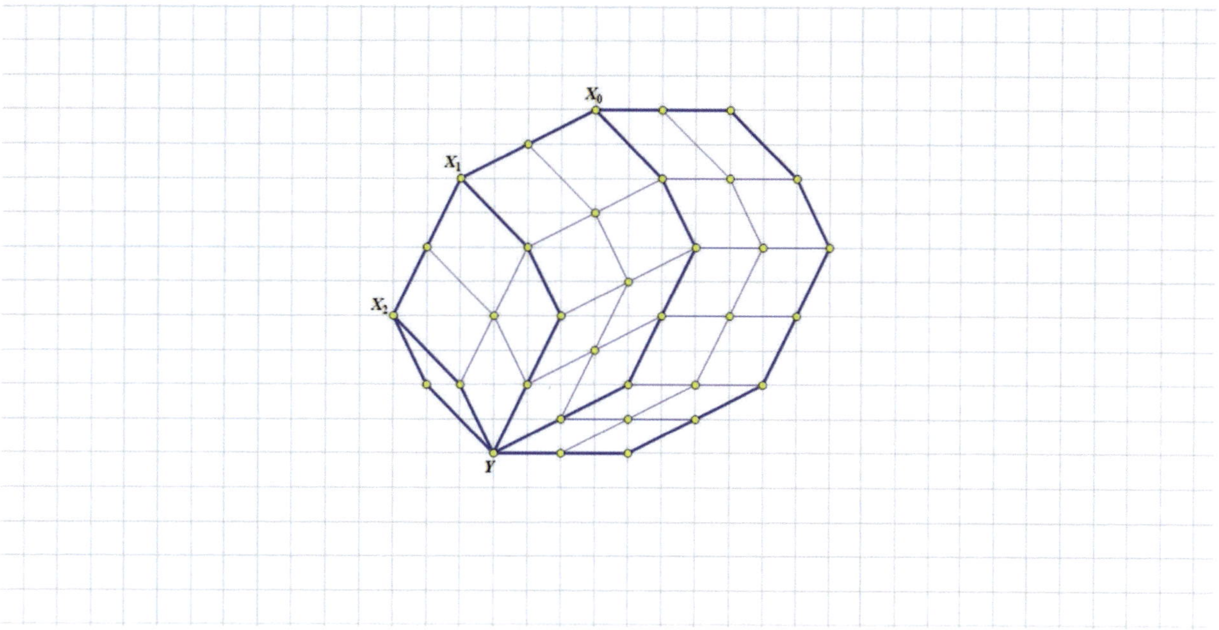

Figure 1.8

Theorem 1.5: Let P be a semiregular V-polygon with two $P_0(V)$ tilings T, U . Then there exists a sequence of hexagon inversions which transform T into U .

Proof. Let P = $(X_0, \mathbf{s}_1, X_1, \mathbf{s}_2, \ldots, X_n, -\mathbf{s}_1, X_{n+1}, \ldots, X_{2n-1}, -\mathbf{s}_n, X_0)$ be a semiregular V-polygon with corner points X_i and sides $\mathbf{s}_i = e_i\mathbf{v}_i$ for i = 1, …, n as in Figure 1.9. We will proceed by induction on the number of tiles which constitute any $P_0(V)$ tiling of a semiregular V- polygon. For small numbers of tiles the Theorem is easily seen to be true. Let P be defined as above and assume inductively that the Theorem holds true for any semiregular V-polygon which can be tiled with fewer tiles than P. Let $\mathbf{v}_1$ denote the boundary vector which extends from W_0 to X_1. If angle $\theta_1 = (W_0, \mathbf{v}_1, X_1, \mathbf{v}_2)$ is not covered by a single tile, then there exists an interval $I_1 = (\theta_1, E_1)$ in T, where E_1 is the tile which occurs at the intersection of the $\mathbf{v}_1$ and $\mathbf{v}_2$ tile streams. By Lemma 1.4, this interval can be inverted to obtain a new tiling T_1 of P in which θ_1 is covered by a single tile of the form $(\mathbf{v}_1, \mathbf{v}_2, -\mathbf{v}_1 - \mathbf{v}_2)$.

For each i ∈ {2, …, n} label the points on side $\mathbf{s}_i$ by $Q_{i,t} = (X_{i-1}, (t-1)\mathbf{v}_i)$ for t = 1, …, e_i as in Figure 1.9. Suppose side $\mathbf{s}_2$ contains more than one edge. In T_1 if angle $\theta_2 = (\mathbf{v}_1, Q_{2,2}, \mathbf{v}_2)$ is not covered by a single tile, then the interval $I_2 = (\theta_2, E_2)$ can be inverted by Lemma 1.4, where E_2 is the tile at the intersection of the $\mathbf{v}_1$ stream and the new $\mathbf{v}_2$ stream. Thus T_1 can be retiled to obtain a new tiling T_2 in which θ_2 is covered by a single tile of the form $(\mathbf{v}_1, \mathbf{v}_2, -\mathbf{v}_1 - \mathbf{v}_2)$. By repeating this retiling procedure for each of the sides $\mathbf{s}_1, \ldots, \mathbf{s}_n$ a tiling T' of P is obtained with the crescent

$$C(W_0, W_1) = (W_0, \mathbf{v}_1, X_1, \mathbf{s}_2, X_2, \mathbf{s}_3, X_3, \ldots, \mathbf{s}_n, X_n, -\mathbf{v}_1, W_1, -\mathbf{s}_n, -\mathbf{s}_{n-1}, \ldots, -\mathbf{s}_2, W_0).$$

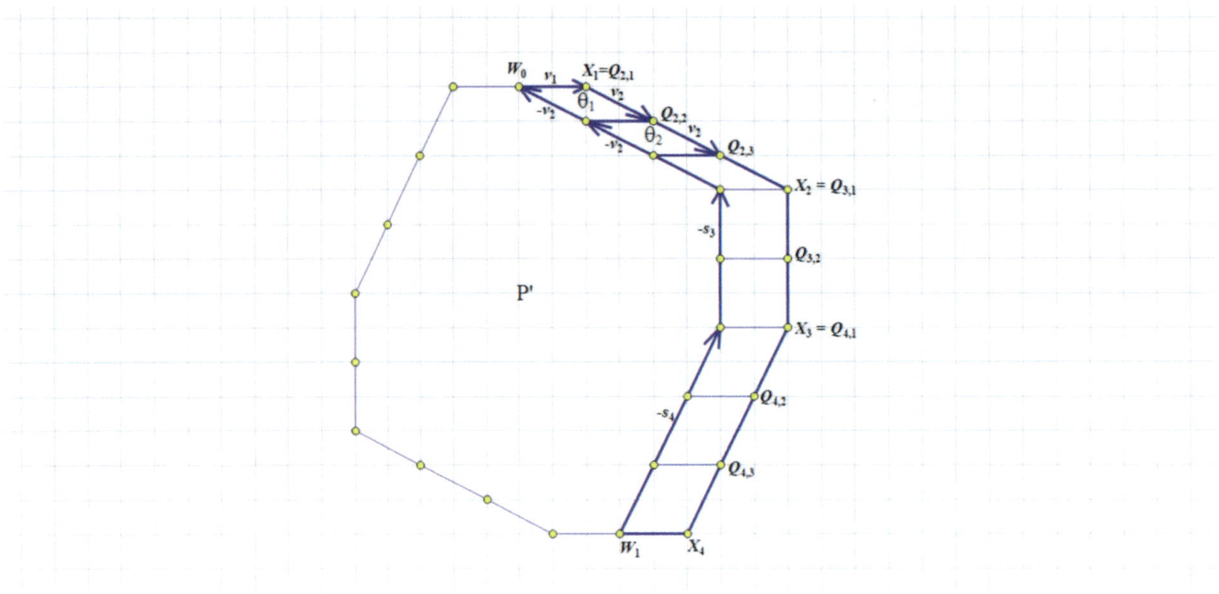

Figure 1.9

When the tiles belonging to this crescent are removed, the remaining tiles of P form a tiling of a semiregular polygon P'. Since P' has fewer tiles than P we can conclude by the inductive hypothesis that P' has a polar crescent tiling . When the tiles belonging to $C(W_0, W_1)$ are reinserted in their original position, we obtain a polar crescent tiling R for our original semiregular polygon P. The same procedure can be applied to the tiling U to obtain precisely the same polar crescent tiling of P. If we then apply the sequence of hexagon inversions which take T to R followed by the reverse of the sequence of hexagon inversions which carry U to R, we obtain the desired sequence of hexagon inversions. □

Note that hexagon inversion preserves the refined composition sequence of a parallelogram tiling, and thus Theorem 1.2 is an immediate consequence of Theorem 1.5.

In Figure 1.10(I- XIII) the complete interval inversion process is depicted for a semiregular octagon P. Starting with a random tiling of P, a polar crescent tiling is ultimately obtained. At each stage the interval to be inverted is outlined in bold. The first crescent is obtained in 1.10(V) and the second in 1.10(IX). Note that in 1.10(IV,VIII, XI) the interval to be inverted is just a simple hexagon.

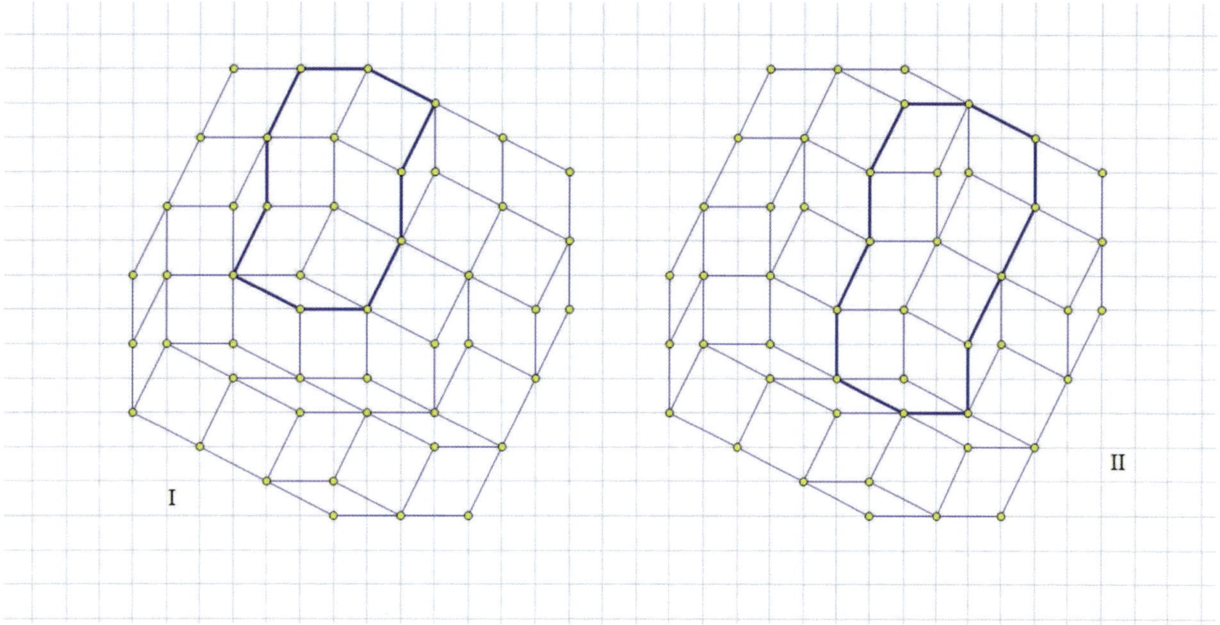

Figure 1.10

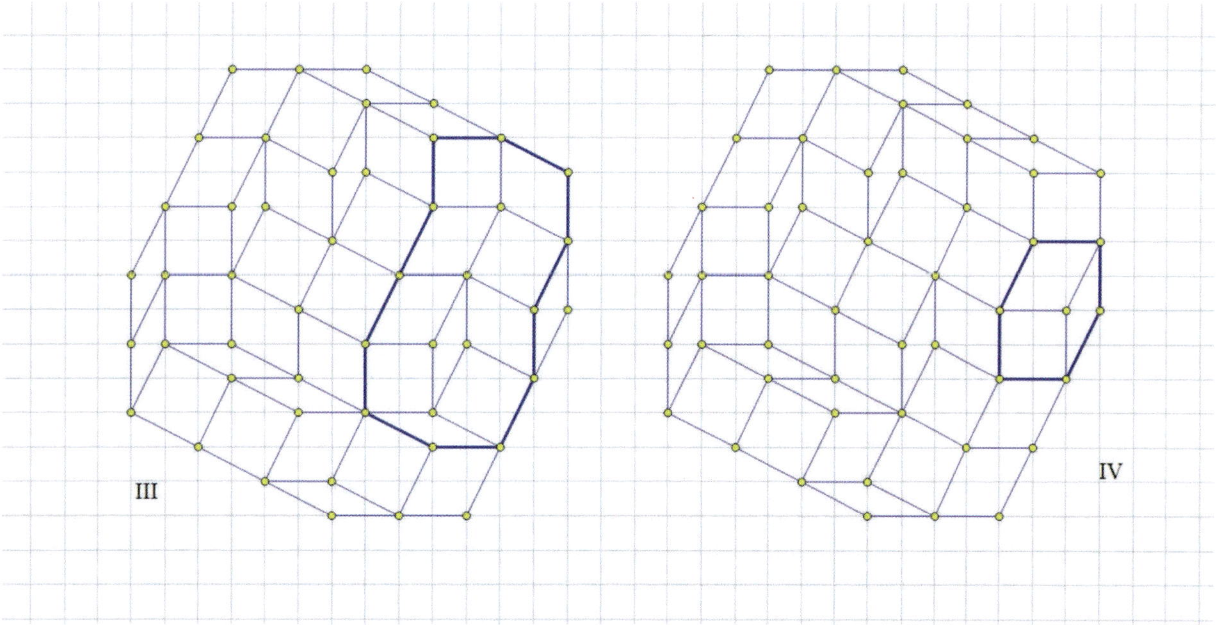

Figure 1.10

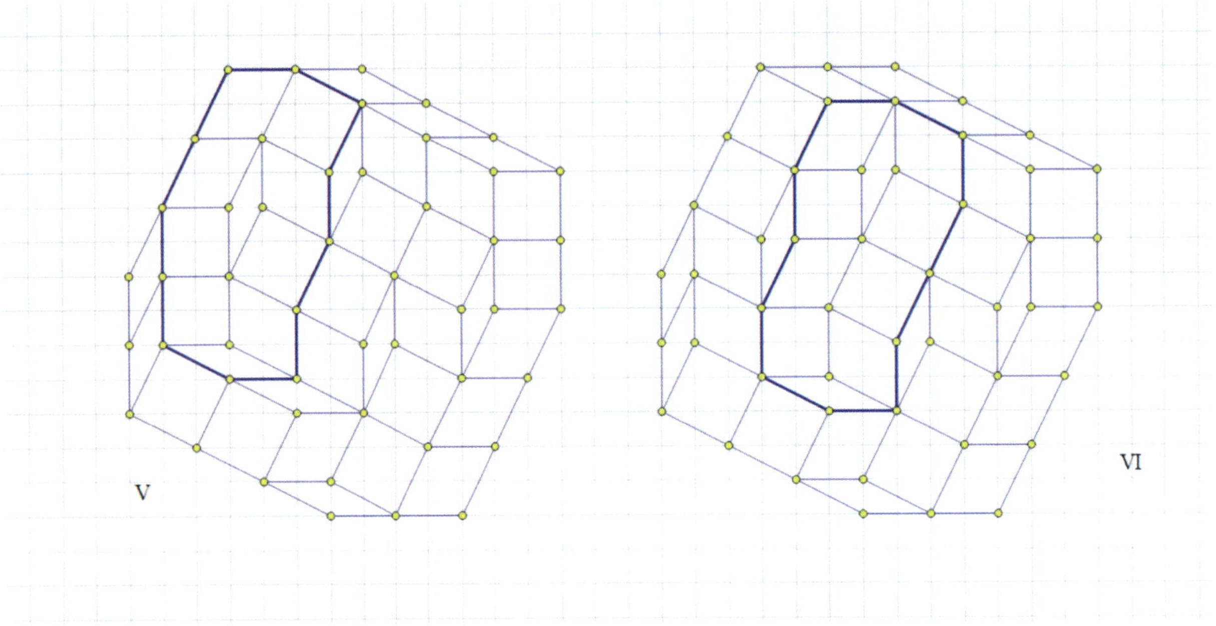

Figure 1.10

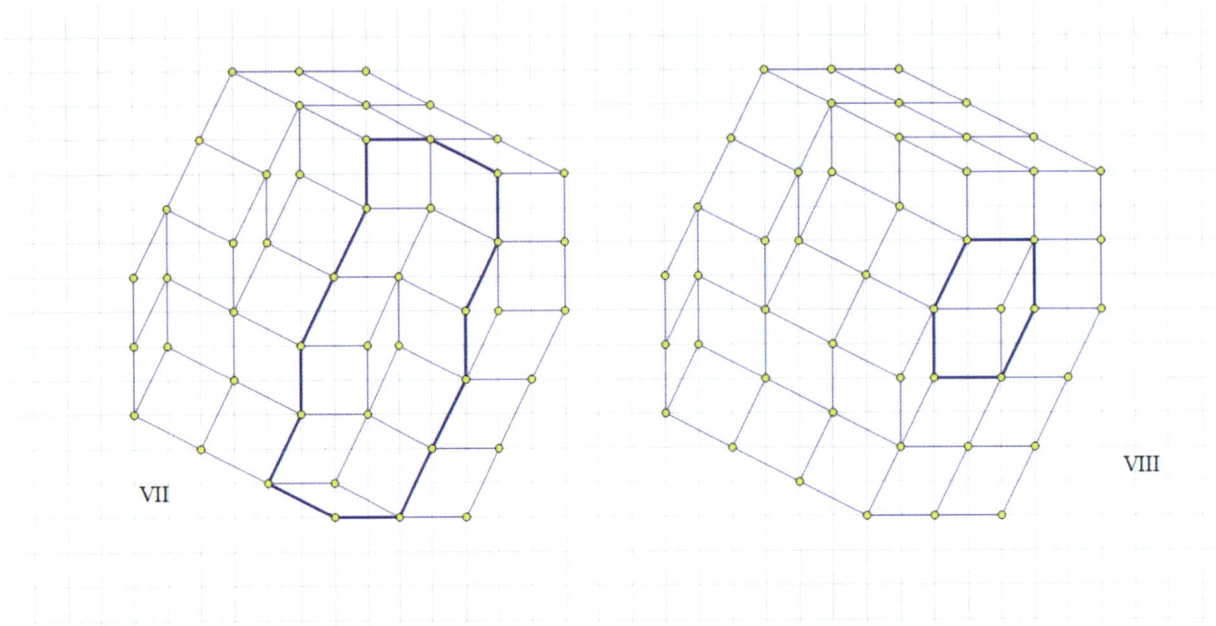

Figure 1.10

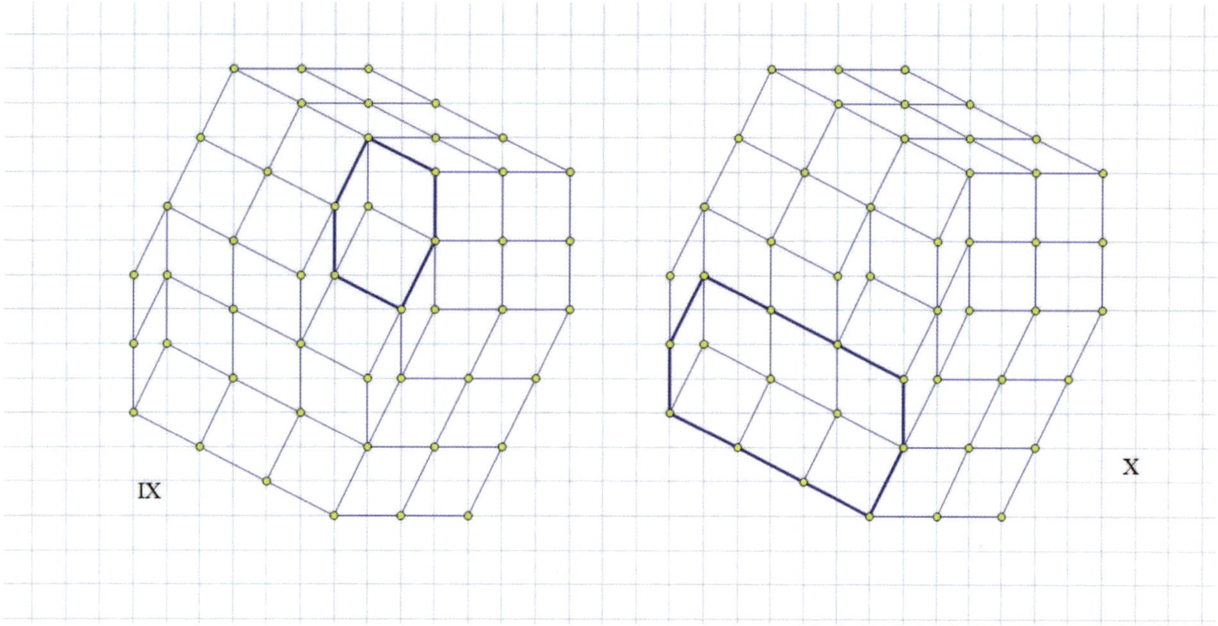

Figure 1.10

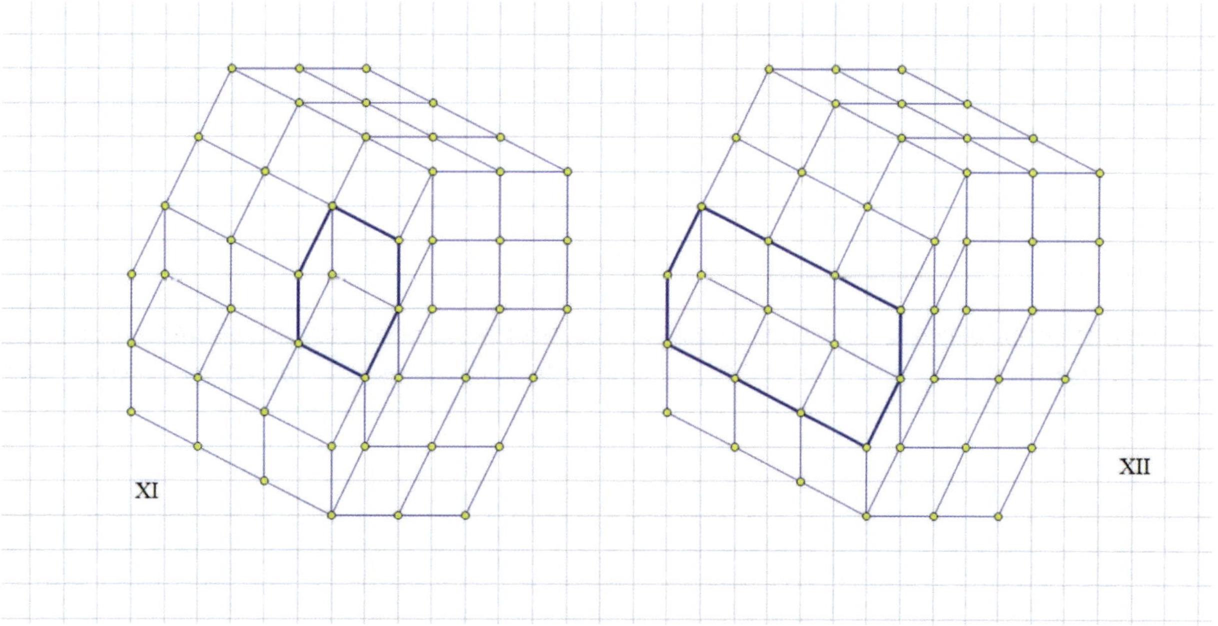

Figure 1.10

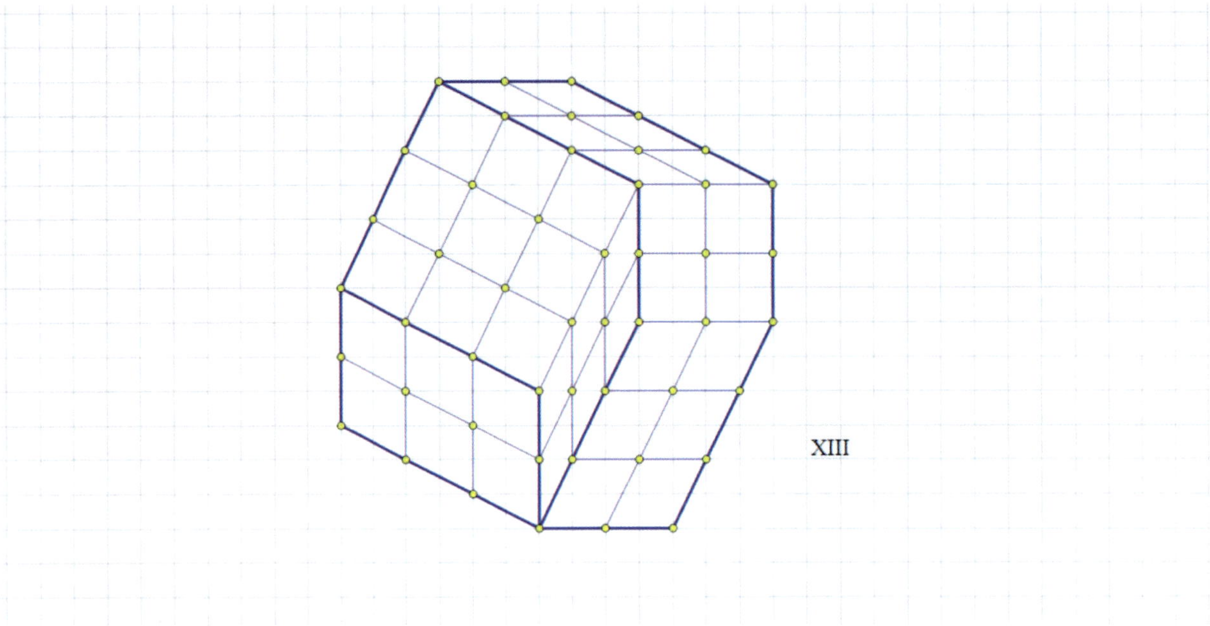

Figure 1.10

CHAPTER 2

LINEAR REPRESENTATION

2.1 The Diameter of a Semiregular Polygon

It is a curious phenomenon that a semiregular polygon with a tiling by parallelograms can equally well be represented by a configuration of lines. A *line configuration* $\Lambda = (V, L)$ *of degree 2* refers to a finite collection L of lines in the Euclidean plane and intersection points V with the property that no more than two lines meet at any given point in V. Each line l contains a sequence W_l of points of V; thus Λ consists of a collection of sequences $\{W_{i},$: l is a line of L}.

Suppose $\{W_1, W_2, \ldots, W_t\}$ and $\{S_1, S_2, \ldots, S_t\}$ are two collections of sequences. Let $W = W_1 \cup \ldots \cup W_t$ and $S = S_1 \cup \ldots \cup S_t$. A bijection φ: $W \to S$ *preserves sequences* in case W_i and S_i have the same number of elements; $\varphi(W_i) = S_i$ and the elements of $W_{i,}$ S_i occur in the same order for each $i \in \{1, \ldots, t\}$. Thus if $W_i = (w_{1,} w_2, \ldots, w_k)$ then $S_i = (\varphi(w_1), \varphi(w_2), \ldots, \varphi(w_t))$.

We will first start with a line configuration and construct the corresponding tiled semiregular polygon P. A *diameter* of P is a concept which occurs in our construction. Let T be a parallelogram tiling of P and suppose X, Y are two vertices of G(T). The distance from X to Y is the length of a shortest path in G(T) which joins X and Y. An X,Y path which achieves this shortest distance is a *distance path.* A *diameter of G(T)* is a distance path which joins two opposite corners of P.

Lemma 2.1: *Let P be a semiregular V-gon with a $P_0(V)$ tiling T and with semiperimeter n. Then every diameter of G(T) has exactly n edges. Moreover, any path in G(T) which joins two boundary vertices of P and contains exactly one edge from each stream must be a diameter.*

Proof. Let X, Y be a pair of opposite corner points of P and let ρ denote an X,Y path in G(T). The edges along the boundary from X to Y and from Y to X determine two *semiperimetric X, Y paths* ρ_1, ρ_2 each with n edges. If **v** is any edge of ρ_1 then the final edge of the **v** - stream lies on ρ_2 . Thus for each edge **v** the **v** - stream must cross ρ. The set $\{\mathbf{v}$ - streams : $\mathbf{v} \in \rho_1\}$ provides a partition of the edge set of G(T), and thus ρ must contain at least n edges. Since the semiperimetric X, Y paths each have n edges we can conclude that an X, Y diameter must have exactly n edges, one edge belonging to each stream in G(T).

Now let ρ be a path in G(T) which joins two boundary points of P at least one of which is not a corner and which contains exactly one edge from each of the n streams. Suppose ρ contains the boundary point W which is not a corner. Then there is a straight angle on the boundary of P of the form ($\mathbf{v}_1$, W, $\mathbf{v}_2$) with $\mathbf{v}_1 = \mathbf{v}_2$. The path ρ is required to contain one edge e_1 belonging to the $\mathbf{v}_1$ stream and one edge e_2 belonging to the $\mathbf{v}_2$ stream. We may suppose without loss of generality that ρ encounters e_1 before e_2. Now since $\mathbf{v}_1 = \mathbf{v}_2$ the corresponding streams are parallel, but then once the $\mathbf{v}_1$ stream is crossed via edge e_1 it would have to be crossed again in order to reach e_2. Since this is not possible, we can conclude that ρ cannot contain a non-corner point of P. □

There are in fact, many diameters for a given pair X, Y of opposite corner points, one of which is illustrated in Figure 2.1(II). To construct a diameter start at X and add edges one by one taking care not to recross or touch twice any stream until the point Y is reached.

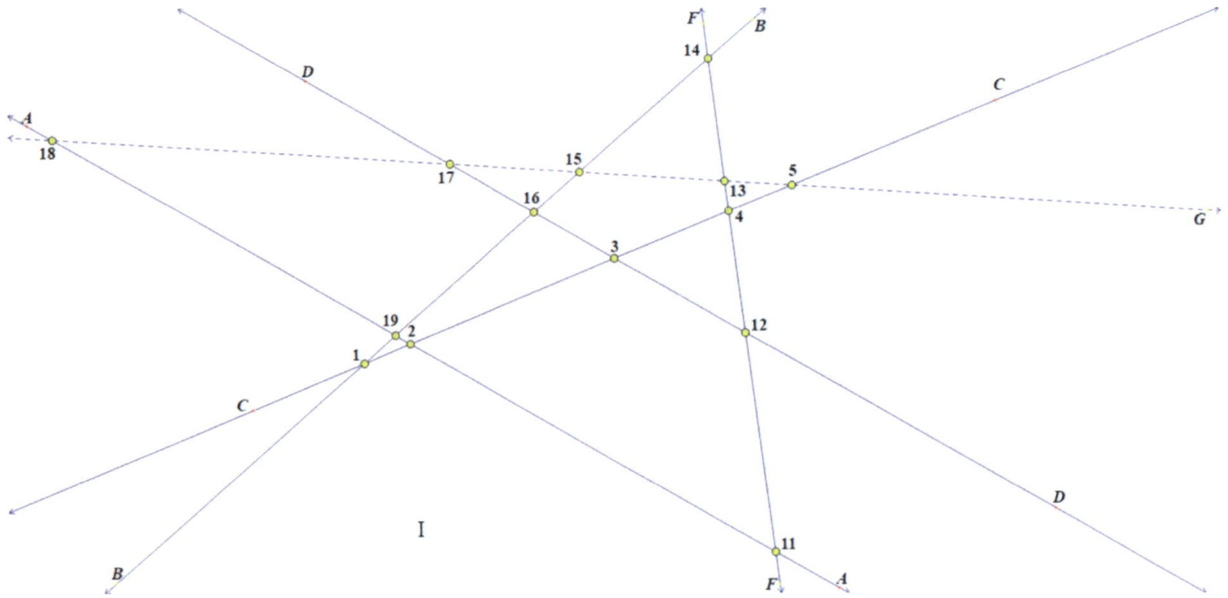

Figure 2.1

2.2 Conversion of Line Configuration to Semiregular Polygon

In Figure 2.1(I) a particular degree 2 line configuration Λ is illustrated with lines A,B,C,D,F,G. In Figure 2.1(II) a semiregular polygon P with tiling T is depicted which corresponds to Λ *before* line G is inserted, and in Figure 2.1(III) the tile stream corresponding to line G has been inserted to obtain a semiregular polygon P' with tiling T' which corresponds to the full line configuration Λ. We will describe first the procedure for inserting the tile stream which corresponds to a line of Λ which is not parallel to any previously inserted tile streams. Line G of Figure 2.1(I) is such a line.

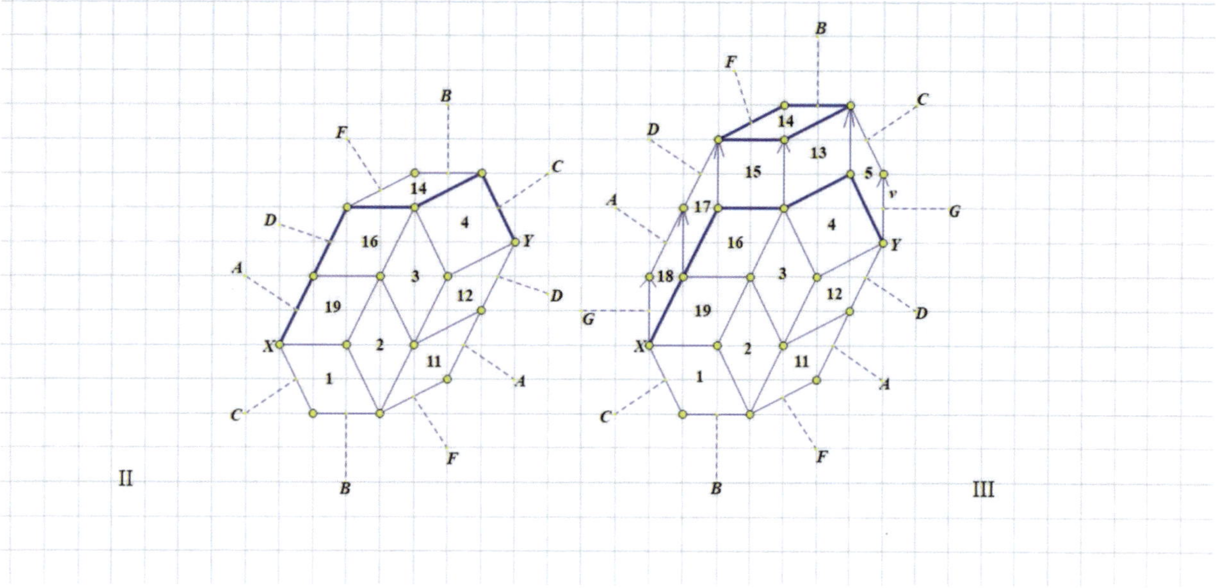

Figure 2.1

The end intersection points of line G, labeled 5, 18 determine where the end tiles of tile stream G will be located, in this example at the end of tile streams C, A respectively. The boundary points X,Y which lie on end edges of these tile streams are opposite corners of P. This must be so since, in accordance with the order of points (18, 17, 15, 13,5) on G which lie on lines A, D, B, F,C respectively, the corresponding tiles must occur on tile streams A,D,B, F, C in the same order. As the corresponding edge streams are disjoint it must be that an X, Y path has 5 edges, one for each intervening stream. Thus an X,Y path has as many edges as a semiperimetric path of P, making it a diameter.

The particular X,Y diameter δ that is chosen is determined by the line G: since, for example, line G meets F at point 13 between points 14, 4, the edge between tiles 14, 4 is included in δ. Similarly line G meets B at point 15 between points 14, 16, so the edge shared by tiles 14, 16 is included in δ. The other intersection points of G occur at endpoints of lines A,D,C so the corresponding edges of δ are boundary edges of P. The resulting X,Y diameter δ is indicated by bold lines in Figure 2.1(II).

The diameter δ is used to dissect and separate P into two parts as in Figure 2.1(III). (In case δ is a semiperimetric path, one of these two parts is empty and the inserted tiles will all be boundary tiles.) An edge v is easily found which serves for the creation of a v tile stream corresponding to line G and which cements the two parts of P together and at the same time produces the semiregular polygon P' as in Figure 2.1(III). The polygon P' is then equipped with a tiling T' which corresponds to the original line configuration Λ.

To the line configuration in Figure 2.1(I) suppose we now insert the line E parallel to C as in Figure 2.2(I).

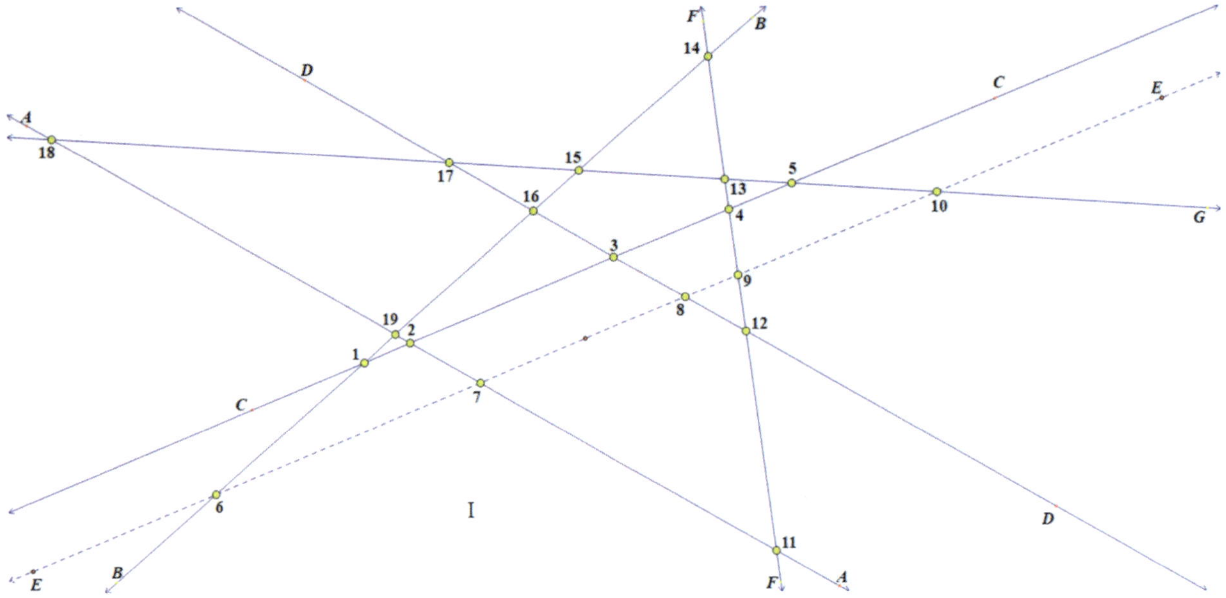

Figure 2.2

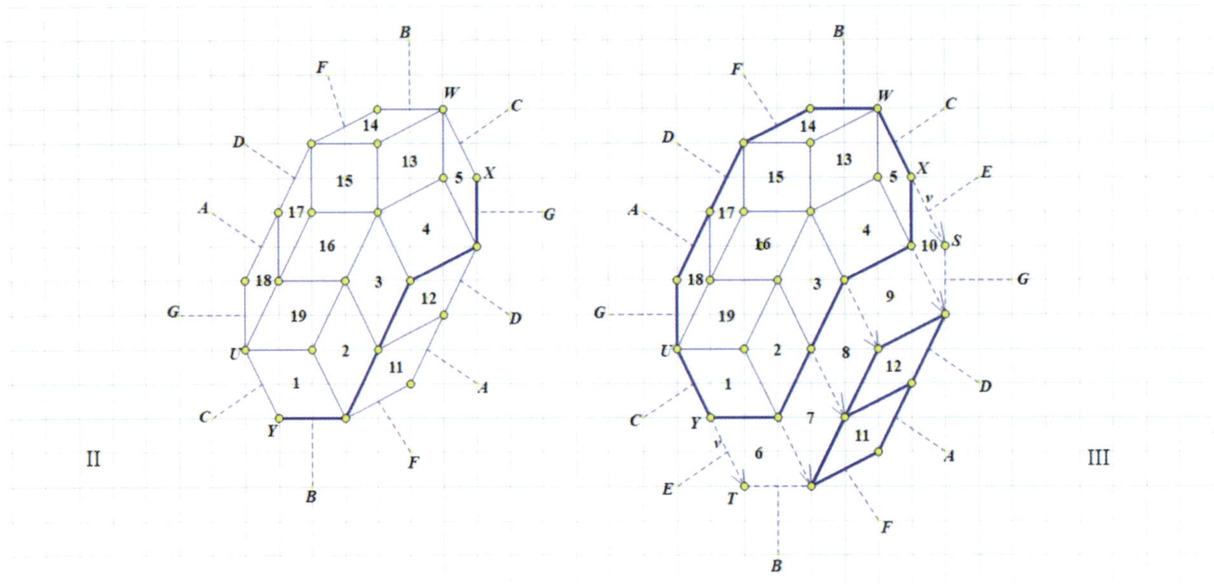

Figure 2.2

In Figure 2.2(II) the same tiled semiregular polygon P' as in Figure 2.1(III) is illustrated but this time an X,Y path which is *not* a diameter is used to separate P' into two parts. These two parts are outlined in bold in Figure 2.2(III). The opposite edges WX and UY of P are the boundary edges of stream C. Since line E is parallel to line C, the boundary edges XS, YT of stream E are chosen to form straight angles with WX and UY respectively, and thus stream E will be parallel to stream C. The separating path XY is chosen according to the locations of intersection points of line E with lines A, B, C, D, F, G.

By repeating the stream insertion operations just described a proof of the following result is obtained.

Theorem 2.1: *Let Λ= (V,L) denote a line configuration of degree 2. Then there exists a semiregular polygon P with a parallelogram tiling T, tile set R, and a sequence preserving bijection φ : V→R which maps linear sequences of points belonging to lines in L to sequences of tiles in corresponding tile streams of T.* □

In Figure 2.3 a degree 2 line configuration Λ is illustrated in which lines H, B are parallel. Positive integers are simply used to label intersection points, and the same set of integers is used in Figures 2.4(I - VI) to name tiles in the tile streams which correspond to lines in Λ. The construction begins by selecting three mutually nonparallel lines A, B, D of Λ and then designing a simple hexagon with streams A, B, D as in Figure 2.4(I). The three tiles are labeled 2, 8, 4 in accordance with the labels on the intersection points of lines A, B, D. Thus we obtain a semiregular polygon which represents the linear configuration consisting of the three lines. The corresponding integer labels indicate the sequence preserving bijection. (Note that if no three mutually nonparallel lines exist then Λ consists of only two parallel classes and thus can be represented by a parallelogram with a unque tiling by parallelograms.)

In Figures 2.4(I-V) the separating diameter is indicated in bold although in Figures 2.4(I,II,III) the separating path consists entirely of boundary edges and so does not really separate the polygon into two parts, but just indicates where the new stream is to be inserted. In 2.4(II) the stream corresponding to line G is inserted; in 2.4(III) the stream corresponding to line C is inserted; In 2.4(IV) the stream corresponding to line E is inserted; in 2.4(V) the stream corresponding to line F is inserted, and finally in 2.4(VI) the stream corresponding to line H is inserted. Since line H is parallel to line B, the separating path indicated in 2.4(V) is *not* a diameter. Each inserted stream is labeled according to the sequence of labeled intersection points in the corresponding line, and thus finally the entire linear configuration Λ is represented by a tiled semiregular polygon.

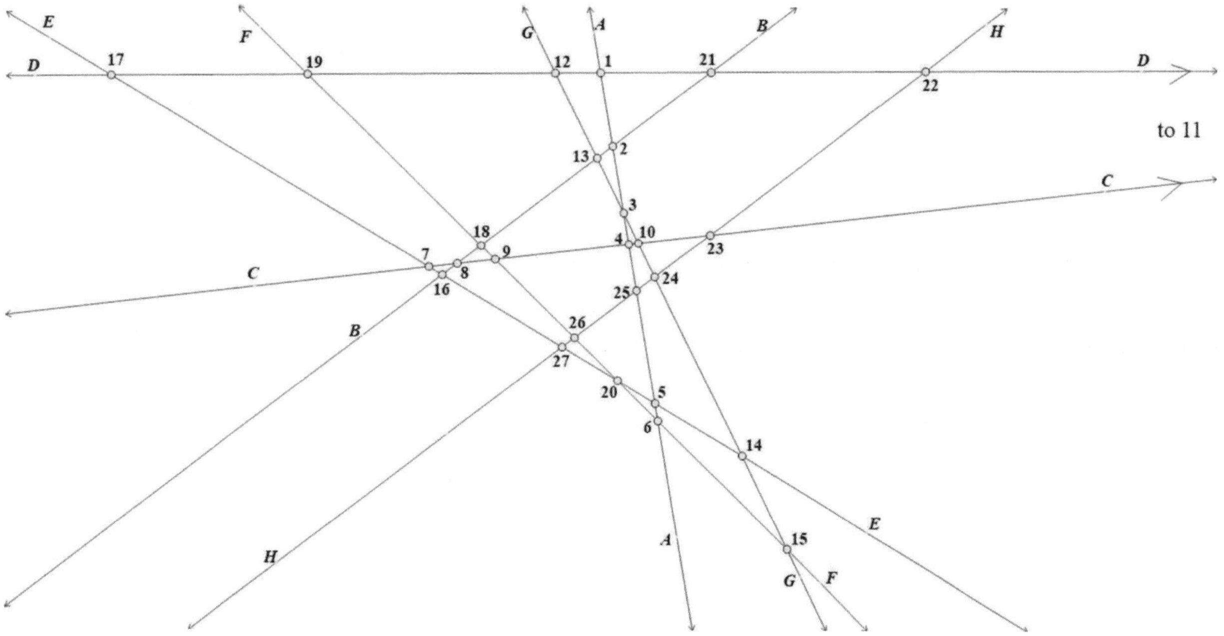

Figure 2.3

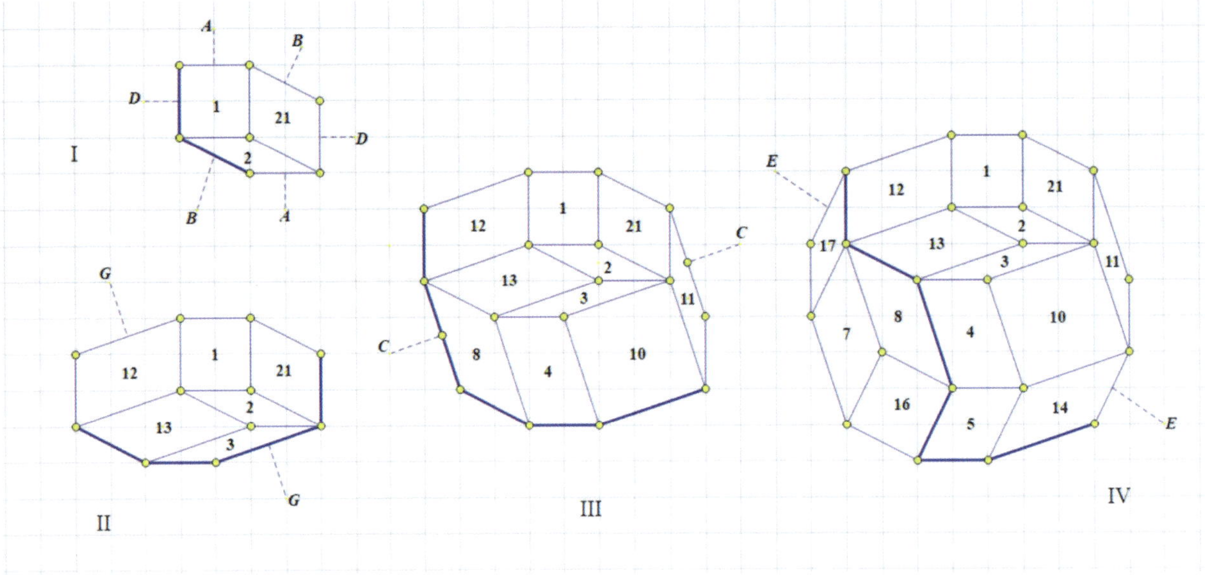

Figure 2.4

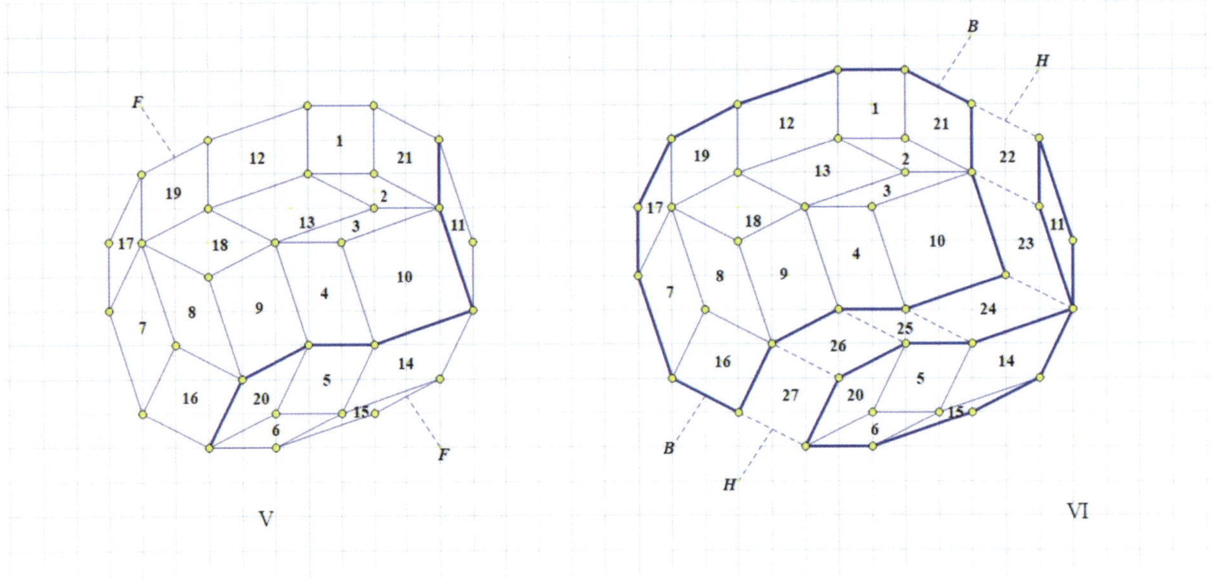

Figure 2.4

2.3 Tiling with Semiregular Polygons

Theorem 2.1 admits an interesting generalization which involves tiling a semiregular polygon with semiregular polygons. A *line configuration Λ of degree n* consists of lines and their intersection points with the requirement that there exists at least one point with exactly n lines passing through it, and all the remaining points have 2 up to n lines passing through them. In a corresponding semiregular polygon P parallelograms correspond to points of Λ which have exactly two lines through them; semiregular hexagons correspond to points of Λ with exactly three lines through them, and in general semiregular m-gons correspond to points of Λ with exactly m lines through them.

One such configuration Λ of degree 4 is depicted in Figure 2.5(I) with the corresponding tiled semiregular polygon P illustrated in Figure 2.5(II). Let T denote the tiling of P and G(T) the corresponding plane graph. The streams of G(T) are determined as in a parallelogram tiling: just start with a boundary edge and proceed from edge to opposite edge until an opposite boundary edge is obtained. The numbered tiles in Figure 2.5(II) correspond in order to the labeled points in each line of Figure 2.5(I), and thus we may say that the tiled semiregular polygon P represents Λ. This example has special significance since the labels on points of Λ are derived from the ancient 3 by 3 magic square called *Lo Shu* which is illustrated in Figure 2.5(III). We have subtracted 1 from each entry of the Lo Shu magic square in order to obtain a labeling for Λ by members of the group Z_9. Thus with these labels each line of Λ has the same sum 12, and each corresponding tile stream of G(T) has sum 12. Traditionally the two main diagonals as well as the rows and columns of a magic square must have the same sum, and this is reflected in the tiling in Figure 2.5(II). The tiled polygon of Figure 2.5(II) is thus a complete representation of the line configuration Λ.

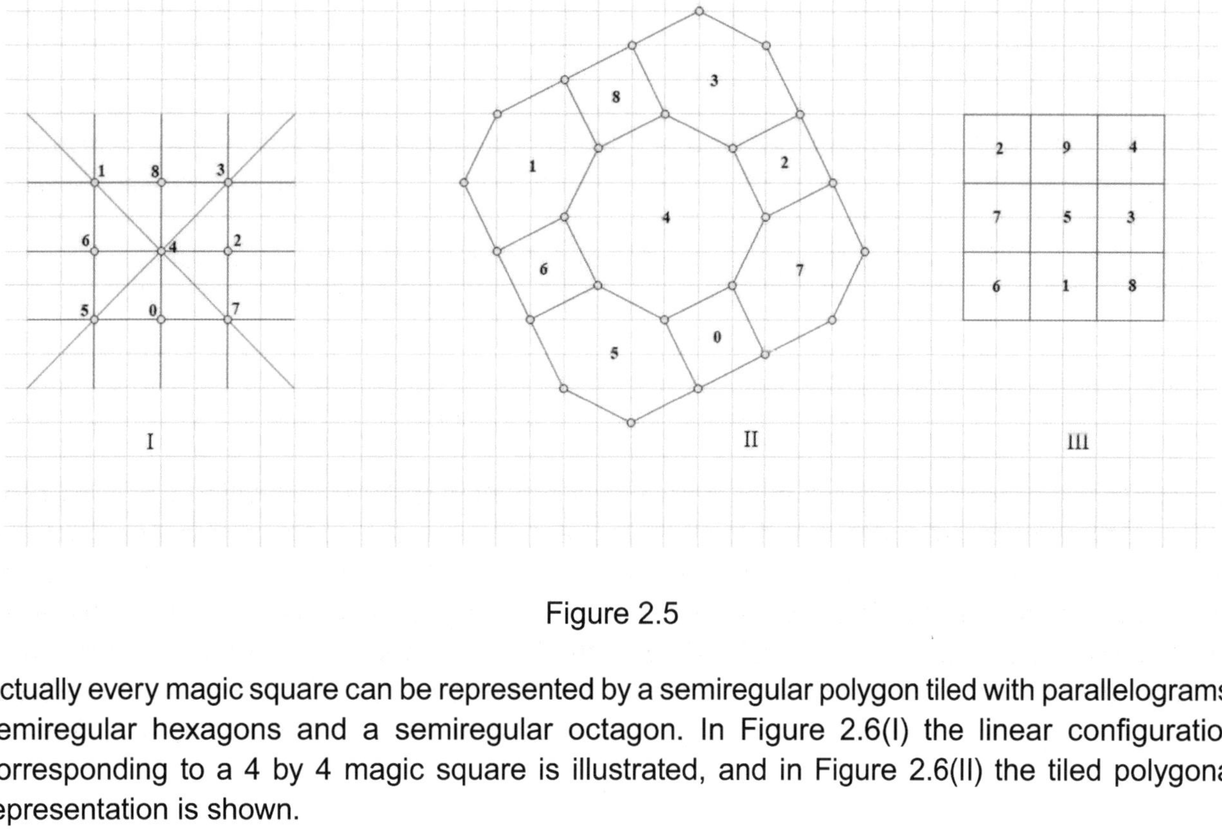

Figure 2.5

Actually every magic square can be represented by a semiregular polygon tiled with parallelograms, semiregular hexagons and a semiregular octagon. In Figure 2.6(I) the linear configuration corresponding to a 4 by 4 magic square is illustrated, and in Figure 2.6(II) the tiled polygonal representation is shown.

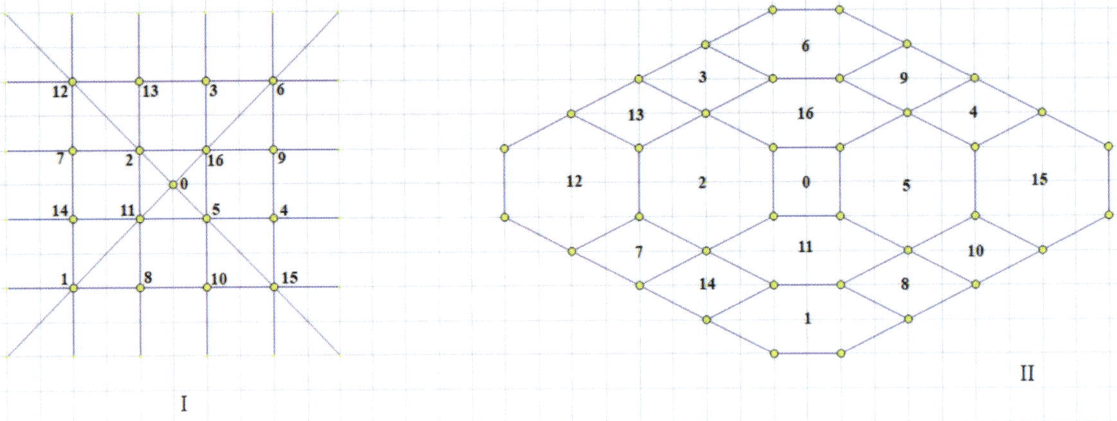

Figure 2.6

For magic squares of even order the diagonal lines of a linear representation do not meet at a point of the original lattice square. But luckily this point can be labeled with 0, thus also achieving our desired labeling with a group. The pattern for the polygonal representation of an even order magic square can easily be extended to any even order square as suggested in Figures 2.7(I, II).

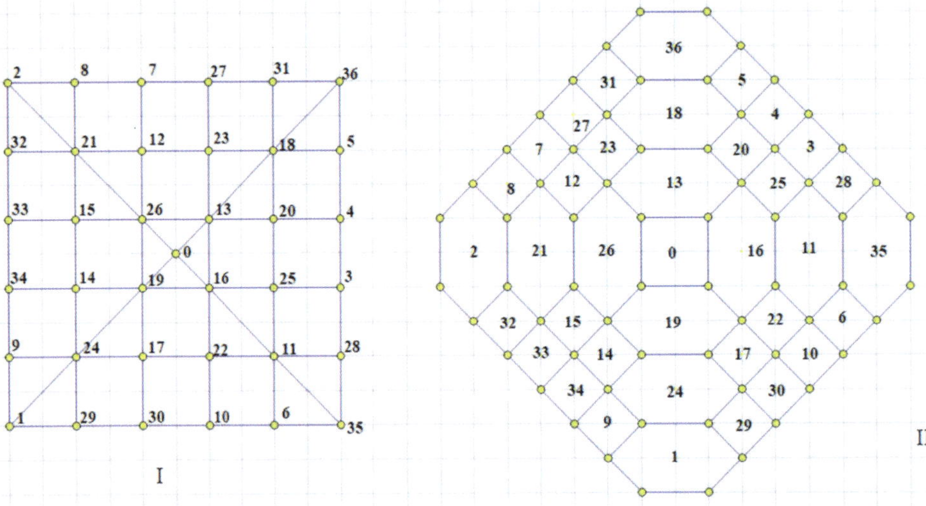

Figure 2.7

The pattern for the polygonal representation of odd order magic squares is also easy to see as in Figures 2.8(I,II).

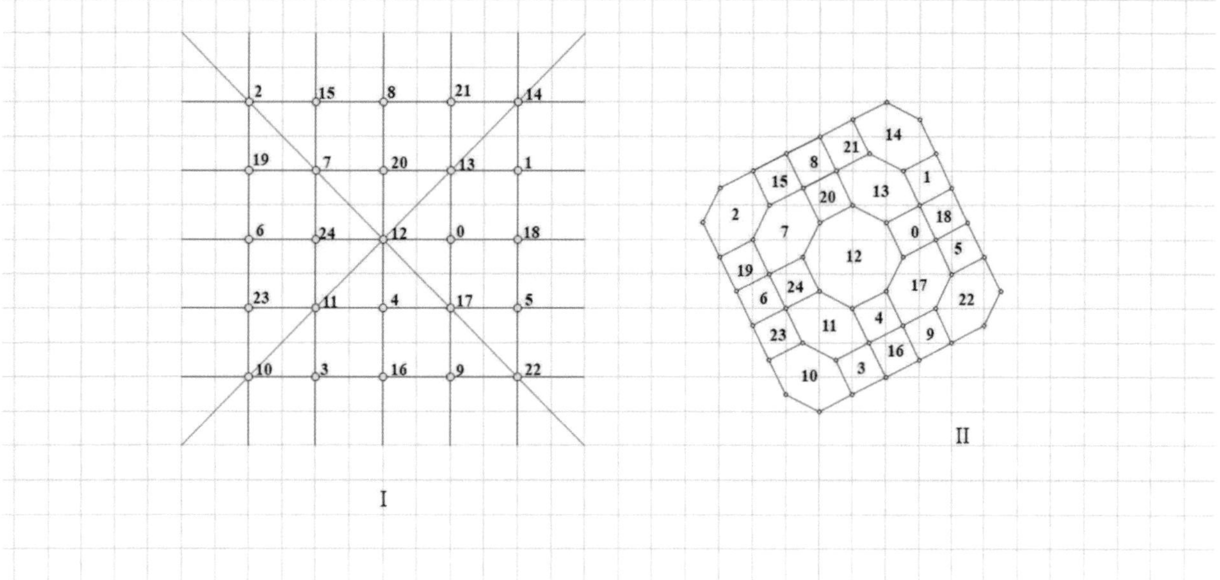

Figure 2.8

The tiled semiregular octagons in Figures 2.5(II), 2.6(II), 2.7(II) and 2.8(II), as well as the infinite class of magic square representations they suggest, furnish our first examples of *magic polygons.* They have bases consisting respectively of the abelian groups Z_9, Z_{17}, Z_{37}, Z_{25}. Since all streams have the same sum when summed as integers, (without regard to the modulus), we call these constructions *natural* . Theorem 2.1 has the following generalization.

Theorem 2.2: *Let Λ be a linear configuration of degree n. Then there exists a corresponding semiregular polygon P with a tiling T consisting of parallelograms and simple semiregular polygons with at most 2n sides, and such that there is a sequence preserving bijection between the lines of Λ and the tile streams of T.*

Proof. The lines of Λ can be *perturbed,* i.e., adjusted ever so slightly, so that no more than two lines meet at any given point, and otherwise the sequencing of intersection points of Λ is not altered. The resulting line configuration Λ' has degree two, and thus by Theorem 2.1 there exists a corresponding semiregular polygon P' with a parallelogram tiling T'. In place of a point of Λ with three incident lines, a simple semiregular hexagon will appear in T'; in place of a point of Λ which is incident with four lines, there will appear in T' a simple semiregular octagon, etc. The tiles in T' which constitute these various semiregular polygons are then removed leaving only the boundaries. The result is a tiling T corresponding to the original line configuration Λ using as prototiles simple semiregular polygons with up to 2n sides. □

In Figure 2.9(I) the line configuration Λ of Figure 2.5(I) has been perturbed so as to obtain a degree 2 line configuration Λ': points 1, 3, 5, 7 where three lines meet in Λ are replaced by the triples of points {1a, 1b, 1c}, {3a, 3b, 3c}, {5a, 5b, 5c}, {7a,7b,7c} respectively, and point 4 of Λ where four lines meet is replaced by the six points {4a, 4b, 4c, 4d, 4e, 4f}. Figure 2.9(II) shows the corresponding semiregular polygon P' with parallelogram tiling T'. The boundaries of the four hexagons and octagon which correspond to tiles 1,3,5,7,4 of Figure 2.5(II) are indicated in bold.

2.4 An Open Question

It is not clear how to prove the Converse of Theorem 2.1 although it appears to be true. We shall leave it as a Conjecture and provide an example.

Conjecture 2.3: *Let P be a semiregular polygon with a parallelogram tiling T. Then there exists a linear configuration Λ of degree 2 and a sequence preserving bijection between the tile streams of T and the lines of Λ.*

Consider the example in Figure 2.10 below in which the tile stream labels of the polygon in 2.10(I) have been extracted in 2.10(II) to form what might be called a *degree 2 system of open curves.* Can this system of curves be transformed into a degree 2 line configuration ? It always seems to be possible but often some extensive redrawing of lines is required. The corresponding line configuration in Figure 2.10(III)

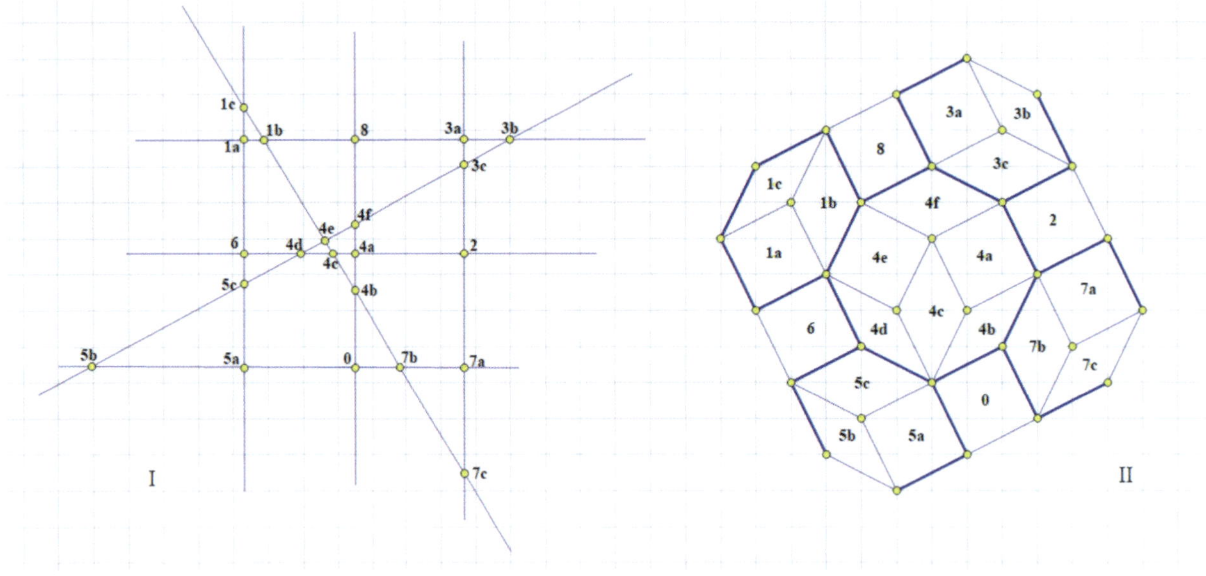

Figure 2.9

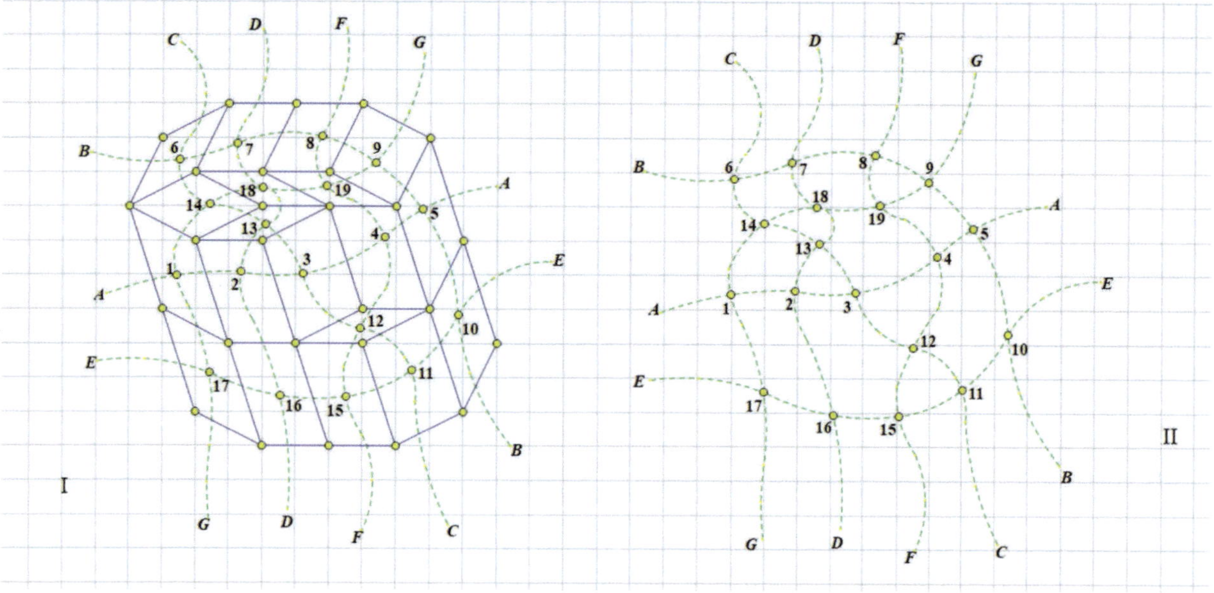

Figure 2.10

was found easily by first inserting the lines A,C,D which form a small triangular region.

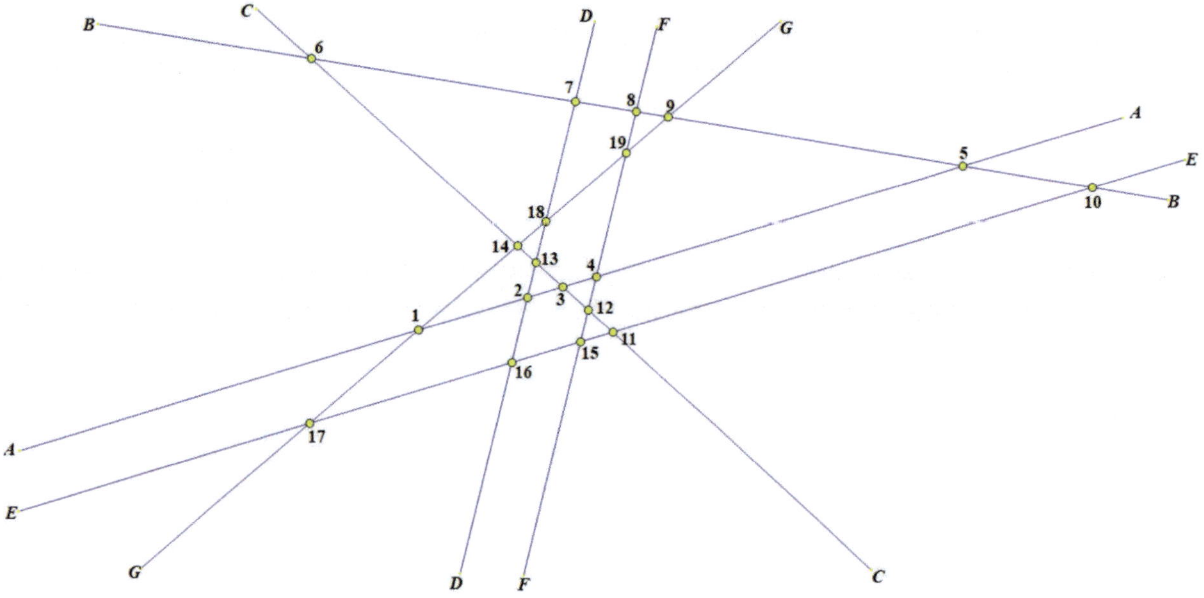

Figure 2.10(III)

CHAPTER 3

MAGIC SQUARES

3.1 Small Squares

Instead of numbers 1, …, n^2 which are usually used to construct a magic square of order n, we shall use instead the elements of various abelian groups. As far as the traditional magic squares are concerned, there is no loss since subtracting one from each cell converts to a square labeled with the elements of the group Z(n). As well, the constructive possibilities are greatly increased due to the variety of abelian groups available for a fixed square order. The ancient magic square "Lo Shu" is the smallest. It is illustrated in Figure 3.1(I) along with the corresponding square with base Z(9) in Figure 3.1(II).

2	9	4
7	5	3
6	1	8

I

1	8	3
6	4	2
5	0	7

II

3	1	5
8	6	4
7	2	0

III

0	2	1
2	1	0
1	0	2

IV

1	2	0
0	1	2
2	0	1

V

(0,1)	(2,2)	(1,0)
(2,0)	(1,1)	(0,2)
(1,2)	(0,0)	(2,1)

VI

Figure 3.1

Figure 3.1(II) illustrates a *natural* square, i.e., all rows columns and corner diagonals have the same sum (12) regardless of the modulus. We will always strive to obtain natural magic polygons as they make the closest analog to the traditional magic square. The square in Figure 3.1(III), which was obtained by adding 2(mod 9) to each cell in Figure 3.1(II), is *basic* but not natural: there are two sums namely 9, 18 if modulus is disregarded. But since $18 \equiv 9 \equiv 0 \pmod 9$ we can regard this square as a *basic magic square* with sum 0.

There are, up to isomorphism, two abelian groups of order 9, namely Z(9) and Z(3) ×Z(3) . A magic square with base Z(3) ×Z(3) can be constructed using two orthogonal Latin squares over Z(3) as in Figures 3.1(IV) and 3.1(V). These are actually *natural magic latin squares* since, without regard to modulus, each row, column and corner diagonal sums to 3. The direct product of these squares is a natural magic square over Z(3)×Z(3) illustrated in Figure 3.1(VI). Recall that two Latin squares over a set S with the same order are *orthogonal* in case their superimposition results in the formation of the full set of ordered pairs in S×S.

0	13	6	11
7	10	1	12
9	4	15	2
14	3	8	5

I

0	1	2	3
3	2	1	0
1	0	3	2
2	3	0	1

II

0	1	2	3
2	3	0	1
3	2	1	0
1	0	3	2

III

(0,0)	(1,1)	(2,2)	(3,3)
(3,2)	(2,3)	(1,0)	(0,1)
(1,3)	(0,2)	(3,1)	(2,0)
(2,1)	(3,0)	(0,3)	(1,2)

IV

2	7	1	4
4	1	7	2
5	0	6	3
3	6	0	5

V

0	1	0	1
0	1	0	1
1	0	1	0
1	0	1	0

VI

(0,2)	(1,7)	(0,1)	(1,4)
(0,4)	(1,1)	(0,7)	(1,2)
(1,5)	(0,0)	(1,6)	(0,3)
(1,3)	(0,6)	(1,0)	(0,5)

VII

Figure 3.2

We shall say that square M is *magic over an abelian group G* in case there exists an assignment of the full set of elements of G to the cells of M such that each row, column and corner diagonal has the same sum in G. If M has n^2 cells then M is *universally magic* in case it has a natural magic solution over every representation of an abelian group of order n^2. We have thus seen that the 3 by 3 square is universally magic.

The 4 by 4 square is also universally magic as we will now demonstrate. In Figure 3.2(I) is depicted a square which is magic over Z(16). It was derived from a magic square given in [6]. The Latin square in Figure 3.2(II) is obtained by applying the homomorphism ϕ:Z(16)⟶Z(4) defined by ϕ(x) = x(mod 4) to the elements in 3.2(I). Figure 3.2(III) is a Latin square orthogonal to 3.2(II), and in 3.2(IV) the superimposition of these squares is illustrated resulting in a magic square with base Z(4) × Z(4) . In Figures 3.2(V), 3.2(VI) orthogonal magic $(Z(8))^2$ and $(Z(2))^8$ squares are given, yielding in Figure 3.2(VII) a magic square with base Z(2)×Z(8). In Figures 3.3(I), 3.3(II) orthogonal magic $(Z(2)\times Z(2))^4$ and $(Z(4))^4$ squares are given, and *superimposed* in 3.3(III) to obtain a square with base Z(2)×Z(2)×Z(4). Finally in Figure 3.3(IV) a magic square with base Z(2)×Z(2)×Z(2)×Z(2) is illustrated.

2	3	0	1
1	0	3	2
3	2	1	0
0	1	2	3

I

(0,0)	(0,1)	(1,1)	(1,0)
(1,1)	(1,0)	(0,0)	(0,1)
(1,0)	(1,1)	(0,1)	(0,0)
(0,1)	(0,0)	(1,0)	(1,1)

II

(0,0,2)	(0,1,3)	(1,1,0)	(1,0,1)
(1,1,1)	(1,0,0)	(0,0,3)	(0,1,2)
(1,0,3)	(1,1,2)	(0,1,1)	(0,0,0)
(0,1,0)	(0,0,1)	(1,0,2)	(1,1,3)

III

(0,0,0,0)	(0,1,0,0)	(1,0,0,0)	(1,1,0,0)
(1,1,1,0)	(1,0,1,0)	(0,1,1,0)	(0,0,1,0)
(1,0,1,1)	(1,1,1,1)	(0,0,1,1)	(0,1,1,1)
(0,1,0,1)	(0,0,0,1)	(1,1,0,1)	(1,0,0,1)

IV

Figure 3.3

Note that the square in Figure 3.3 (IV) is basic but not natural since, for example, the first row has the natural sum (2,2,0,0) but the other rows and columns have sum (2,2,2,2). This defect is remedied if the reader will look ahead to Figure 3.26. We can thus conclude that the 4 by 4 square is universally magic.

3.2 Magic Orthogonal Latin Squares

Magic orthogonal Latin squares were used in Figures 3.2(I, II) to construct the Z(3)×Z(3) magic square in Figure 3.2(III), and also in Figures 3.3(II, III) to construct the Z(4)×Z(4) magic square in Figure 3.3(IV). The following general construction can be used to create Z(n)×Z(n) magic squares for odd values of n.

Theorem 3.1 *A natural magic square with base Z(n)×Z(n) can be constructed for every odd positive integer n≥3. (See Figure 3.4).*

Proof. It suffices to exhibit two orthogonal magic latin squares over Z(n). Let N denote the sequence (0, 1, 2, ..., n-1) consisting of the elements of Z(n) and let $\rho^t(N)$ denote a right circulation of N by t units. Let A_1, ..., A_n denote the rows of an n by n square A with $A_1 = \rho^{(n+1)/2}(N)$, $A_2 = \rho(A_1)$, $A_3 = \rho(A_2)$, ..., $A_n = \rho(A_{n-1})$. Then A is a *cyclic* latin square. The left - right corner diagonal of A consists of the constant value (n-1)/2 and the opposite corner diagonal consists of the full set of elements belonging to Z(n). As a consequence both diagonals sum to n(n-1)/2, which is also the sum of each row and column of A. Thus A is a magic latin square.

Let B denote the square obtained by rotating A 90° clockwise. Then, of course, B is latin, and since the 90° rotation has the effect of interchanging the corner diagonals, B is also magic. Finally, to see that A is orthogonal to B, observe that the broken diagonals of A are constant and the corresponding broken diagonals of B consist of the full set of elements of Z(n). A set of cells in a latin square, exactly one in each row and one in each column, whose entries are distinct is called a *transversal* of the latin square. Thus the broken diagonals of B are either transversals or constant, as are the broken diagonals of A, and the square obtained by superimposing A, B is magic with base Z(n) ×Z(n). □

3	4	5	6	0	1	2
2	3	4	5	6	0	1
1	2	3	4	5	6	0
0	1	2	3	4	5	6
6	0	1	2	3	4	5
5	6	0	1	2	3	4
4	5	6	0	1	2	3

A

4	5	6	0	1	2	3
5	6	0	1	2	3	4
6	0	1	2	3	4	5
0	1	2	3	4	5	6
1	2	3	4	5	6	0
2	3	4	5	6	0	1
3	4	5	6	0	1	2

B

Figure 3.4

In case n is even, magic orthogonal latin squares are harder to obtain. In 1960 Bose, Shrikande, and Parker, [3] constructed orthogonal Latin squares of order n for each n≡2(mod 4), n>6. In 1900 Tarry [4] showed that no Latin square of order 6 can have an orthogonal mate. Many other constructions of orthogonal Latin squares can be found in [5]. For now we will just exhibit the following example which involves two orthogonal Latin squares L_1, L_2 of order 8 found in [5] and which we have duplicated below.

a b c d e f g h	a b c d e f g h
b c d a f g h e	f e h g d c b a
c d a b g h e f	h g f e b a d c
d a b c h e f g	c d a b g h e f
e h g f a d c b	g f e h c b a d
f e h g b a d c	b c d a h e f g
g f e h c b a d	d a b c f g h e
h g f e d c b a	e h g f a d c b
L_1	L_2

3 1 5 0 6 2 7 4	3 1 5 0 6 2 7 4
1 5 0 3 2 7 4 6	2 6 4 7 0 5 1 3
5 0 3 1 7 4 6 2	4 7 2 6 1 3 0 5
0 3 1 5 4 6 2 7	5 0 3 1 7 4 6 2
6 4 7 2 3 0 5 1	7 2 6 4 5 1 3 0
2 6 4 7 1 3 0 5	1 5 0 3 4 6 2 7
7 2 6 4 5 1 3 0	0 3 1 5 2 7 4 6
4 7 2 6 0 5 1 3	6 4 7 2 3 0 5 1
L_3	L_4

By making the substitutions: a =3, b = 1, c = 5, d = 0, e = 6, f = 2, g = 7, h = 4 in L_1 and L_2, the orthogonal Latin squares L_3, L_4 are obtained, which are each magic over Z(8). We will let $L_3 \times L_4$ denote the square obtained by superimposing the squares L_3, L_4 . Then $L_3 \times L_4$ is a natural magic square with base Z(8)×Z(8).

3.3 Squares Constructed from equipartitions of Z(n)

If A is a matrix with entries from an abelian group G, we use the notation {A} to indicate the set, or multiset, consisting of the full set of entries in A. If the positive integer n = pq is composite, then an *equipartition of Z(n)* is a partition of Z(n) into p subsets each with q elements and each with sum q(pq-1)/2. Such a partition is useful in the construction of a number of magic squares. In what follows let m = pq-1.

Lemma 3.2 *If n = pq is a composite positive integer then (i) if p, q have the same parity then Z(n) has an equipartition into p subsets as well as an equipartition into q subsets. (ii) If one of p, q, (say p) is odd and q is even, then Z(n) has an equipartition into p subsets, but no equipartition into q subsets.*

Proof. Suppose first that at least one of p, q, (say q), is even and let A denote the 2 by pq/2 matrix:

A =

m	*m*-1	*m*-2	...	*m*-(*pq*/2 - 1)
0	1	2	...	*pq*/2 - 1

and let A = [A_1, A_2, ..., $A_{q/2}$] where the A_i are consecutive 2 by p blocks of A. Note that {A} = Z(pq) and thus for M = [A_1^T, A_2^T, ..., $A_{q/2}^T$] we also have {M} = Z(pq). The rows of M yield a partition of Z(pq) into p subsets each with q elements and each with sum mq/2 . Note that if p is odd then an equipartition into q subsets is not possible since mp/2 is not an integer.

Now suppose p, q are both odd, (p≥3, q≥5), and express q in the form q = 2s + 3. Let A denote the 2 by ps matrix:

$A =$

m	m-1	m-2	...	m - (ps-1)
0	1	2	...	ps-1

and express A in block form: A = [A_1, A_2, ..., A_s] where each A_i is a 2 by p matrix. Let matrices E, F be defined as below. Note that {A}∪{E}∪{F} = Z(pq) and (m/2)+p = m - ps - (p-1)/2. Also each row of E and F has sum 3m/2. Now define the matrix Q as in Figure 3.5. Then the rows of Q determine a partition of Z(pq) each with q points and each with sum mq/2.

$E =$

m-ps	ps	m/2
m-ps-1	ps+2	m/2 - 1
m-ps-2	ps+4	m/2 - 2
...	...	...
m/2 + p	ps+p-1	m/2 - (p-1)/2

$F =$

m/2+p-1	ps+1	m-ps-p
m/2+p-2	ps+3	m-ps-p-1
m/2+p-3	ps+5	m-ps-p-2
...	...	...
m-ps-p+1	ps+p-2	m/2 + 1

Q =

A_1^T	A_2^T	A_3^T	...	...	E
					F

Figure 3.5

The last case to consider is p, q odd with p≥3 and q = 3. To obtain an equipartition of Z(pq) into p subsets each with 3 elements, define matrices E, F as above with s = 0, and leaving out the matrices A_i, and let the matrx Q be defined as in Figure 3.6.

Q =

E
F

Figure 3.6

The rows of Q determine the desired equipartition. We shall refer to the equipartition constructed in this Lemma as the *natural equipartition of Z(n)* . □

Let S be a set with |S| = n ≥3. A *latin square over S* is a square matrix L(S) of order n in which each element of S occurs exactly once in each row and column. If S is a subset of an abelian group G, then ΣS is the sum of elements in each row and column. In case the two corner diagonals also have sum ΣS, then L(S) is a *magic latin square over S.* If only the diagonal which extends between the upper left and lower right corners has sum ΣS then L(S) is *left semimagic*, and if only the diagonal which extends between the upper right and lower left corners has sum ΣS, then L(S) is *right semimagic*. These types of latin squares will play an important role in the construction of magic squares with bases formed by direct products of cyclic groups.

If S, T are two sets with the same cardinality n then (i) the latin squares L(S), L(T) are *isotopes* in case there exists a bijection ϕ: S⟶T such that $L_{ij}(T) = \phi(L_{ij}(S))$ for each i,j∈{1, ..., n}, and (ii) L(S) and L(T) are *orthogonal*, denoted L(S)⊥L(T), in case for each element (s,t) of S×T there is exactly one set of indices i,j such that $L_{ij}(S)$ = s and $L_{ij}(T)$ = t. We will sometimes state this more simply by saying that when the matrices L(S), L(T) are *superimposed,* (and thus a pair (s,t) of elements is associated with each cell), then the entire set of elements in S×T appears. With these definitions in hand the following Theorem can be stated.

Theorem 3.3 *Let p, q be odd integers with p≥3 and q≥3. Then there exists a natural magic square of order pq with base* $Z(p)\times Z(pq^2)$.

Proof. By Lemma 3.2, $Z(pq^2)$ has the natural equipartition Π into q^2 subsets, each with p elements and each with sum $p(pq^2-1)/2$. Let S(t) denote the member of Π which contains t, and express q in the form q = 2k+1 for the appropriate positive integer k. Let m = $(pq^2-1)/2$ and let Π_0 denote the subcollection of Π defined by

Π_0 = {S(m-k), ..., S(m-1), S(m), S(m+1), ..., S(m+k)}.

The subsets in Π_0 are distinct and will be used to construct Latin squares of order p. Let A, B denote orthogonal magic Latin squares of order p, each with base Z(p), as constructed in the proof of Theorem 3.1.

For p = 5 the squares A, B are illustrated in Figure 3.7. Now for each of the q^2 subsets S∈Π construct a latin square L(S) which is isotopic to A. Construct a square matrix Q of order pq using {L(S): S∈Π} as blocks with q of these blocks in each row and with the main diagonal consisting of the latin squares {L(S): S∈Π_0}. This latter class of latin squares is constructed so that the elements m-k, ..., m-1, m, m+1, ..., m+k occur as the constant value of the main diagonals in S(m-k), ..., S(m-1), S(m), S(m+1), ..., S(m+k) respectively. For p = 5 and q = 3 the natural equipartition of Z(45) into q^2 subsets each with p elements and sum $p(pq^2-1)/2$ is given below Figure 3.7 and in Figure 3.8(I) the corresponding matrix Q is given.

2	3	4	0	1
1	2	3	4	0
0	1	2	3	4
4	0	1	2	3
3	4	0	1	2

A

3	4	0	1	2
4	0	1	2	3
0	1	2	3	4
1	2	3	4	0
2	3	4	0	1

B

Figure 3.7

44	0	35	9	22
43	1	34	11	21
42	2	33	13	20
41	3	32	15	19
40	4	31	17	18
39	5	30	10	26
38	6	29	12	25
37	7	28	14	24
36	8	27	16	23

Construct a square matrix M of order pq using q^2 copies of latin square B as in Figure 3.8(II). Since L(S) is isotopic to A for each S∈Π, and B⊥A, it follows that B⊥L(S) for each S∈Π. As a consequence M⊥Q and the square M×Q is magic over $Z(p)\times Z(pq^2)$. Since M, Q are naturally magic, so is M×Q. □

I

21	1	11	34	43	20	2	13	33	42	24	7	14	28	37
43	21	1	11	34	42	20	2	13	33	37	24	7	14	28
34	43	21	1	11	33	42	20	2	13	28	37	24	7	14
11	34	43	21	1	13	33	42	20	2	14	28	37	24	7
1	11	34	43	21	2	13	33	42	20	7	14	28	37	24
19	3	15	32	41	22	0	9	35	44	26	5	10	30	39
41	19	3	15	32	44	22	0	9	35	39	26	5	10	30
32	41	19	3	15	35	44	22	0	9	30	39	26	5	10
15	32	41	19	3	9	35	44	22	0	10	30	39	26	5
3	15	32	41	19	0	9	35	44	22	5	10	30	39	26
25	6	12	29	38	18	4	17	31	40	23	8	16	27	36
38	25	6	12	29	40	18	4	17	31	36	23	8	16	27
29	38	25	6	12	31	40	18	4	17	27	36	23	8	16
12	29	38	25	6	17	31	40	18	4	16	27	36	23	8
6	12	29	38	25	4	17	31	40	18	8	16	27	36	23

II

3	4	0	1	2	3	4	0	1	2	3	4	0	1	2
4	0	1	2	3	4	0	1	2	3	4	0	1	2	3
0	1	2	3	4	0	1	2	3	4	0	1	2	3	4
1	2	3	4	0	1	2	3	4	0	1	2	3	4	0
2	3	4	0	1	2	3	4	0	1	2	3	4	0	1
3	4	0	1	2	3	4	0	1	2	3	4	0	1	2
4	0	1	2	3	4	0	1	2	3	4	0	1	2	3
0	1	2	3	4	0	1	2	3	4	0	1	2	3	4
1	2	3	4	0	1	2	3	4	0	1	2	3	4	0
2	3	4	0	1	2	3	4	0	1	2	3	4	0	1
3	4	0	1	2	3	4	0	1	2	3	4	0	1	2
4	0	1	2	3	4	0	1	2	3	4	0	1	2	3
0	1	2	3	4	0	1	2	3	4	0	1	2	3	4
1	2	3	4	0	1	2	3	4	0	1	2	3	4	0
2	3	4	0	1	2	3	4	0	1	2	3	4	0	1

Figure 3.8

In case p is even and two magic orthogonal Latin squares of order p can be found then an order pq natural magic square with base $Z(p)\times Z(pq^2)$ can be constructed in similar fashion. As an example let p = 4 and q = 3 and let A, B denote the two magic orthogonal latin squares over Z(4) illustrated in Figure 3.9. The elements of $Z(pq^2)$ = Z(36) have a natural equipartition into 9 subsets each with 4 elements and each with sum 70 as given below Figure 3.9.

0	1	2	3
2	3	0	1
1	0	3	2
3	2	1	0

A

0	3	2	1
3	0	1	2
1	2	3	0
2	1	0	3

B

Figure 3.9

These sets are used to construct 9 order 4 Latin squares isotopic to A. These are arranged in three rows and three columns to form an order 12 square Q illustrated below.

35	0	26	9
34	1	25	10
33	2	24	11
32	3	23	12
31	4	22	13
30	5	21	14
29	6	20	15
28	7	19	16
27	8	18	17

These sets are used to construct 9 order 4 Latin squares isotopic to A. These are arranged in three rows and three columns to form an order 12 square Q illustrated below.

0	9	26	35
26	35	0	9
9	0	35	26
35	26	9	0

5	14	21	30
21	30	5	14
14	5	30	21
30	21	14	5

3	12	23	32
23	32	3	12
12	3	32	23
32	23	12	3

8	17	18	27
18	27	8	17
17	8	27	18
27	18	17	8

1	10	25	34
25	34	1	10
10	1	34	25
34	25	10	1

6	15	20	29
20	29	6	15
15	6	29	20
29	20	15	6

4	13	22	31
22	31	4	13
13	4	31	22
31	22	13	4

7	16	19	28
19	28	7	16
16	7	28	19
28	19	16	7

2	11	24	33
24	33	2	11
11	2	33	24
33	24	11	2

A second square matrix M of order 12 is constructed with 9 copies of the magic Latin square B. Since B⊥A, we also have that B is orthogonal to every isotope of A. It then follows that M⊥Q and consequently M×Q is a magic square of order 12 with base Z(4)×Z(36). Since M, Q are natural squares, so is M×Q.

3.4 Magic squares with base $G_1 \times G_2$

If M is a square matrix then we will use the notation J(t,M) to denote a t by t block matrix in which each block is identically the matrix M.

Theorem 3.4 *Let M_1, M_2 be magic squares of order t with bases the abelian groups G_1, G_2 respectively. Then there exists a magic square with base $G_1 \times G_2$.*

Proof. Replace each value x in M_1 using the substitution x⟶J(t, x) and let A denote the resulting magic square of order t^2. Let B = J(t, M_2). The square A is magic over the multiple group $(G_1)^t$ and B is magic over the multiple group $(G_2)^t$. The square A×B has order t^2 and is magic over $G_1 \times G_2$. □

An example of this construction is given in Figure 3.10 where the squares M_1, M_2 are two magic squares over Z(9) and Z(3)×Z(3) respectively.

1	8	3
6	4	2
5	0	7

M_1

(1,0)	(2,2)	(0,1)
(0,2)	(1,1)	(2,0)
(2,1)	(0,0)	(1,2)

M_2

1	1	1	8	8	8	3	3	3
1	1	1	8	8	8	3	3	3
1	1	1	8	8	8	3	3	3
6	6	6	4	4	4	2	2	2
6	6	6	4	4	4	2	2	2
6	6	6	4	4	4	2	2	2
5	5	5	0	0	0	7	7	7
5	5	5	0	0	0	7	7	7
5	5	5	0	0	0	7	7	7

A

M_2	M_2	M_2
M_2	M_2	M_2
M_2	M_2	M_2

B

Figure 3.10

Note that Theorem 3.4 can be used to create natural magic squares of order 9 with bases Z(9)×Z(9), Z(9)×Z(3)×Z(3) and Z(3)×Z(3)×Z(3)×Z(3). Theorem 3.3 can be used to construct a natural square with base Z(3)×Z(27) and several examples of order 9 squares with base Z(81) can be constructed based on those found in [6]. Thus we can conclude that the square of order 9 is naturally universally magic.

3.5 Magic Squares with Base Z(n)×Z(n), n even

In Theorem 3.1 Magic squares with base Z(n)×Z(n), n odd were constructed. Now we will consider the case when n is even; in particular when n is a multiple of 4. A *special* Latin square over an abelian group G is one in which both corner diagonals, (as well as all rows and columns), contain the full set of elements of G.

Theorem 3.5 *If a pair of special orthogonal Latin squares exist of even order s, then a magic square with base Z(st)×Z(st) can be constructed for every odd integer t ≥3.*

Proof. Let L_1 and L_2 be special orthogonal Latin squares of even order s over Z(s) as in Figure 3.11, and let L_3, L_4 be orthogonal magic Latin squares of odd order t, (t≥3), over Z(t). Note that L_3, L_4 are provided by Theorem 3.1. By Lemma 3.2 there exists a natural equipartition of Z(st) into t subsets $\{S_0, \ldots, S_{t-1}\}$ each with s elements, an each with sum (s/2)(st-1). For each $i \in \{0, \ldots, t-1\}$ let $L_1(S_i)$ be an isotope of L_1 with elements from S_i, Let A denote the square of order st obtained by replacing each value i, (from Z(t)), in L_3 according to the rule $i \longrightarrow L_1(S_i)$. For each $i \in \{0, \ldots, t-1\}$ let $L_2(S_i)$ be an isotope of L_2 with elements from S_i, and let B denote the square of order st obtained by replacing each value i in L_4 by the rule $i \longrightarrow L_2(S_i)$. Each of the Latin squares $L_1(S_i)$, $L_2(S_i)$ are magic due to the fact that L_1, L_2 are special. The squares A, B are each magic Latin squares over Z(st) since they are composed of the order s Latin magic squares $L_1(S_i)$, $L_2(S_i)$. Also A⊥B due to the orthogonality of $L_1(S_i)$, $L_2(S_i)$ for each i. As a consequence A×B is a magic square with base Z(st)×Z(st). □

Note that the proof of Theorem 3.5 provides, in the squares A, B orthogonal Latin magic squares of any order which is a multiple of s. In Figure 3.11 an example of the construction of Theorem 3.5 is given with s = 4 and t = 3. The equipartition of Z(12) is given by S_0 = {0,11,1,10}, S_1={4,7,5,6}, S_2 = {2,9,8,3}. In Theorem 3.5, any pair of even order special orthogonal Latin squares could be used for L_1, L_2 and t need not be odd. For example by taking L_1, L_2 as in Figure 3.11, and then taking $L_3 = L_1$, and $L_4 = L_2$ two special orthogonal Latin squares A, B of order 16 are obtained. The equipartition of Z(16) and the squares A, B are illustrated in Figure 3.12. Using these squares in place of L_1, L_2 in Theorem 3.5 makes possible the construction of magic squares with base Z(16t)×Z(16t) for any odd positive integer t.

The construction of special Latin squares is an interesting problem. Besides the even order ones which can be constructed as in Figure 3.12, two pairs of orthogonal special Latin squares of order n = 5 and n = 7 are illustrated in Figure 3.13. Note that in both constructions the main corner diagonal contains the elements in numerical order, and the opposite corner diagonal is chosen so that opposite elements have the same sum (n). The orthogonal mate is obtained by simply taking the transpose. We call a Latin square which is orthogonal to its transpose, *reflective.*

L_1

0	1	2	3
2	3	0	1
3	2	1	0
1	0	3	2

L_2

0	1	2	3
3	2	1	0
1	0	3	2
2	3	0	1

L_3

0	2	1
2	1	0
1	0	2

L_4

1	0	2
2	1	0
0	2	1

$L_1(S_0)$

0	1	10	11
10	11	0	1
11	10	1	0
1	0	11	10

$L_1(S_1)$

4	5	6	7
6	7	4	5
7	6	5	4
5	4	7	6

$L_1(S_2)$

2	3	8	9
8	9	2	3
9	8	3	2
3	2	9	8

$L_2(S_0)$

0	1	10	11
11	10	1	0
1	0	11	10
10	11	0	1

$L_2(S_1)$

4	5	6	7
7	6	5	4
5	4	7	6
6	7	4	5

$L_2(S_2)$

2	3	8	9
9	8	3	2
3	2	9	8
8	9	2	3

A

$L_1(S_0)$	$L_1(S_2)$	$L_1(S_1)$
$L_1(S_2)$	$L_1(S_1)$	$L_1(S_0)$
$L_1(S_1)$	$L_1(S_0)$	$L_1(S_2)$

B

$L_2(S_1)$	$L_2(S_0)$	$L_2(S_2)$
$L_2(S_2)$	$L_2(S_1)$	$L_2(S_0)$
$L_2(S_0)$	$L_2(S_2)$	$L_2(S_1)$

Figure 3.11

S_0 = {0, 15, 1, 14}
S_1 = {2, 13, 3, 12}
S_2 = {4, 11, 5 ,10}
S_3 = {6, 9, 7, 8}

$L_1(S_0)$	$L_1(S_2)$	$L_1(S_3)$	$L_1(S_1)$
$L_1(S_3)$	$L_1(S_1)$	$L_1(S_0)$	$L_1(S_2)$
$L_1(S_1)$	$L_1(S_3)$	$L_1(S_2)$	$L_1(S_0)$
$L_1(S_2)$	$L_1(S_0)$	$L_1(S_1)$	$L_1(S_3)$

A

$L_2(S_0)$	$L_2(S_2)$	$L_2(S_3)$	$L_2(S_1)$
$L_2(S_1)$	$L_2(S_3)$	$L_2(S_2)$	$L_2(S_0)$
$L_2(S_2)$	$L_2(S_0)$	$L_2(S_1)$	$L_2(S_3)$
$L_2(S_3)$	$L_2(S_1)$	$L_2(S_0)$	$L_2(S_2)$

B

Figure 3.12

0	3	4	2	1
2	1	3	4	0
1	4	2	0	3
4	0	1	3	2
3	2	0	1	4

0	6	3	2	1	4	5
4	1	0	6	5	2	3
5	3	2	1	0	6	4
6	5	4	3	2	1	0
2	0	6	5	4	3	1
3	4	1	0	6	5	2
1	2	5	4	3	0	6

0	4	5	6	2	3	1
6	1	3	5	0	4	2
3	0	2	4	6	1	5
2	6	1	3	5	0	4
1	5	0	2	4	6	3
4	2	6	1	3	5	0
5	3	4	0	1	2	6

0	2	1	4	3
3	1	4	0	2
4	3	2	1	0
2	4	0	3	1
1	0	3	2	4

Figure 3.13

3.6 Constructions Using Semilatin Squares

An n by n square M is *row semilatin over Z(n)* in case (i) each of its rows contains the full set of elements of Z(n); (ii) the n^2 cells of M contain exactly n of each of the n values of Z(n); (iii) All rows and columns of M have the same sum namely n((n-1)/2). A *column semilatin square* is defined similarly except that “rows” in (i) is replaced by “columns”. If M is row semilatin them M^T is column semilatin. We will be especially interested in the case in which M is special and M is reflective, i.e., $M \perp M^T$. Then $M \times M^T$ will yield an n by n magic square with base $Z(n) \times Z(n)$. Some examples of special reflective row semilatin squares of even order are provided in Figure 3.14. These examples have the additional properties: (i) values which are reflections of each other across the vertical axis of symmetry all have sum n-1, and (ii) each column consists of n/2 equal pairs which sum to n-1. The 4 by 4 and the 8 by 8 square in Figure 3.14 suggest a recursive method for constructing magic squares with base $Z(2^n) \times Z(2^n)$.

Let $m = 2^n$ and let $k = m/4$. If M is any matrix whose cells are populated with elements from Z(n), let M^R denote the matrix obtained from M by reversing the order of each of its rows, and let k + M denote the matrix obtained by adding the group element k to each cell of M. Let B(k) denote the square of order k over Z(k) in which each row consists of the sequence (0, 1, …, k-1). Let D(k) = k + B(k); E(k) = 2k + B(k) and C(k) = 3k + B(k). Let $A(2^2)$ denote the 4 by 4 square given in Figure 3.14 and for n>2 define $A(2^n) = A(m)$ recursively as in Figure 3.15.

0	2	1	3
3	1	2	0
3	1	2	0
0	2	1	3

0	1	2	3	4	5	6	7
0	1	2	3	4	5	6	7
7	6	2	4	3	5	1	0
7	6	5	3	4	2	1	0
7	6	5	3	4	2	1	0
7	6	2	4	3	5	1	0
0	1	5	4	3	2	6	7
0	1	5	4	3	2	6	7

0	8	7	6	5	4	3	2	1	9
9	1	7	6	5	4	3	2	8	0
9	8	2	6	4	5	3	7	1	0
0	8	2	3	5	4	6	7	1	9
9	1	2	3	4	5	6	7	8	0
0	1	7	3	4	5	6	2	8	9
0	8	7	3	5	4	6	2	1	9
9	8	2	6	4	5	3	7	1	0
9	1	2	6	4	5	3	7	8	0
0	1	7	3	5	4	6	2	8	9

0	4	3	2	1	5
5	1	3	2	4	0
5	4	2	3	1	0
0	4	2	3	1	5
5	1	2	3	4	0
0	1	3	2	4	5

Figure 3.14

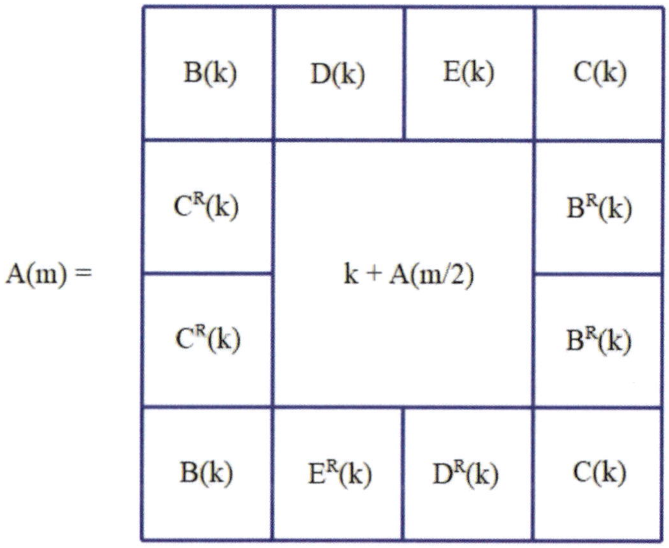

Figure 3.15

It is easily seen that A(m) is row semilatin and special. To show that A(m) is reflexive we must compute $A(m) \times A^T(m)$. We proceed by induction on n. The basis n = 2, ($m = 2^2$), is provided by the 4 by 4 square in Figure 3.14. We assume inductively that the square A(m/2) is reflexive and thus the square k + A(m/2), (k = m/4) is reflexive. As a consequence:

$$(k+A(m/2)) \times (k+A(m/2))^T = \{(k+Z(k)) \cup (2k+Z(k)\} \times \{(k+Z(k)) \cup (2k+Z(k)\}$$

From Figure 3.15 it can be seen that

$$A(m)\times A^T(m) = (B(k)\times B^T(k))\cup(D(k)\times C^{RT}(k))\cup (C^{RT}(k)\times D(k))\cup$$
$$(E(k)\times C^{RT}(k))\cup(C^{RT}(k)\times E(k)\cup(C(k)\times B^T(k))\cup(B^T(k)\times C(k))\cup$$
$$(B^R(k)\times E^{RT}(k))\cup(E^{RT}(k)\times B^R(k)) \cup(B^R(k)\times D^{RT}(k))\cup(D^{RT}(k)\times B^R(k))\cup$$
$$(C(k)\times C^T(k)) \cup (k+A(m/2))\times(k+A(m/2))^T.$$

Now consider

$$B(k)\times B^T(k) = Z(k)\times Z(k);\ D(k)\times C^{RT}(k)= (k+Z(k))\times(3k+Z(k));$$
$$C^{RT}(k)\times D(k) = (3k+Z(k))\times(k+Z(k));$$
$$E(k)\times C^{RT}(k) = (2k+Z(k))\times(3k+Z(k));\ (C^{RT}(k)\times E(k)) = (3k+Z(k))\times(2k+Z(k))$$
$$C(k)\times B^T(k) = (3k+Z(k))\times Z(k);\ (B^T(k)\times C(k) = Z(k)\times(3k+Z(k));$$
$$B^R(k)\times E^{RT}(k) = Z(k)\times(2k+Z(k));\ E^{RT}(k)\times B^R(k) = (2k+Z(k))\times Z(k)$$
$$B^R(k)\times D^{RT}(k) = Z(k)\times(k+Z(k));\ D^{RT}(k)\times B^R(k) = (k+Z(k))\times Z(k)$$
$$C(k)\times C^T(k) = (3k+Z(k))\times(3k+Z(k));$$
$$(k+A(m/2))\times(k+A(m/2)^T) = \{(k+Z(k))\cup(2k+Z(k)\}\times \{(k+Z(k))\cup(2k+Z(k)\}.$$

From the above series of equations it is seen that $A(m)\times A^T(m) = Z(m)\times Z(m)$ and thus we have proved the following.

Theorem 3.6 *For each positive integer $n\geq 2$ there exists a magic square M of order 2^n with base $Z(2^n)\times Z(2^n)$. Moreover there exists a special reflexive row semilatin square A such that $A\times A^T = M$.* □

In Figure 3.16 the square $A(2^4)$ is illustrated. Combining Theorems 3.5, 3.6 yields the following Corollary.

Corollary 3.7 *If n is a multiple of 4, then a magic square with base $Z(n)\times Z(n)$ can be constructed.*

Proof. Express n in the form $n = 2^m t$ where $m\geq 2$ and t is odd. By Theorem 3.6 there exists a special reflective row semilatin square A of order 2^m. Then $A\times A^T$

0	1	2	3	4	5	6	7	8	9	10	11	12	13	14	15
0	1	2	3	4	5	6	7	8	9	10	11	12	13	14	15
0	1	2	3	4	5	6	7	8	9	10	11	12	13	14	15
0	1	2	3	4	5	6	7	8	9	10	11	12	13	14	15
15	14	13	12	4	5	6	7	8	9	10	11	3	2	1	0
15	14	13	12	4	5	6	7	8	9	10	11	3	2	1	0
15	14	13	12	11	10	6	8	7	9	5	4	3	2	1	0
15	14	13	12	11	10	9	7	8	6	5	4	3	2	1	0
15	14	13	12	11	10	9	7	8	6	5	4	3	2	1	0
15	14	13	12	11	10	6	8	7	9	5	4	3	2	1	0
15	14	13	12	4	5	9	8	7	6	10	11	3	2	1	0
15	14	13	12	4	5	9	8	7	6	10	11	3	2	1	0
0	1	2	3	11	10	9	8	7	6	5	4	12	13	14	15
0	1	2	3	11	10	9	8	7	6	5	4	12	13	14	15
0	1	2	3	11	10	9	8	7	6	5	4	12	13	14	15
0	1	2	3	11	10	9	8	7	6	5	4	12	13	14	15

Figure 3.16

yields a magic square with base $Z(2^m)\times Z(2^m)$, so the Corollary is proved in case t = 1. If $t\geq3$ then in Theorem 3.5 take $L_1 = A$ and $L_2 = A^T$ to obtain a magic square with base $Z(n)\times Z(n)$. □

In Light of Theorem 3.1 and Corollary 3.7, a magic square with base $Z(n)\times Z(n)$ can be constructed whenever n is odd, ($n\geq3$), and whenever n is a multiple of 4. It remains only to consider the case in which n is twice an odd number. A general procedure for constructing such squares is not clear. In Figure 3.17 we provide two special reflective row semilatin squares A, B of orders 6, 10 respectively. Then $A\times A^T$ and $B\times B^T$ will yield magic squares with bases $Z(6)\times Z(6)$ and $Z(10)\times Z(10)$.

3.7 Special Reflective Latin Squares of Odd Order

A well known standard procedure can be used to construct special reflective Latin squares of odd order. Let M be a square of odd order $n\geq5$ and populate the cells of M with the elements of Z(n) according to the following procedure. Place 0 in the middle cell of the top row and use a sequence of *knight moves*, (over two squares and down 1) to place the remaining n-1 zeros. If a boundary on the right side of M is encountered then count two squares over as if M was scrolled into a vertical cylinder thus jumping from the right to the left side of M. If the bottom boundary of M is encountered then make the one square knight move from bottom to top as if M was scrolled into a horizontal cylinder. This procedure was used to construct the Latin squares of order 5, 7 and 11 in Figure 3.18.

0	4	3	2	1	5
5	1	3	2	4	0
5	4	2	3	1	0
0	4	2	3	1	5
5	1	2	3	4	0
0	1	3	2	4	5

A

0	8	7	6	5	4	3	2	1	9
9	1	7	6	5	4	3	2	8	0
9	8	2	6	4	5	3	7	1	0
0	8	2	3	5	4	6	7	1	9
9	1	2	3	4	5	6	7	8	0
0	1	7	3	4	5	6	2	8	9
0	8	7	3	5	4	6	2	1	9
9	8	2	6	4	5	3	7	1	0
9	1	2	6	4	5	3	7	8	0
0	1	7	3	5	4	6	2	8	9

B

Figure 3.17

3	4	0	1	2
1	2	3	4	0
4	0	1	2	3
2	3	4	0	1
0	1	2	3	4

4	5	6	0	1	2	3
2	3	4	5	6	0	1
0	1	2	3	4	5	6
5	6	0	1	2	3	4
3	4	5	6	0	1	2
1	2	3	4	5	6	0
6	0	1	2	3	4	5

6	7	8	9	10	0	1	2	3	4	5
4	5	6	7	8	9	10	0	1	2	3
2	3	4	5	6	7	8	9	10	0	1
0	1	2	3	4	5	6	7	8	9	10
9	10	0	1	2	3	4	5	6	7	8
7	8	9	10	0	1	2	3	4	5	6
5	6	7	8	9	10	0	1	2	3	4
3	4	5	6	7	8	9	10	0	1	2
1	2	3	4	5	6	7	8	9	10	0
10	0	1	2	3	4	5	6	7	8	9
8	9	10	0	1	2	3	4	5	6	7

Figure 3.18

Note that the *knight move construction* is completed by placing the values in numerical sequence, (cyclically) in each row. The following Theorem tells precisely when the knight move construction yields a special reflexive Latin square.

Theorem 3.8 *The knight move construction yields a special reflexive Latin square over Z(n) iff $n \geq 5$ is odd and not a multiple of 3.*

Proof. If $2|n$ then after n/2 knight moves a second occurrence of the same value results in the same column, and thus a Latin square is not obtained. If n = 3 then the knight move construction results in a square over Z(3) which is symmetric with respect to the main corner diagonal, hence nonreflective. So suppose n is odd and larger than or equal to 5. Suppose $x \in Z(n)$ lies in the top row of an n by n square matrix M. Then x will be placed in every row since a knight move goes down only one square. Since gcd(n,2)=1 the knight moves will place an x in every column as well. No more than one x will be placed in each column since Z(n) is generated by 1 and no more than one x will be placed in each row since Z(n) is generated by 2. Thus the knight move construction yields a Latin square.

If x lies in the upper left corner of M, then the successive values in the main corner diagonal are obtained by repeatedly subtracting 1. This results in a diagonal which contains every value of Z(n). The successive values in the opposite corner diagonal are obtained by repeatedly adding 3 (mod n). Since gcd(3,n) = 1 by stipulation in the Theorem statement, Z(n) is generated by 3 and thus all values of Z(n) will occur in this diagonal as well. As a consequence M is a special Latin square. If 3|n then repeated value will occur in this diagonal and so M would not be special. M is in fact, *pandiagonal* since even the broken diagonals contain all the elements of Z(n).

To show that M is reflexive we must show that the reflection of each constant cell set is a transversal. The reflection of the two over and one down knight move is a one over and two down knight move, so it suffices to show that every cell set constructed by a series of one over and two down knight moves, is a transversal. If x is the value in any cell of M, then the effect of moving over one and down 2 results in a cell which contains the value x-3. Again, since gcd(n,3) = 1, repeated subtraction of 3 produces every element of Z(n). □

Theorem 3.8 provides a second method along with Theorem 3.1 for constructing magic squares with base Z(n)×Z(n) for n odd and n≥5.

3.8 Universally Magic Squares

Recall that a square of order n is universally magic in case it is naturally magic over every representation of an abelian group of order n^2. In Section 3.1 it was shown that the squares of order 3 and 4 are universally magic. A magic square over Z(n) is known to exist for every integer n≥3. Various constructions of these can be found in [7]. There are only two groups of order 25, namely, Z(25) and Z(5)×Z(5). By Theorem 3.1 a natural magic square with base Z(5)×Z(5) exists, and thus the 5 by 5 square is universally magic. Similarly the 7 by 7 square is universally magic.

Now consider a square of order 6. A Z(6)×Z(6) square is obtained via the special reflective row semilatin square of order 6 illustrated in Figure 3.17. A Z(2)×Z(18) square is obtained via the orthogonal $(Z(18))^2$ and $(Z(2))^{18}$ squares illustrated in Figure 3.19 (I, II), and a Z(3)×Z(12) square is obtained via the orthogonal $(Z(12))^3$ and $(Z(3))^{12}$ squares given in Figure 3.19(III, IV).

Note that the square in Figure 3.19(II) is not natural since one of the diagonals has the natural sum 4, and the rows and columns all have sum 3. Thus our Z(2)×Z(18) construction is not natural. Also, although Z(4)×Z(9) is isomorphic to Z(36) and the existence of the Z(36) square implies the existence of a Z(4)×Z(9) *basic* square via the isomorphism, the natural character of the Z(36) square *does not* imply that the corresponding Z(4)×Z(9) square is natural, only that it is basic. To show that the 6 by 6 square is universally magic, natural solutions with bases Z(2)×Z(18), Z(2)×Z(2)×Z(9), Z(2)Z(2)×Z(3)×Z(3), etc., would need to be constructed.

I

0	4	14	15	13	5
11	7	9	8	10	6
17	4	14	15	1	0
5	13	2	3	16	12
6	7	9	8	10	11
12	16	3	2	1	17

III

5	2	10	1	11	4
10	5	2	11	4	1
2	10	5	4	1	11
9	0	7	6	8	3
0	7	9	3	6	8
7	9	0	8	3	6

II

0	1	0	1	1	0
0	1	0	1	1	0
0	0	1	0	1	1
1	0	1	1	0	0
1	0	1	0	0	1
1	1	0	0	0	1

IV

0	2	1	1	0	2
2	1	0	2	1	0
1	0	2	0	2	1
1	0	2	0	2	1
2	1	0	2	1	0
0	2	1	1	0	2

Figure 3.19

The $(Z(12))^3$ square in Figure 3.19(III) has a structure which suggests the following construction.

Theorem 3.19 *A natural Z(p)×Z(4p) magic square can be constructed for every odd integer p≥3.*

Proof. Let n=4p and note that by Lemma 3.2, Z(n) has no equipartition into 4 subsets each with p elements. However a *2-sum partition* into 4 subsets two of which have sum (1/2)(p(n-1)-1) and two of which have sum (1/2)(p(n-1)+1) is possible as we shall demonstrate.

Form a matrix A with 4 columns and p rows by listing the elements of Z(n) in consecutive order 4 elements per row and weaving back and forth between rows. See the example with p = 5 and n = 20 in Figure 3.20. In consecutive pairs of rows the sum of the 2 elements in each column is the same: the constant sum is 7 in the first pair of rows; in the second pair of rows the constant sum is 7 + 16; in the third pair of rows the sum is 7 + 2(16), etc. As a consequence, the column sums in A excluding values in the last row will all be the same, and in fact the expression (1/2)(p-1)(7) + 2(p-1)(p-3) yields exactly this common sum.

A =

0	1	2	3
7	6	5	4
8	9	10	11
15	14	13	12
16	17	18	19
46	47	48	49

A′ =

3	0	1	2
6	7	5	4
8	9	10	11
15	14	13	12
16	17	18	19
48	47	47	48

9	0	7	14	17	3	6	15	16	8
17	9	0	7	14	6	15	16	8	3
14	17	9	0	7	15	16	8	3	6
7	14	17	9	0	16	8	3	6	15
0	7	14	17	9	8	3	6	15	16
2	4	12	19	11	10	1	5	13	18
4	12	19	11	2	18	10	1	5	13
12	19	11	2	4	13	18	10	1	5
19	11	2	4	12	5	13	18	10	1
11	2	4	12	19	1	5	13	18	10

M

0	3	1	4	2	2	4	1	3	0
3	1	4	2	0	0	2	4	1	3
1	4	2	0	3	3	0	2	4	1
4	2	0	3	1	1	3	0	2	4
2	0	3	1	4	4	1	3	0	2
2	4	1	3	0	0	3	1	4	2
0	2	4	1	3	3	1	4	2	0
3	0	2	4	1	1	4	2	0	3
1	3	0	2	4	4	2	0	3	1
4	1	3	0	2	2	0	3	1	4

Q

Figure 3.20

The last row of A is given by [n-4, n-3, n-2, n-1] = [4p-4. 4p-3, 4p-2, 4p-1]. When these values are added to the sum (1/2)(p-1)(7) + 2(p-1)(p-3) of the first p rows of A, the column sums of A are given in order by

$$(1/2)(4p^2 - p - 3),\ (1/2)(4p^2 - p - 1),\ (1/2)(4p^2 - p + 1),\ (1/2)(4p^2 - p + 3).$$

These column sums are adjusted as in Figure 3.20 by replacing the first two rows of A by the rows

3	0	1	2
6	7	5	4

As indicated in the matrix A′ of Figure 3.20. The column sums of A′ are given in order by

$$(1/2)(4p^2 - p + 1),\ (1/2)(4p^2 - p - 1),\ (1/2)(4p^2 - p - 1),\ (1/2)(4p^2 - p + 1) =$$
$$(1/2)(p(n-1)+1),\ (1/2)(p(n-1)-1),\ (1/2)(p(n-1)-1),\ (1/2)(p(n-1)+1).$$

We thus obtain the stipulated 2-sum partition of Z(n).

Now let A, B denote two orthogonal magic Latin Squares of order p over Z(p), as constructed in Theorem 3.1. We may choose A so that it is *left-special,* i.e., the main diagonal contains every element of Z(p), and then B will be *right-special, i.e.,* the diagonal which extends from the bottom left corner of B to the top right corner, contains every element of Z(p). The opposite corner diagonals of A, B will contain the constant value (1/2)(p-1).

One of the rows of A′ will contain the values {n/2, (n-2)/2, (n-4)/2, (n+2)/2}; Let S(n/2) denote the values in the column of A′ which contains n/2, and define S((n-2)/2), S((n-4)/2) and S((n+2)/2) similarly. Let L(S((n-2)/2)) be a p by p magic Latin square isotopic to B with constant main diagonal occupied by (n-2)/2; let L(S(n/2)) be a p by p magic Latin square also isotopic to B with constant main diagonal occupied by n/2. Let L(S((n-4)/2)) be a p by p magic Latin square isotopic to A whose the lower left to top right diagonal has the constant value (n-4)/2, and let L(S((n+2)/2)) be a p by p magic Latin square isotopic to A and with the constant value (n+2)/2 in the lower left to top right diagonal . Let M and Q denote the 2p by 2p matrices defined as in Figure 3.21.

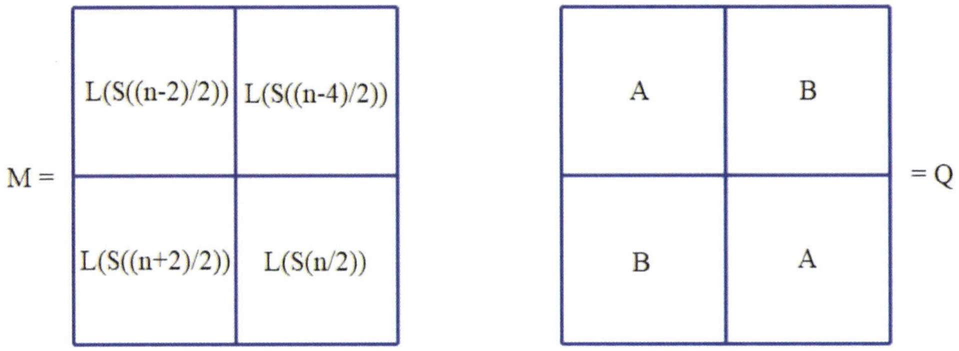

Figure 3.21

The matrix M is constructed so that all rows columns and corner diagonals have sum p(n-1), and thus M is magic over $(Z(n))^p$. The matrix Q is magic over $(Z(p))^n$. Also corresponding blocks in these two matrices are chosen so that M⊥Q, and thus Q×M is a magic square with base Z(p)×Z(n). Since M, Q are natural squares, so is Q×M. □

In Figure 3.20 an example of the construction of Theorem 3.19 is given with p = 5 and n = 20.

We have seen equipartitions (*1-sum partitions*) of Z(n) in Lemma 3.2 and 2-sum partitions of Z(n) in Theorem 3.19. We will now present a 4-sum partition of Z(48) which will be useful in constructing a magic square with base Z(3)×Z(48). The partition, indicated below, consists of 16 sets each with 3 elements. These sets are organized in groups A, B, C, D each with four sets . The sets in group A all have sum 69; in group B the common sum is 70; in group C the common sum is 71, and in group D the common sum is 72.

$A_1 = \{0, 47, 22\}$	$B_1 = \{1, 46, 23\}$	$C_1 = \{2, 43, 26\}$	$D_1 = \{3, 42, 27\}$
$A_2 = \{4, 30, 35\}$	$B_2 = \{5, 32, 33\}$	$C_2 = \{7, 25, 39\}$	$D_2 = \{6, 28, 38\}$
$A_3 = \{11, 24, 34\}$	$B_3 = \{10, 29, 31\}$	$C_3 = \{9, 44, 18\}$	$D_3 = \{8, 45, 19\}$
$A_4 = \{12, 21, 36\}$	$B_4 = \{13, 20, 37\}$	$C_4 = \{14, 17, 40\}$	$D_4 = \{15, 16, 41\}$
A	B	C	D

Let T, U denote the orthogonal magic Latin squares of order 3 illustrated in Figure 3.22.

T =

1	0	2
2	1	0
0	2	1

U =

2	0	1
0	1	2
1	2	0

T	U	U	U
U	T	U	U
U	U	T	U
U	U	U	T

Q

$U(A_1)$	$T(D_3)$	$T(B_3)$	$T(C_2)$
$T(C_3)$	$U(B_1)$	$T(D_2)$	$T(A_2)$
$T(D_4)$	$T(A_3)$	$U(C_1)$	$T(B_4)$
$T(B_2)$	$T(C_4)$	$T(A_4)$	$U(D_1)$

M

Figure 3.22

Let Q denote the block matrix whose blocks consist of copies of U, T as indicated in Figure 3.22. For each i ∈ {1,2,3,4}, let $T(A_i)$, $T(B_i)$, $T(C_i)$, $T(D_i)$ denote 3 by 3 squares over A_i, B_i, C_i, D_i respectively which are isotopic to T, and let $U(A_1)$, $U(B_1)$, $U(C_1)$, $U(D_1)$ be squares over A_1, B_1, C_1, D_1 respectively which are isotopic to U. Form the matrix M as in Figure 3.22. Note that M is so structured that each column, row and corner diagonal contains one each from the set groups A, B, C, D. Also $M \perp Q$, and thus $M \times Q$ is a 12 by 12 magic square with base $Z(3) \times Z(48)$.

The 16 set 4-sum partition {A,B,C,D} was difficult to find and no pattern could be found for constructing 4-sum partitions of Z(16m) for other odd values of m.

3.9 Base Conversions

Let M be an n by n magic square over $Z(n^2)$ and let $\Gamma_n(M)$ represent the n by n matrix obtained by converting each value in every cell of M to its equivalent base n representation. If g is a value in a cell of M, then the value in the corresponding cell of $\Gamma_n(M)$ will be represented by the ordered pair (g(div n), g(mod n)) where g(div n) represents the quotient when g is divided by n, and g(mod n) represents, as usual, the remainder when g is divided by n. In case $\Gamma_n(M)$ is a magic square with base $Z(n) \times Z(n)$ then we say that M is *base n regular.* Some of the standard constructions of n by n magic squares are base n regular. In Figure 3.23 is illustrated a 4 by 4, 5 by 5 and 7 by 7 magic squares with their respective base 4 and base 5 and base 7 conversions.

15	1	2	12
4	10	9	7
8	6	5	11
3	13	14	0

9	17	0	13	21
10	23	6	19	2
16	4	12	20	8
22	5	18	1	14
3	11	24	7	15

29	38	47	0	9	18	27
37	46	6	8	17	26	28
45	5	7	16	25	34	36
4	13	15	24	33	35	44
12	14	23	32	41	43	3
20	22	31	40	42	2	11
21	30	39	48	1	10	19

(3,3)	(0,1)	(0,2)	(3,0)
(1,0)	(2,2)	(2,1)	(1,3)
(2,0)	(1,2)	(1,1)	(2,3)
(0,3)	(3,1)	(3,2)	(0,0)

(1,4)	(3,2)	(0,0)	(2,3)	(4,1)
(2,0)	(4,3)	(1,1)	(3,4)	(0,2)
(3,1)	(0,4)	(2,2)	(4,0)	(1,3)
(4,2)	(1,0)	(3,3)	(0,1)	(2,4)
(0,3)	(2,1)	(4,4)	(1,2)	(3,0)

(4,1)	(5,3)	(6,5)	(0,0)	(1,2)	(2,4)	(3,6)
(5,2)	(6,4)	(0,6)	(1,1)	(2,3)	(3,5)	(4,0)
(6,3)	(0,5)	(1,0)	(2,2)	(3,4)	(4,6)	(5,1)
(0,4)	(1,6)	(2,1)	(3,3)	(4,5)	(5,0)	(6,2)
(1,5)	(2,0)	(3,2)	(4,4)	(5,6)	(6,1)	(0,3)
(2,6)	(3,1)	(4,3)	(5,5)	(6,0)	(0,2)	(1,4)
(3,0)	(4,2)	(5,4)	(6,6)	(0,1)	(1,3)	(2,5)

Figure 3.23

The 5 by 5 magic square in Figure 3.23 features a knight move construction, 2 squares up and one square to the right as described in Theorem 3.8, and the 7 by 7 magic square in Figure 3.23 features a *diagonal construction.* (In a typical diagonal construction 0 is placed in the middle cell of the top row and then values are inserted sequentially by moving one cell to the right and one up. In case a right boundary is encountered at the end of a row, then the right move is accomplished by moving to the first cell in the row, and if the top boundary is encountered then the up move is accomplished by moving to the bottom cell in the column. Whenever the move leads to an occupied square, then move one square down.) In Figure 3.23 it can be seen that both of these standard magic square constructions yield base conversions which can be dissected into direct products of orthogonal Latin squares.

Another standard construction of magic squares is based on a nesting of subsquares each of which is magic over a subset of $Z(n^2)$. In Figure 3.24 three examples of such squares are illustrated. The reader can check that base conversions *do not* yield squares which are magic over $Z(n)\times Z(n)$. In Chapter 6, a similar nesting construction will be used to construct integral magic octagons.

22	0	1	19	18
21	15	8	13	3
4	10	12	14	20
7	11	16	9	17
6	24	23	5	2

45	0	1	2	41	40	39
44	34	12	13	31	30	4
43	33	27	20	25	15	5
6	16	22	24	26	32	42
10	19	23	28	21	29	38
11	18	36	35	17	14	37
9	48	47	46	7	8	3

0	62	61	3	4	58	57	7
55	14	48	47	18	43	19	8
54	46	24	38	37	27	17	9
10	21	35	29	30	32	42	53
52	41	31	33	34	28	22	11
12	23	36	26	25	39	40	51
13	44	15	16	45	20	49	50
56	1	2	60	59	5	6	63

Figure 3.24

Most of the 6 by 6 magic squares are not base 6 regular, but a base 6 regular magic square of order 6 can be constructed using the special reflective square A in Figure 3.17. The square $A \times A^T$ consists of ordered pairs (a,b). By replacing each such ordered pair by the single value 6a+b a 6 by 6 magic square M is obtained which is base 6 regular.

3.10 Coset Constructions

Yet another method of obtaining Magic squares with base $Z(n) \times Z(n)$ involves the natural partition of Z(n) afforded by cosets. We shall require an integer $p \geq 3$ such that $p|n$ and such that $n/p \geq 3$. These requirements allow the creation of two pairs of orthogonal Latin squares $\{L_1, L_2\}$ of order p, and two pairs of orthogonal Latin squares $\{L_3, L_4\}$ of order n/p. Let $\{A_1, \ldots, A_p\}$ denote the cosets of the subgroup ⟨p⟩ generated by p, each with n/p elements. For each $i \in \{1, \ldots, p\}$ let $L_3(A_i)$ denote a Latin square over A_i isotopic to L_3 and let $L_4(A_i)$ denote a Latin square over A_i isotopic to L_4.

Let M denote a square of order n consisting of p^2 blocks, each block consisting of the Latin squares just constructed and arranged in a pattern isotopic to L_1. Thus each coset will be repeated exactly p times such that all p cosets occur in each row and column of M. This will insure that all rows and columns of M will have the same sum equal to the sum n(n-1)/2 of all the elements of Z(n). See the example in Figure 3.25 where n = 12 and p = 4. The Latin squares $L_3(A_i)$ and $L_4(A_i)$ which occupy diagonal positions in M must be chosen carefully so that the corner diagonals of M also have sum n(n-1)/2.

A second square Q of order n is designed similarly: it will consist of p^2 blocks, each block consisting of the Latin squares $L_3(A_i)$ and $L_4(A_i)$ but now arranged in a pattern isotopic to L_2. Wherever an $L_3(X)$ occurs in M, an $L_4(X)$ must occur in the corresponding location of Q, and wherever an $L_4(X)$ occurs in M, an $L_3(X)$ must occur in the corresponding position in Q. Special attention must again be given to the blocks in diagonal locations to insure that both corner diagonals of Q have the sum n(n-1)/2. The resulting squares M, Q will be orthogonal and magic over (Z(n)). Them $M \times Q$ will be a magic square with base $Z(n) \times Z(n)$.

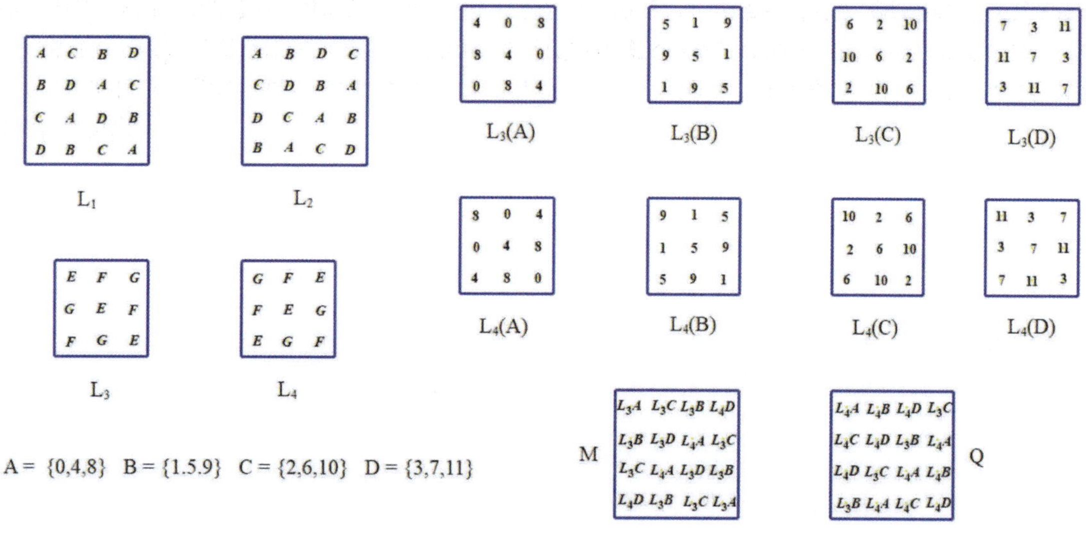

Figure 3.25

3.11 Magic squares of order 8

The square in Figure 3.3(IV) over Z(2)×Z(2)×Z(2)×Z(2) is not natural, but a natural solution is available as indicated in Figure 3.26. Thus the 4 by 4 square, as well as the 3 by 3 square are universally magic.

0 0 0 0	1 0 0 1	1 1 1 0	0 1 1 1
1 1 1 1	0 1 1 0	0 0 0 1	1 0 0 0
0 1 0 1	1 1 0 0	1 0 1 1	0 0 1 0
1 0 1 0	0 0 1 1	0 1 0 0	1 1 0 1

Figure 3.26

Several standard magic squares of order 8 can be found in [7]. These can be used to construct natural magic squares of order 8 with base Z(64). The orthogonal Latin squares L_3, L_4 listed in Section 3.2 can be used to create a natural magic square of order 8 with base Z(8)×Z(8).

To show that the square of order 8 is universally magic, it remains to construct magic squares over the groups Z(2)×Z(32), Z(4)×Z(16), Z(2)×Z(2)×Z(16), Z(2)×Z(2)×Z(2)×Z(8), Z(2)×Z(2)×Z(2)×Z(2)×Z(4), Z(4)×Z(4)×Z(4), Z(2)×Z(2)×Z(4)×Z(4), Z(2)×Z(2)×Z(2)×Z(2)×Z(2)×Z(2), and Z(2)×Z(4)×Z(8). The orthogonal squares I, II in Figure 3.27 can be superimposed to create a natural magic square with base Z(2)×Z(32), and the orthogonal squares III, IV in Figure 3.27 can be superimposed to create a natural magic square with base Z(4)×Z(16).

By converting the entries in square 3.27(III) to base 2 and superimposing the result with the $(Z(16))^4$ square 3.27(IV), an order 8 magic square with base Z(2)×Z(2)×Z(16) is obtained. When the entries in 3.27(IV) are converted to base 4, the $(Z(4)\times Z(4))^4$ square in Figure 3.28(I) is obtained. If this square is then superimposed on the square in Figure 3.27(III), the order 8 magic square in Figure 3.28(II) with base Z(4)×Z(4)×Z(4) is obtained. If each of the digits {0,1,2,3} which occur in 3.28(II) is converted to base 2, a magic square with base Z(2)×Z(2)×Z(2)×Z(2)×Z(2)×Z(2) is obtained. Convert each of the digits {0,1,2,3} in the $(Z(4))^{16}$ square in Figure 3.27(III) to base 2 to obtain a $(Z(2)\times Z(2))^{16}$ square. This square consists of four mutually orthogonal Latin squares over Z(2)×Z(2), each

I

0	0	1	1	0	0	1	1
1	1	0	0	1	1	0	0
1	1	0	0	1	1	0	0
0	0	1	1	0	0	1	1
1	1	0	0	1	1	0	0
0	0	1	1	0	0	1	1
0	0	1	1	0	0	1	1
1	1	0	0	1	1	0	0

III

3	1	2	0	2	0	3	1
2	0	3	1	3	1	2	0
1	3	0	2	0	2	1	3
0	2	1	3	1	3	0	2
1	3	0	2	0	2	1	3
0	2	1	3	1	3	0	2
3	1	2	0	2	0	3	1
2	0	3	1	3	1	2	0

II

1	12	19	30	6	8	23	25
18	31	0	13	21	27	4	10
14	3	28	17	9	7	24	22
29	16	15	2	26	20	11	5
6	8	23	25	1	12	19	30
21	27	4	10	18	31	0	13
9	7	24	22	14	3	28	17
26	20	11	5	29	16	15	2

IV

0	13	6	11	0	13	6	11
7	10	1	12	7	10	1	12
9	4	15	2	9	4	15	2
14	3	8	5	14	3	8	5
0	13	6	11	0	13	6	11
7	10	1	12	7	10	1	12
9	4	15	2	9	4	15	2
14	3	8	5	14	3	8	5

Figure 3.27

occupying a quarter of an 8 by 8 square. When superimposed over the$(Z(4)\times Z(4))^4$ square in Figure 3.28(I) a Z(2)×Z(2)×Z(4)×Z(4) square is obtained, and when superimposed over an 8 by 8 square M whose quarters consist of copies of the Z(2)×Z(8) square in Figure 3.2(VII), a magic square with base Z(2)×Z(2)×Z(2)×Z(8) is obtained. If the square M is superimposed over the $(Z(4))^{16}$ square in 3.27(III), a magic square with base Z(2)×Z(4)×Z(8) is obtained. Finally to obtain a magic square with base Z(2)×Z(2)×Z(2)×Z(2)×Z(4), construct an 8 by 8 square each of whose quarters consist of copies of the Z(2)×Z(2)×Z(2)×Z(2) square in Figure 3.2, and. superimpose this square over the $(Z(4))^{16}$ square in 3.27(III). With these constructions we have shown that the square of order 8 is universally magic.

(0,0)	(3,1)	(1,2)	(2,3)	(0,0)	(3,1)	(1,2)	(2,3)
(1,3)	(2,2)	(0,1)	(3,0)	(1,3)	(2,2)	(0,1)	(3,0)
(2,1)	(1,0)	(3,3)	(0,2)	(2,1)	(1,0)	(3,3)	(0,2)
(3,2)	(0,3)	(2,0)	(1,1)	(3,2)	(0,3)	(2,0)	(1,1)
(0,0)	(3,1)	(1,2)	(2,3)	(0,0)	(3,1)	(1,2)	(2,3)
(1,3)	(2,2)	(0,1)	(3,0)	(1,3)	(2,2)	(0,1)	(3,0)
(2,1)	(1,0)	(3,3)	(0,2)	(2,1)	(1,0)	(3,3)	(0,2)
(3,2)	(0,3)	(2,0)	(1,1)	(3,2)	(0,3)	(2,0)	(1,1)

0 0 3	3 1 1	1 2 2	2 3 0	0 0 2	3 1 0	1 2 3	2 3 1
1 3 2	2 2 0	0 1 3	3 0 1	1 3 3	2 2 1	0 1 2	3 0 0
2 1 1	1 0 3	3 3 0	0 2 2	2 1 0	1 0 2	3 3 1	0 2 3
3 2 0	0 3 2	2 0 1	1 1 3	3 2 1	0 3 3	2 0 0	1 1 2
0 0 1	3 1 3	1 2 0	2 3 2	0 0 0	3 1 2	1 2 1	2 3 3
1 3 0	2 2 2	0 1 1	3 0 3	1 3 1	2 2 3	0 1 0	3 0 2
2 1 3	1 0 1	3 3 2	0 2 0	2 1 2	1 0 0	3 3 3	0 2 1
3 2 2	0 3 0	2 0 3	1 1 1	3 2 3	0 3 1	2 0 2	1 1 0

Figure 3.28

Recall that the square of order 9 was shown to be universally magic in Section 3.4. Thus we have shown that squares of order 3,4,5,7,8,9 are universally magic. Only the universality of the order 6 square is in question. Based on this, admittedly scanty, evidence we make the following Conjecture.

Conjecture 3.20 *Every square of order n with n ≥3 is universally magic.*

3.12 Integral Magic Squares

In this Section we will construct magic squares over the integral representation $Z^*(n^2)$. The notation M_n will be used to represent an integral square of order n and F_n will be used to denote a *frame* consisting of four linear polysquares attached at their ends to form a square. A nesting of integral squares will be constructed by successively attaching frames; thus in general $M_n = M_{n-2} \cup F_n$. The case n even and n odd will be dealt with separately. In Figure 3.29 the first several integral squares of each type are illustrated. The smallest squares M_3, M_4 we call the *kernel* of the nesting. All the squares in the nesting are called *inclusions*. If P is any integral polygon, then any proper subpolygon of P which is integral in its own right will be called an *inclusion.*

-19	14	18	13	-21	-22	17
-15	-8	5	9	-12	6	15
-16	-7	-1	4	-3	7	16
20	10	-2	0	2	-10	-20
23	11	3	-4	1	-11	-23
24	-6	-5	-9	12	8	-24
-17	-14	-18	-13	21	22	19

30	-30	27	-27	25	-25	22	-18
-31	16	-16	13	-13	11	-8	32
-23	-17	1	-2	7	-4	18	24
-22	-11	8	-5	2	-3	12	23
-19	-9	-6	5	0	3	10	20
21	15	-1	4	-7	6	-14	-20
29	9	17	-12	14	-10	-15	-28
19	31	-26	28	-24	26	-21	-29

Figure 3.29

Note that for odd n the sum of M_n remains at 0. Values at opposite ends of each row, each column and each corner diagonal are inverses. For even n, M_4 has sum 2, and with the addition of each frame the sum increases by 1; thus the sum of M_n is n/2. The procedure for generalizing these nested constructions is given in the following Theorem.

Theorem 3.21 *For each integer n ≥3 an integral magic square can be constructed. Moreover, these can be constructed in nested fashion.*

Proof. Suppose first that n is odd, then $F_n = \{\mp((n-2)^2+1))/2), \ldots, \mp(n^2-1)/2\}$. With

$p = ((n-2)^2+1))/2)$; $q = (n^2-3n)/2$; $m = (n^2-2n-1)/2$; $r = (n^2-n)/2$; $s = (n^2-1)/2$,

the elements of F_n are organized into the sections indicated in Figure 3.30.

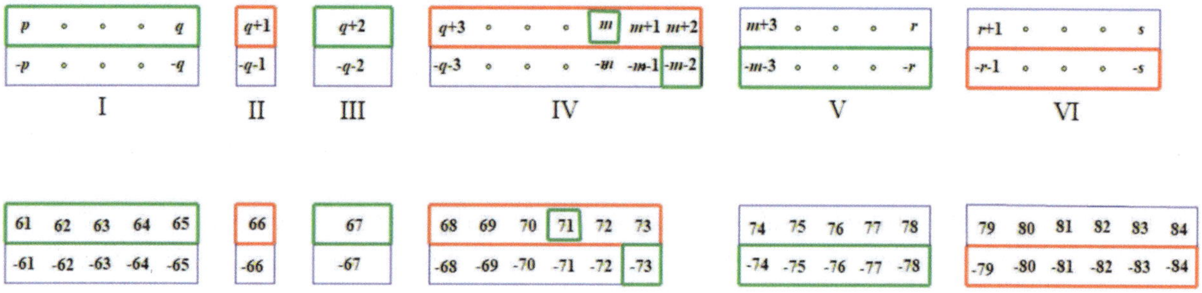

Figure 3.30

In this Figure the values to be placed in the top side of F_n are enclosed in red, and the values to be placed in the right side of F_n are enclosed in green. The corner value, m which is shared by the top and right side is enclosed in both colors. The first value of the top side and the bottom value of the right side, m+2, -m-2 respectively are opposite values with respect to the main diagonal of M_n . The values not enclosed in red or green are placed in the left and bottom sides of the frame F_n. Each of these values is placed at the opposite end of the row, column or corner diagonal of its inverse so that the sum in each of these rows, columns and diagonals remains at 0. So it is necessary only to show that the sum of the red as well as the green values is also 0. The sum of the red values is given by:

$$(q+1) + (1/2)(q+m+5)(m-q) - (1/2)(s+r+1)(s-r) =$$
$$(1/8)(4n^2-12n+8) + (1/8)(2n^2 -5n+9)(n-1) - (1/8)(2n^2-n+1)(n-1) = 0,$$

and the sum of the green values is given by:

$$(1/2)(p+q)(q-p+1) + q - (1/2)(m+r+3)(r-m-2) =$$
$$(1/8)(2n^2-7n+5)(n-3) + (1/8)(4n^2 - 12n) - (1/8)(2n^2 - 3n +5)(n-3) = 0.$$

Thus all four sides of the frame F_n have sum 0. It follows by a simple induction that each square M_n for n odd and n ≥3 is a a magic integral square with sum 0. In Figure 3.30 a sample construction of F_{13} is provided.

Now suppose that n is even and n≥4. In Figure 3.29 it is seen that the sum of the order four integral square M_4 is 2; the sum of M_6 is 3, and the sum of M_8 is 4. In general M_n will be constructed so that it has sum n/2. F_n will denote the frame which encloses M_{n-2} ; thus $M_n = M_{n-2} \cup F_n$ as in the case n odd, but now the sides of F_n must be constructed with sum n/2. If n is twice an odd number then the general form for F_n is given in Figure 3.31 with:

$$p = 2((n/2)-1)^2 + 1;\ q = p+(n/2)-3;\ r = p+n-5;\ s = (n^2/2)-4.$$

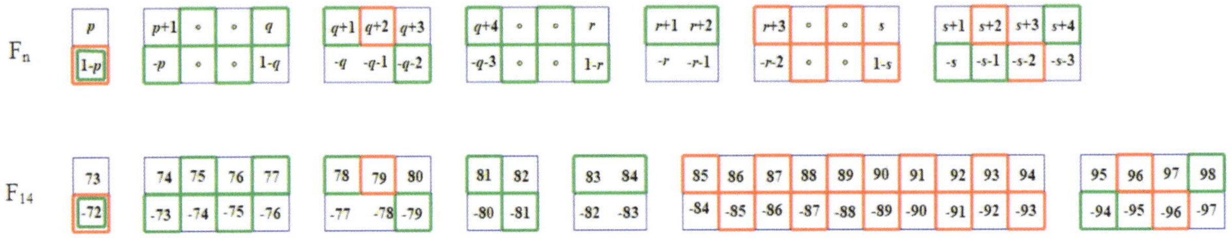

Figure 3.31

In this Figure the cells enclosed in red represent, as usual, the values belonging to the top side of F_n and the cells enclosed in green contain the values assigned to the right side of F_n. The remaining, (unenclosed values) will be placed on the left and bottom sides of F_n so that values at opposite ends of every row column and corner diagonal of M_n have sum 1. The sum of the red values is:

$$(1-p) + (q+2) = (1-p) + (p + (n/2)-1) = n/2,$$

And the sum of the green values is:

$$(1-p) + (q-p) -1 + 0 + (2r+3) + (3-s) =$$

$$q - 2p + 2r - s + 6 = (p+(n/2) - 3) -2p +(2p +2n - (n^2/2)) =$$

$$p+(n/2) - 3 +2n -(n^2/2) = 2((n/2)-1)^2 + (n/2) -2 + 2n -(n^2/2) = n/2.$$

As a consequence of the values assigned in F_n, and under the inductive assumption that M_{n-2} is an integral magic square with sum (n-2)/2, each row, column and corner diagonal which extends the corresponding row, column and corner diagonal of M_{n-2} now has sum 1 + (n-2)/2 = (n/2) as desired. In addition we have just shown that the top and right side of the frame F_n has sum n/2. It remains only to check that the left and bottom side of F_n has the same sum. Let $\{x_1, x_2, \ldots, x_n\}$ represent the n red values which comprise the top side of F_n. We have shown that the sum of these values is n/2. The values in the bottom row are then given by $\{1-x_1, \ldots, 1-x_n\}$ which have sum n - (n/2) = n/2. Similarly the sum of the values in the left side of F_n have sum n/2. We have now shown that M_n is an integral magic square with sum n/2.

Finally, suppose that n is a multiple of 4. In this case the frame F_n is constructed a little differently. With:

$$p = 2((n/2)-1)^2+1;\ q = p+(n/2)-2;\ r = q+(n/2)-3;\ s = (n^2/2) -4,$$

Let F_n be represented as in Figure 3.32.

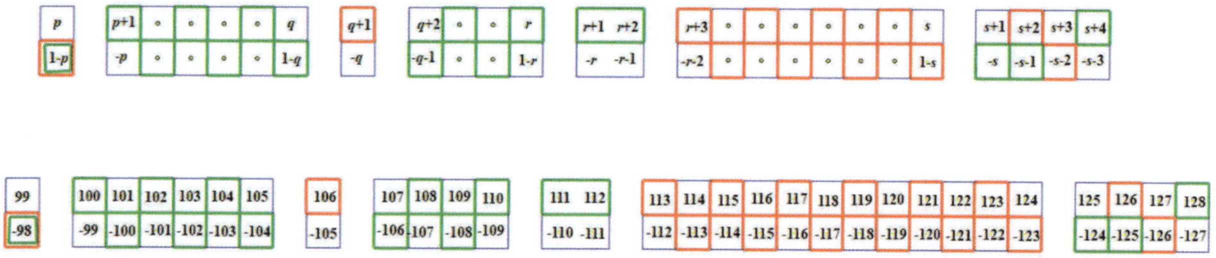

F_{16}

Figure 3.32

The sum of the red values is: (1-p) + (q+1) = n/2, and the sum of the green values is:

$(1\text{-}p) + ((n/2)\text{-}4) + (2r+3) + (n^2/2) - n^2 + 7 =$

$-p + (n/2) + 2r - (n^2/2) + 7 = -p + (n/2) + (2p+n-4) + n + 1 - (n^2/2) =$

$p + (n/2) + 2n - 3 - (n^2/2) = n/2.$

Thus both the top and right side of F_n have the desired sum. As in the previously considered case, the left and bottom sides of F_n also have sum n/2, and M_n is an integral magic square with sum n/2. □

Note that ΣZ*(n) = 0 if n is odd and ΣZ*(n) = n/2 if n is even. As a consequence the sum of an integral magic square of odd order *must* be 0, and the sum of an integral magic square of even order n *must* be n/2. The integral magic squares M_3 and M_4 are *primary*, that is, they do not occur as extensions of any smaller integral magic squares. A *composite* integral magic square is one which is constructed as an extension of a smaller integral magic square. Theorem 3.21 shows that M_3 and M_4 are the only primary integral magic squares; every other integral magic square is composite.

It is also possible to convert a given standard magic square M over $Z(n^2)$ to an integral near magic square over $Z^*(n^2)$ by the following method. First suppose n is odd and let M* denote the square obtained from M by replacing an element x by x-n if x>(n-1)/2, and leaving x unchanged in case x ≤(n-1)/2. M* is thus a square over $Z^*(n^2)$ but it is not magic as its column sums and row sums can take on various values. It is, however, often the case that there exists an integer k such that M** = k+M* is a 0-sum integral near magic square. We call M** the *integral mate of M,* and k the *conversion factor.* In case a magic square M has an integral mate, we say that M is *convertible.* In Figure 3.33 an example of a convertible magic square of order 9 is illustrated.

10	17	12	73	80	75	28	35	30
15	13	11	78	76	74	33	31	29
14	9	16	77	72	79	32	27	34
55	62	57	37	44	39	19	26	21
60	58	56	42	40	38	24	22	20
59	54	61	41	36	43	23	18	25
46	53	48	1	8	3	64	71	66
51	49	47	6	4	2	69	67	65
50	45	52	5	0	7	68	63	70

M

10	17	12	-8	-1	-6	28	35	30	117
15	13	11	-3	-5	-7	33	31	29	117
14	9	16	-4	-9	-2	32	27	34	117
-26	-19	-24	37	-37	39	19	26	21	36
-21	-23	-25	-39	40	38	24	22	20	36
-22	-27	-20	-40	36	-38	23	18	25	-45
-35	-28	-33	1	8	3	-17	-10	-15	-126
-30	-32	-34	6	4	2	-12	-14	-16	-126
-31	-36	-29	5	0	7	-13	-18	-11	-126
-126	-126	-126	-45	36	36	117	117	117	

M*

24	31	26	6	13	8	-39	-32	-37	0
29	27	25	11	9	7	-34	-36	-38	0
28	23	30	10	5	12	-35	-40	-33	0
-12	-5	-10	-30	-23	-28	33	40	35	0
-7	-9	-11	-25	-27	-29	38	36	34	0
-8	-13	-6	-26	-31	-24	37	32	39	0
-21	-14	-19	15	22	17	-3	4	-1	0
-16	-18	-20	20	18	16	2	0	-2	0
-17	-22	-15	19	14	21	1	-4	3	0
0	0	0	0	0	0	0	0	0	

M** = M* + 14

Figure 3.33

The row (column) sums in a standard magic square M of order n over $Z(n^2)$ are given by $(1/2)n(n^2-1)$ and the row and column sums in M* must be congruent to this number mod (n^2). The number $(1/2)n(n-1)$ satisfies this congruence and so the column/row sums of M* all must have the form $(1/2)n(n-1) + tn^2$ for some integer t. The possible conversion factors are obtained by dividing this expression by n or -n, i.e., $(1/2)(n-1) + tn$ and $(-1/2)(n-1) + tn$. None of these serve as conversion factors for the square given in Figure 3.34, so we may conclude that this square is not convertible. We could find no smaller examples of odd order nonconvertible magic squares.

70	0	50	31	49	1	79	2	78
20	40	60	55	25	12	68	9	71
30	80	10	19	61	64	16	62	18
33	39	59	42	27	63	17	54	26
47	41	21	53	38	74	6	24	56
67	52	14	32	15	43	57	76	4
13	28	66	48	65	23	37	51	29
75	3	69	72	7	29	35	36	34
5	77	11	8	73	51	45	46	44

M

-11	0	-31	31	-32	1	-2	2	-3	-45
20	40	-21	-26	25	12	-13	9	-10	36
30	-1	10	19	-20	-17	16	-19	18	36
33	39	-22	-39	27	-18	17	-27	26	36
-34	-40	21	-28	38	-7	6	24	-25	-45
-14	-29	14	32	15	-38	-24	-5	4	-45
13	28	-15	-33	-16	23	37	-30	29	36
-6	3	-12	-9	7	29	35	36	34	117
5	-4	11	8	-8	-30	-36	-35	-37	-126
36	36	-45	-45	36	-45	36	-45	36	

M*

Figure 3.34

The square M in Figure 3.33 is the product L⊕L where L is the 3 by 3 magic square of Lo Shu. The square M in Figure 3.34 is due to Frierson and was found in [7]. It features a variation on the construction of nested magic subsquares.

Now suppose n is even and M is a standard magic square of order n over $Z(n^2)$. An integral version of M can also be obtained using a variation of the above procedure. Let M* be defined as above and suppose there exists an integer k such that the row/column sums of k+M* consist of n-1 sums with value -n/2 and one sum with value (1/2)n(n-1), and such that the cell which contains $n^2/2$ lies precisely at the intersection of the row and column with sum (1/2)n(n-1). (This seems like a lot to require, but it is amazing how often this condition arises.) The value in this cell is then replaced with $-n^2/2$, resulting in a square in which each row and column has sum -n/2. Finally, every value is replaced by its inverse to obtain an integral near magic square over $Z^*(n^2)$. The first two steps in this process are illustrated by the order 10 magic square in Figure 3.35. We have not been able to find any even order magic square which is *not* convertible.

0	2	88	90	76	77	47	46	35	34
3	1	91	89	79	78	44	45	32	33
56	58	27	26	12	13	83	82	68	70
59	57	24	25	15	14	80	81	71	69
95	94	63	62	48	49	36	38	4	6
92	93	60	61	51	50	39	37	7	5
28	30	19	18	84	85	75	74	40	42
31	29	16	17	87	86	72	73	43	41
64	66	52	54	20	21	11	10	99	98
67	65	55	53	23	22	8	9	96	97

M

0	2	-12	-10	-24	-23	47	46	35	34	95
3	1	-9	-11	-21	-22	44	45	32	33	95
-44	-42	27	26	12	13	-17	-18	-32	-30	-105
-41	-43	24	25	15	14	-20	-19	-29	-31	-105
-5	-6	-37	-38	48	49	36	38	4	6	95
-8	-7	-40	-39	-49	50	39	37	7	5	-5
28	30	19	18	-16	-15	-25	-26	40	42	95
31	29	16	17	-13	-14	-28	-27	43	41	95
-36	-34	-48	-46	20	21	11	10	-1	-2	-105
-33	-35	-45	-47	23	22	8	9	-4	-3	-105
-105	-105	-105	-105	-5	95	95	95	95	95	

M*

10	12	-2	0	-14	-13	-43	-44	45	44	-5
13	11	1	-1	-11	-12	-46	-45	42	43	-5
-34	-32	37	36	22	23	-7	-8	-22	-20	-5
-31	-33	34	35	25	24	-10	-9	-19	-21	-5
5	4	-27	-28	-42	-41	46	48	14	16	-5
2	3	-30	-29	-39	-40	49	47	17	15	-5
38	40	29	28	-6	-5	-15	-16	50	-48	95
41	39	26	27	-3	-4	-18	-17	-47	-49	-5
-26	-24	-38	-36	30	31	21	20	9	8	-5
-23	-25	-35	-37	33	32	18	19	6	7	-5
-5	-5	-5	-5	-5	-5	-5	-5	95	-5	

10+M*

Figure 3.35

In Figure 3.36 a standard procedure, found in [6], is illustrated which can be used to create magic squares of any odd order. A square W consisting of consecutive values from 0 to n^2-1 is constructed at a 45 degree tilt. We shall refer to such a square as a *seed square.* Inside this square a second square M is constructed with its corners at the midpoints of the sides of S. M is partially populated with the values of S which lie inside M. The values in S which lie outside of M are translated to the opposite side of M and inserted in the blank cells; thus M contains every value of S. The horizontal diagonals, (including broken diagonals), of S become rows of M, and the vertical diagonals of S become columns of M. In Chapter 4 we show that the diagonals, including broken diagonals, of a seed square all have the same sum. So, in this case M is a magic square over $Z(n^2)$. We will refer to odd order squares constructed in this way as *diamonds.* In the following we outline a proof that every diamond is convertible.

Theorem 3.22 *Every diamond of odd order q over $Z(q^2)$ is convertible with conversion factor (1/2)(q+1).*

Proof. Let M denote a diamond over $Z(q^2)$ and let M* denote the corresponding square over $Z^*(q^2)$. Partition M and M* into two squares S_1, S_2 with orders (1/2)(q+1) and (1/2)(q-1) respectively, and two rectangles R_1 with (1/2)(q-1)rows and (1/2)(q+1) columns, and R_2 with (1/2)(q+1) rows and (1/2)(q-1) columns as in the example in Figure 3.37. Call the dividing lines which determine these rectangles the *horizontal* and *vertical axes* of M or M*.

The first thing to notice is that pairs of values in the first column of M which are opposite each other with respect to the horizontal axis all have sum q^2. There remains only the unpaired value (1/2)q(q-1) in the lower left corner of M. Similarly the pairs of values in the bottom row of M which are opposite each other with respect to the vertical axis, have sum q^2. Again, only the value (1/2) q(q-1) in the lower left corner of M is unpaired. Two values in M which sum to q^2 are inverses, and in M* are represented in the form x and -x. As a consequence the first column and bottom row of M* sum to the one unpaired value (1/2)q(q-1).

The next thing to observe is that adjacent columns in the section of M* occupied by R_1 and S_1 have the same sum, This is due to the special structure of M*: suppose C_1 and C_2 are adjacent columns in this section with C_1 preceding C_2 and note that, with one exception, the value in C_2 which lies in the cell diagonally one cell up from a value in C_1 exceeds it in value by exactly 1. Even the bottom value in C_2 exceeds the top value in C_1 by 1. These plus ones improve the sum of C_2 over C_1 by q-1. There is, however one remaining pair of diagonally opposite values which reduce the sum of C_2 by q-1. As a consequence the sum of all columns in the section of M* occupied by R_1 and C_1 is the same, and thus must equal the sum (1/2)q(q-1) of the first column. Similarly, the sum of all rows in the section of M* occupied by S_1 and R_2 must be the same as the sum of the bottom row, namely (1/2)q(q-1). From the same considerations the sums of the columns which occupy S_2 and R_2 and the sums of the rows which occupy R_1 and S_2 must all be the same. Since $\Sigma Z^*(n^2) = 0$ we must have

$$(1/2)(q+1)(1/2)q(q-1) + (1/2)(q-1)s = 0,$$

Where s denotes the common sum of the columns of M* which occupy S_2 and R_2. Solving for s we obtain s = (-q/2)(q+1). Similarly the common sum of the rows which lie in R_1 and S_2 is (-q/2)(q+1).

In the rightmost column, pairs of values opposite each other with respect to the horizontal axis sum to q^2-q-1. When (1/2)(q+1) is added to each element belonging to these pairs, the sum of each pair equals q^2. Thus in M** = M*+(1/2)(q+1) these are inverse pairs with sum 0. The bottom element of this column in M*, which belongs to none of these pairs, is (-1/2)(q+1), thus in M** this entire column has sum 0. The equivalence of column sums in S_2 and R_2 in M* is not affected by the addition of (1/2)(q+1) to each value in the columns, and thus all these columns have sum 0 in M**. We have seen that the columns in R_1 and S_1 all have the same sum in M*, and thus in M**. Since $\Sigma Z^*(n^2) = 0$, and the columns in S_2 and R_2 in M** have total sum 0, it must be that the total sum of the columns in R_1 and S_1 is also 0. Putting these facts together we can deduce that each column in S_2 and R_2 in M** must have sum 0. We have shown that every column in M** has sum 0.

Next, to show that the rows of M** have sum 0, consider the bottom row of M. With w = (1/2)(q+1); t = q+1 and m = w+(1/2)(q-3)t, the elements in this row lie in the set

$$\{w, w+t, w+2t, \ldots, m, m+1, m+1+t, m+1+2t\ldots, m+1+(1/2)(q-3)t, m+1+(1/2)(q-1)t\}.$$

When w is added to each element of this set we obtain

$$\{2w, 2w+t, 2w+2t, \ldots, w+m, w+r, w+r+t, \ldots, w+r+(1/2)(q-3)t, w+r+(1/2)(q-1)t\}$$

where r = m+1. All but the last element in this set form pairs which sum to q^2. For example, (w+m)+(w+r) = 2w+2m+1 = (q+1)+2w+(q-3)(q+1)+1 = q^2. The last element has the value (w+r)+(1/2)(q-1)t = (1/2)(q^2+1) + (1/2)(q^2-1) = q^2. Pairs which sum to q^2 sum to 0 in $Z^*(n^2)$ and thus the bottom row of M** has sum 0. By a similar argument to the above, we can conclude that every row of M** has sum 0, and thus M** is a near magic integral 0-sum square. □

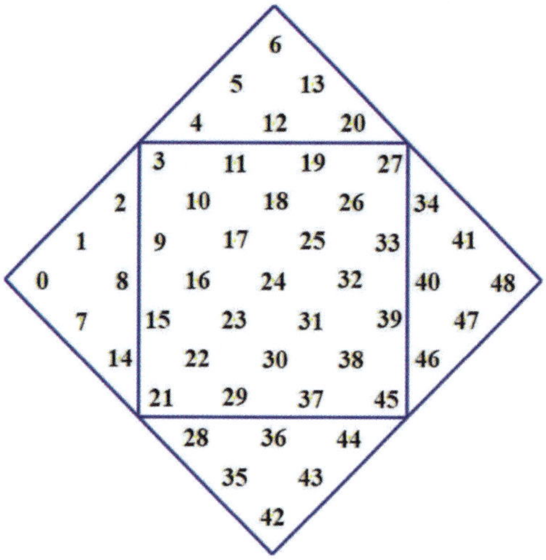

3	28	11	36	19	44	27
34	10	35	18	43	26	2
9	41	17	42	25	1	33
40	16	48	24	0	32	8
15	47	23	6	31	7	39
46	22	5	30	13	38	14
21	4	29	12	37	20	45

M

Figure 3.36

							S_2	
R_1	3	-21	11	-13	19	-5	-22	-28
	-15	10	-14	18	-6	-23	2	-28
	9	-8	17	-7	-24	1	-16	-28
	-9	16	-1	24	0	-17	8	21
	15	-2	23	6	-18	7	-10	21
	-3	22	5	-19	13	-11	14	21
S_1	21	4	-20	12	-12	20	-4	21
	21	21	21	21	-28	-28	-28	R_2

M*

7	-17	15	-9	23	-1	-18	0
-11	14	-10	22	-2	-19	6	0
13	-4	21	-3	-20	5	-12	0
-5	20	3	-21	4	-13	12	0
19	2	-20	10	-14	11	-6	0
1	-23	9	-15	17	-7	18	0
-24	8	-16	16	-8	24	0	0
0	0	0	0	0	0	0	

4+M*

Figure 3.37

CHAPTER 4

MAGIC RECTANGLES

4.1 2-sum Rectangles

The geometric base for the constructions in this Chapter will be a p by q rectangular polysquare R over Z(pq) with q>p≥2. We will consider two types of constructions, one in which R is a magic integral rectangle, and one in which all rows have the natural sum S_1 and all columns have the natural sum S_2. The latter construction will be called a *2-sum rectangle*. It is not possible, using natural sums, for S_1 to equal S_2 since we must have $pS_1= qS_2= qS_1$ which would imply p=q. We consider 2-sum constructions first. The first basic result is the following.

Theorem 4.1 *Let R be a 2-sum p by q rectangular polysquare over Z(pq). Then p,q must have the same parity. Moreover $S_1= (1/2)q(pq-1)$ and $S_2= (1/2)p(pq-1)$.*

Proof. Consider: ΣZ(pq) = (1/2)pq(pq-1), and since there are p rows each with the same sum S_1, we must have $S_1= (1/2)q(pq-1)$. The determination of S_2 is obtained similarly. If p is odd and q is even then S_2 is not an integer, and if p is even and q is odd then S_1 is not an integer. □

4.2 Rectangles with Even Dimensions

For the simplest case, p=2, (q≥4, q even), we will employ a *zig-zag* construction by first putting 0 in the first column of R in the second row, and then putting 1 in the second column in the top row and alternating the positions of consecutive integers until the last column of R is obtained. The remaining elements of Z(2q) are placed in each column so that every column has sum 2q-1. Examples with q=10 and q=12 are provided in Figure 4.1.

I

		55					45			
19	1	17	3	15	5	13	7	11	9	100
0	18	2	16	4	14	6	12	8	10	90
		40					50			

II

		55					40			
19	1	17	3	15	5	6	12	8	9	95
0	18	2	16	4	14	13	7	11	10	95
		40					55			

III

			72						72			
23	1	21	3	19	5	17	7	15	9	13	11	144
0	22	2	20	4	18	6	16	8	14	10	12	132
			66						66			

IV

			72						66			
23	1	21	3	19	5	6	16	8	14	10	12	138
0	22	2	20	4	18	17	7	15	9	13	11	138
			66						72			

Figure 4.1

Figures 4.1(I,III) feature our standard zig-zag pattern. The result is that all columns will have the same sum 2q-1. To equalize the row sums, values in some of the columns are interchanged. The type of interchanges made vary for different values of q. To go from 4.1(I) to 4.1(II), the middle three values in the second block of 4.1(I) are interchanged, and to move from 4.1(III) to 4.1(IV), the entire second block rows are interchanged. In the following Theorem the general procedure is outlined for obtaining 2-sum rectangles with 2 rows and q columns, (q≥4, q even).

Theorem 4.2 *Every rectangle with 2 rows and q columns, (q≥4, q even), has a 2-sum solution over Z(2q).*

Proof. Suppose first that 4|q. Let A denote a 2 by q rectangular polysquare with a zig-zag assignment of the elements of Z(2q) to its cells as in the example in figure 4.1(iii). Divide A into two 2 by q/2 sections and name the rows in each of these sections A_1, A_2, A_3, A_4 where A_1 and A_2 constitute the top row of A and A_3, A_4 constitute the bottom row of A. For i = 1,2,3,4, let M_i denote the sum of the elements in A_i. A_1 contains the odd values 1, 3, 5, ..., ((q/2) - 1), and also the odd numbers 2q-1, 2q-2, 2q-4, ..., ((3q/2)+1). The sum of the first set of odd numbers is $q^2/16$, and the sum of the second set is $7q^2/16$, yielding $M_1 = q^2/2$. The values in A_3 are obtained by subtracting 1 from each of the q/2 values in A_1, and thus $M_3 = (q^2-q)/2$.

A_4 contains the consecutive even values q/2, (q/2)+2, (q/2)+4, ..., (q/2)+ q-2 which sum to $(q^2-q)/2$, thus $M_4 =(q^2-q)/2$. Note that M_2 is obtained by adding 1 to each of the q/2 values in A_4, and thus $M_2 = q^2/2$. If we now interchange the values in A_2, A_4 the resulting polysquare is a 2-sum rectangle with the common sum of $q^2 - (q/2)$ for the rows, and the common sum of 2q-1 for the columns.

Now suppose q is twice an odd number. There are two possibilities which we consider separately: (i) q = 8k+6, and (ii) q= 8k+2. Let A be a 2 by q rectangular polysquare whose cells are assigned values from Z(2q) in zig-zag fashion. The top row of A has sum S_1 = 1+ 3 + 5 + ... + 2q-1 = q^2 and the bottom row of A has the sum S_2 = 0 + 2 + 4 + 6 + ... + (2q-2) = q^2 - q. In Case (i) m_1 = (5q-2)/4 is an odd integer which lies between 1 and 2q-1 in the top row of A. The value in the second row of A which lies beneath (5q-2)/4 is m_2 = (2q-1) - (5q-2)/4 = (3q-2)/4. The difference between these two values is q/2. Thus if m_1 and m_2 are interchanged, the sum of the top row is reduced to q^2 - (q/2), and the sum of the bottom row is increased to q^2 - (q/2). So by this single interchange A has been converted to a 2-sum rectangle.

In Case (ii) the odd integers s_1 = (3q-2)/4, s_2 = (5q+2)/4 and s_3 = (5q-6)/4 all lie in the top row of A. The values which lie beneath these in the second row of A are t_1 = (2q-1) - s_1 = (5q-2)/4; t_2 = (2q-1) - s_2 = (3q-6)/4 and t_3 = (2q-1)- s_3 = (3q+2)/4. Then (s_1 + s_2 + s_3) - (t_1 + t_2 + t_3) = (1/4)(13q-6) - (1/4)(11q-6) = q/2. Thus if the pairs of values (s_1, t_1), (s_2, t_2) and (s_3, t_3) are interchanged then q/2 is subtracted from the top row sum of A and q/2 is added to the bottom row sum of A with the result that both rows of A have the sum q^2 - (q/2). □

The general construction of a 2-sum rectangle over Z(pq) in which both p and q are even is also not difficult.

Theorem 4.3 *Every p by q rectangle in which q>p≥2 with both p, q even, has a 2-sum solution over Z(pq).*

Proof. Let R be a p by q rectangular polysquare with p,q even and q>p≥2. The elements of Z(pq) are assigned row by row in zig-zag fashion as in the example in Figure 4.2.

I

47	1	45	3	43	5	41	7	192
0	46	2	44	4	42	6	40	184
39	9	37	11	35	13	33	15	192
8	38	10	36	12	34	14	32	184
31	17	29	19	27	21	25	23	192
16	30	18	28	20	26	22	24	184

III

59	1	57	3	55	5	53	7	51	9	300
0	58	2	56	4	54	6	52	8	50	290
49	11	47	13	45	15	43	17	41	19	300
10	48	12	46	14	44	16	42	18	40	290
39	21	37	23	35	25	33	27	31	29	300
20	38	22	36	24	34	26	32	28	30	290

II

47	1	45	3	4	42	6	40	188
0	46	2	44	43	5	41	7	188
39	9	37	11	12	34	14	32	188
8	38	10	36	35	13	33	15	188
31	17	29	19	20	26	22	24	188
16	30	18	28	27	21	25	23	188

IV

59	1	57	3	55	5	53	7	46	9	295
0	58	2	56	4	54	6	52	13	50	295
49	11	47	8	45	15	43	17	41	19	295
10	48	12	51	14	44	16	42	18	40	295
39	21	37	23	35	25	26	32	28	29	295
20	38	22	36	24	34	33	27	31	30	295

Figure 4.2

The zig-zag assignment results in all columns having sum (pq-1)(p/2) and the row sums alternating between (pq)(q/2) and (pq-2)(q/2). There are two cases to consider:

CASE 1: *One of p,q is a multiple of 4.* We can suppose without loss of generality that 4|q. Let R be a p by q polysquare whose cells are assigned the elements of Z(pq) in zig-zag fashion as in the example in Figure 4.2(I). Let R_1, R_2, ..., R_p denote the rows of R from top to bottom. For each i∈{1, ..., p} let A_i denote the left half of row R_i, and let B_i denote the right half. In case i is odd we have $\Sigma A_i = \Sigma B_i$ = (pq)(q/4), and if i is even we have $\Sigma A_i = \Sigma B_i$ = (pq-2)(q/4). Interchange each of the pairs half-rows {B_1, B_2}, {B_3, B_4}, ..., {B_{p-1}, B_p}. This interchange will decrease ΣR_i by (pq)(q/4) - (pq-2)(q/4) = q/2 in case i is odd, and increase ΣR_i in case i is even. As a consequence ΣR_i

= (pq)(q/2)-(q/2) = (pq-1)(q/2) if i is odd, and ΣR_i = (pq-2)(q/2)+(q/2) = (pq-1)(q/2) if i is even. All the columns in the revised polysquare have sum (pq-1)(p/2) and all the rows have sum (pq-1)(q/2), so R is now a 2-sum rectangle.

We shall refer to any pair of values which lie in the same column of a rectangular polysquare as a *σ-pair.* Our constructions will involve interchanging the elements of a σ-pair or interchanging two σ-pairs.

CASE 2: *Neither p nor q is a multiple of 4.* Let R be a p by q rectangular polysquare whose cells have been assigned the elements of Z(pq) in zigzag fashion as in Figure 4.2(III). As usual the column sums are all the same and the row sums alternate between (pq)(q/2) and (pq-2)(q/2). Note that the σ-pairs in each of the row pairs $\{R_1, R_2\}$, $\{R_3, R_4\}$, ..., $\{R_{p-1}, R_p\}$ all have sum pq-1 and thus are interchangeable without altering any of the column sums of R. The σ-pair {pq-q+1, q-2} occupies R_1 and R_2, and the σ-pair {(3q/2)-2, pq - (3q/2) + 1} occupies rows R_3, R_4. These σ-pairs are interchanged so that the R_4 value pq - (3q/2) + 1 interchanges places with the R_1 value pq-q+1, and the R_3 value (3q/2)-2 interchanges positions with the R_2 value q-2. The consequence is that ΣR_1 and ΣR_3 are reduced by q/2, and ΣR_2, ΣR_4 are increased by q/2. Then $\Sigma R_1 = \Sigma R_2 = \Sigma R_3 = \Sigma R_4$ = (pq-1)(q/2).

In case p = 4k+2 then there are k groups of four rows which follow each other in succession. In each group σ-pairs in analogous positions can be interchanged as in Figure 4.3(I) with the result that all rows involved will have the same sum, (pq-1)(q/2). This leaves only the rows R_{p-1} and R_p whose sums must be adjusted. In these rows we require a σ-pair {A, B} whose values differ by q/2. Then we have A+B = pq-1 and A-B = q/2 which yields A = (1/2)(pq-1+(q/2)) and B = (1/2)(pq-1-(q/2)). The first value of row R_p is (1/2)(p-2)q and the last value in this row is pq/2. It is easy to check that (1/2)(p-2)q < B < pq/2, and thus the σ-pair {A,B} does indeed lie in rows R_{p-1} and R_p.

Now there are two possibilities: (i) the larger value, (A), lies in R_{p-1}. In this case we just interchange A and B with the result that $\Sigma R_{p-1} = \Sigma R_p$ = (pq-1)(q/2) as desired. This interchange is illustrated in Figure 4.3(II). (ii) the larger value, (A), lies in R_p. In this case, to obtain the desired equivalence between ΣR_{p-1} and ΣR_p, we interchange not only the σ-pair {A,B} but also the two adjacent σ-pairs which we will denote by {C,D} and {E,F}. Putting {C,D} before {A,B} and {E,F} after {A,B} yields: C = A+1; E = A-1; D = pq-1-C and F = pq-1-E. Then:

$$(C+B+E) - (D+A+F) = (2A+B) - (D+A+F) = (A+B) - (D + F) =$$
$$(pq-1)-(2pq-2-2A) = q/2.$$

Thus the interchange of these three σ-pairs reduces R_{p-1} by q/2 and increases R_p by q/2, again resulting in $\Sigma R_{p-1} = \Sigma R_p$ = (pq-1)(q/2). In Figure 4.2(III) this sort of interchange is illustrated. □

139	1	137	3	135	5	133	7	131	9	129	11	127	13	980
0	138	2	136	4	134	6	132	8	130	10	128	12	126	966
125	15	123	17	121	19	119	21	117	23	115	25	113	27	980
14	124	16	122	18	120	20	118	22	116	24	114	26	112	966
111	29	109	31	107	33	105	35	103	37	101	39	99	41	980
28	110	30	108	32	106	34	104	36	102	38	100	40	98	966
97	43	95	45	93	47	91	49	89	51	87	53	85	55	980
42	96	44	94	46	92	48	90	50	88	52	86	54	84	966
83	57	81	59	79	61	77	63	75	65	73	67	71	69	980
56	82	58	80	60	78	62	76	64	74	66	72	68	70	966

I

139	1	137	3	135	5	133	7	131	9	129	11	120	13	973
0	138	2	136	4	134	6	132	8	130	10	128	19	126	973
125	15	123	17	121	12	119	21	117	23	115	25	113	27	973
14	124	16	122	18	127	20	118	22	116	24	114	26	112	973
111	29	109	31	107	33	105	35	103	37	101	39	92	41	973
28	110	30	108	32	106	34	104	36	102	38	100	47	98	973
97	43	95	45	93	40	91	49	89	51	87	53	85	55	973
42	96	44	94	46	99	48	90	50	88	52	86	54	84	973
83	57	81	59	79	61	77	63	75	65	66	67	71	69	973
56	82	58	80	60	78	62	76	64	74	73	72	68	70	973

II

Figure 4.3

4.3 Rectangles with two Odd Dimensions

With two odd dimensions the constructions are more varied and hard to generalize. In case the number p of rows is 3 we can obtain the following general construction.

Theorem 4.4 *Let R be a 3 by q rectangular polysquare with q ≥ 5 and q odd . Then the elements of Z(3q) can be assigned to the cells of R so that each column has sum 3((3q-1)/2), and each row has sum q((3q-1)/2).*

Proof. There are two cases to consider: (1) q = 5+4t, and (2) q = 7+4t for t a nonnegative integer. Let R_1, R_2, R_3 denote the top, middle and bottom row of R respectively. First the cells of R are assigned the elements from Z(3q) according to an equipartition of Z(3q) into q three element subsets each with sum 3((3q-1)/2). The even values 0, 2, ..., q-1 are assigned in zig-zag fashion to R_1, R_2. Opposite these values the consecutive elements 3q-1, 3q-2, ..., (1/2)(5q-1) are inserted. In the remainder of R_1, R_2 the odd values 1, 3, ..., q-2 are assigned in zig-zag fashion, and opposite these values the consecutive elements (1/2)(5q-3), ..., 2q are placed. (See the examples in Figure 4.4 with q = 5, 9, 3.) The elements for R_3 are chosen so that each column has the desired sum, 3((3q-1)/2). This row will contain all the values from q up to 2q-1. The sum of the values in the third row then has the desired value of q((3q-1)/2).

Consequently, it is only necessary to adjust the elements in R_1, R_2 by means of interchanges so that these rows also have the desired sum. Let Δ denote the desired common row sum q((3q-1)/2).

CASE 1 *(q = 5+4t)*. In this case $\Sigma R_1 = \Delta + \varepsilon$, and $\Sigma R_2 = \Delta - \varepsilon$, where ε = (1/4)(3q+1). To equalize the row sums it is thus necessary to transfer a value of ε from R_1 to R_2. The first step in this process involves interchanging the first two values {0, 3q-2} in R_2 with the two values {(1/2)(3q-1), (1/2)(3q-3)} beneath them in R_3 . This interchange will not affect row sums since the two pairs of values have the same sum. The further interchange of the first value in R_1 with the new first value in R_2 results in a reduction of the row sum in R_1 by (1/2)(3q-1). Now it should be observed that consecutive σ-pairs which lie in the first two rows, with one exception, when interchanged either add 3 or subtract 3 from the sum of R_1. We will focus on the consecutive σ-pairs whose interchange increases the sum of R_1 by 3. To this end we must find the integer k such that:

(1) $(-1/2)(3q-1) + 3k = -\varepsilon = (-1/4)(3q+1)$.

This equation yields $q = 4k+1$, and thus $k = (1/4)(q-1) = (1/4)(4+4t) = 1+t$. It is easy to find a total of 1+t σ-pairs each of which improve the sum of the top row by 3. The sum total of these interchanges reduce the sum of R_1 by ε, and thus increase the sum of R_2 by ε. The result is that all three rows now have sum Δ and thus a 2-sum rectangle is created. □

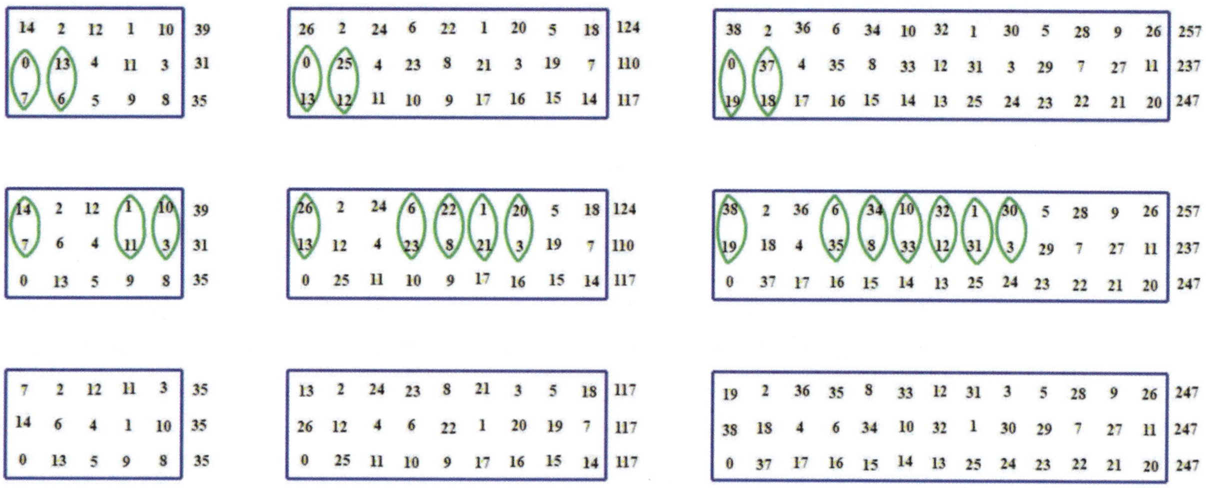

Figure 4.4

CASE 2 *(q = 7+4t)*. This case breaks further into three subcases: (CASE 2.1 (q=7+12t); CASE 2.2 (q = 11+12t), and CASE 2.3 (q=15+12t). For this CASE $\Sigma R_1 = \Delta+\varepsilon$; $\Sigma R_2 = \Delta-\varepsilon$, and $\Sigma R_3 = \Delta$ where $\varepsilon = (1/4)(5q+1)$.

CASE 2.1 *(q = 7+12t)*. The last σ-pair in R_1, R_2 is {2q, q-2} with 2q the last element of R_1. The interchange of this pair reduces ΣR_1 by q+2. We will use the fact again that adjacent σ-pairs can be found in R_1, R_2 which either increase or decrease ΣR_1 by 3. Now observe that the equation:

$$-q - 2 - 3k = -\varepsilon = -(1/4)(5q+1)$$

Has the solution $k = (1/12)(q-7) = t$. Thus by interchanging t σ-pairs along with the σ-pair {2q, q-2}, ΣR_1 is reduced by ε and ΣR_2 is increased by ε, with the consequence that all three rows have sum Δ. Examples with q = 7, 19 are illustrated in Figure 4.5.

20	2	18	6	16	3	14	79
0	19	4	17	1	15	5	61
10	9	8	7	13	12	11	70

56	2	54	6	52	10	50	14	48	18	46	3	44	7	42	11	40	15	38	556
0	55	4	53	8	51	12	49	16	47	1	45	5	43	9	41	13	39	17	508
28	27	26	25	24	23	22	21	20	19	37	36	35	34	33	32	31	30	29	532

20	2	18	6	16	3	5	70
0	19	4	17	1	15	14	70
10	9	8	7	13	12	11	70

56	2	54	6	52	10	50	14	48	18	46	3	44	7	42	11	13	39	17	532
0	55	4	53	8	51	12	49	16	47	1	45	5	43	9	41	40	15	38	532
28	27	26	25	24	23	22	21	20	19	37	36	35	34	33	32	31	30	29	532

Figure 4.5

CASE 2.2 *(q = 11 + 12t).* In this case first interchange the adjacent σ-pairs {q-1, (1/2)(5q-1)} and {(1/2)(5q-3), 1} which lie in R_1, R_2 near the middle of R. These interchanges reduce ΣR_1 by q-3. The equation

$$-(q-3) - 3k = -\varepsilon = -(1/4)(5q+1)$$

Has solution k = (1/12)(q+13) = t+2. Thus by interchanging t+2 σ-pairs in R_1, R_2 each of which reduce ΣR_1 by 3, a net reduction of ΣR_1 by ε is obtained and an increase of ΣR_2 by ε is obtained. Consequently all three rows of R now have the same sum Δ. In Figure 4.6 an example with q = 23 is illustrated.

CASE 2.3 *(q = 15+12t).* First interchange the first two σ-pairs in R_2, R_3. The R_2 components and the R_3 components have the same sum 3q-2, so their interchange will not affect ΣR_2 and ΣR_3. Then interchange the first element of R_1 with the new first element of R_2. These interchanges result in a diminution of ΣR_1 by (1/2)(3q-1). The equation

$$-(1/2)(3q-1) + 3k = -\varepsilon = -(1/4)(5q+1)$$

has solution k = (1/12)(q-3) = (1/12)(12 + 12t) = t+1. Thus the interchange of three adjacent σ-pairs each of which add 3 to ΣR_1 has the effect of reducing ΣR_1 by ε and increasing ΣR_2 by ε thus equalizing the three row sums. In Figure 4.7 an example with q = 27 is illustrated. □

With 5 or more rows the constructions are more difficult. We will just exhibit a few examples in Figures 4.8 and 4.9. In Figure 4.8 we use a method which is standard for the construction of odd order magic squares: values are assigned consecutively by moving one cell to the right and one cell down. In case a right boundary is encountered "one cell to the right" moves to the first cell in the same row as if the rectangle was rolled into a cylinder horizontally. In case the bottom of the rectangle is encountered, then "one cell down" is interpreted to mean a jump to the top of the column, as if the rectangle were rolled into a cylinder vertically. This technique works very nicely for the 5 by 7 and 5 by 9 rectangles, but unfortunately does not work so neatly for larger rectangles. In

68	2	66	6	64	10	62	14	60	18	58	22	56	3	54	7	52	11	50	15	48	19	46	420
0	67	4	65	8	63	12	61	16	59	20	57	1	55	5	53	9	51	13	49	17	47	21	362
34	33	32	31	30	29	28	27	26	25	24	23	45	44	43	42	41	40	39	38	37	36	35	391

0	67	4	65	8	63	62	14	60	18	58	57	1	3	54	7	52	11	50	15	48	19	46	391
68	2	66	6	64	10	12	61	16	59	20	22	56	55	5	53	9	51	13	49	17	47	21	391
34	33	32	31	30	29	28	27	26	25	24	23	45	44	43	42	41	40	39	38	37	36	35	391

Figure 4.6

80	2	78	6	76	10	74	14	72	18	70	22	68	26	66	3	64	7	62	11	60	15	58	19	56	23	54	1114
0	79	4	77	8	75	12	73	16	71	20	69	24	67	1	65	5	63	9	61	13	59	17	57	21	55	25	1046
40	39	38	37	36	35	34	33	32	31	30	29	28	27	53	52	51	50	49	48	47	46	45	44	43	42	41	1080

80	2	78	6	76	10	74	14	72	18	70	22	68	26	66	3	64	7	62	11	60	15	58	19	56	23	54	1114
40	39	4	77	8	75	12	73	16	71	20	69	24	67	1	65	5	63	9	61	13	59	17	57	21	55	25	1046
0	79	38	37	36	35	34	33	32	31	30	29	28	27	53	52	51	50	49	48	47	46	45	44	43	42	41	1080

40	2	78	77	8	75	12	14	72	18	70	22	68	26	66	3	64	7	62	11	60	15	58	19	56	23	54	1080
80	79	4	6	76	10	74	73	16	71	20	69	24	67	1	65	5	63	9	61	13	59	17	57	21	55	25	1080
0	39	38	37	36	35	34	33	32	31	30	29	28	27	53	52	51	50	49	48	47	46	45	44	43	42	41	1080

Figure 4.7

Figure 4.9 a 2-sum 7 by 9 rectangle is constructed using our usual method of using the sets in an equipartition to populate the columns and then applying interchanges within the columns. Although the sequence of required interchanges is more complex, still the method is reliable and so we make the following Conjecture.

Conjecture 4.5 *If p, q have the same parity and q>p≥2, then a 2-sum solution over Z(pq) exists for any p by q rectangular polysquare.*

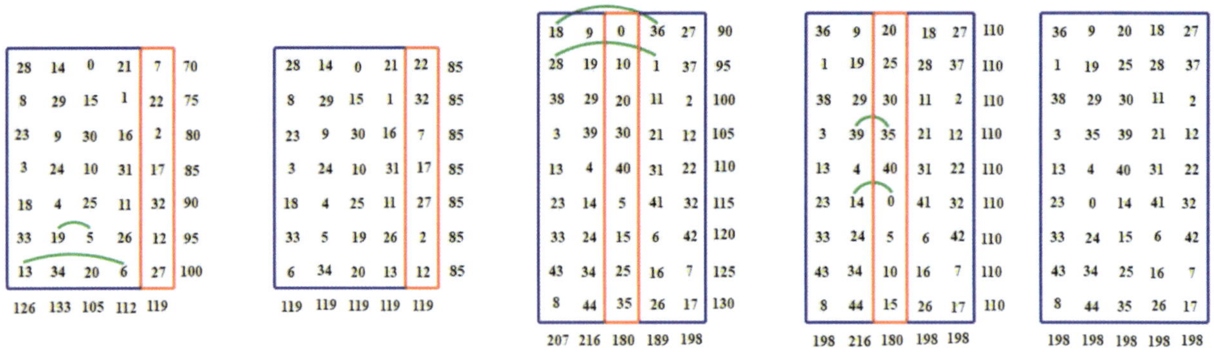

Figure 4.8

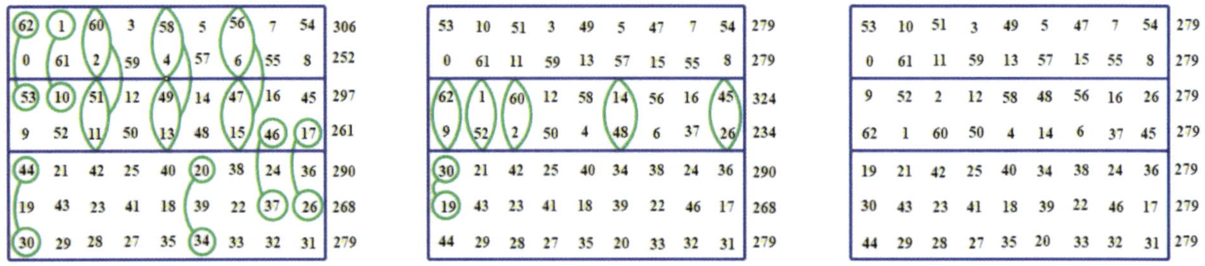

Figure 4.9

In Figure 4.8, the values in the columns with red boundary are simply permuted with the immediate result that all row sums are equalized. It may be that this can be accomplished only in the 5 by 7 and 5 by 9 rectangles. These rectangles can also be solved using equipartitions; an exercise we leave for the reader.

4.4 Factorizations

Let A be a natural p by q 2-sum rectangle over {0, ..., n-1}, (n = pq), and let t be a positive integer. Let t + A denote the p by q rectangle obtained by adding t to the elements in each cell of A. We will call t+A an *associate of A*. If the rows of A have sum S_A and the columns of A have sum $T_{A,}$ then the rows of t+A have sum qt + S_A and the columns of t+A have sum pt+T_A. The sets {n+A}, {2n+A}, ..., {kn+A} form a partition of {0, ..., kn+n-1}.

Let B be a natural r by s 2-sum rectangle over {0, ..., m-1}, (m = rs) with row sum S_B and column sum T_B. Then the sets {n+A}, {2n+A}, ..., {(rs-1)n+A} form a partition of {0, ..., rsn-1} which is the full set of elements required to form a pr by qs 2-sum rectangle. Let M denote the rectangle with these dimensions formed by replacing each element x of B by the the associate nx+A. If (x_1, ..., x_s) is a row of B, then the row of p by q blocks (nx_1+A, ..., nx_s+A) occurs in M. Any row of single elements in this row of blocks will have sum qn(x_1 + ... +x_s) + s(S_A) = qnS_B+sS_A, and, in fact every row of M will have this sum. If (y_1, ..., y_r) is a column of B, and (ny_1+A, ..., ny_r+A) is the corresponding column of blocks in M, then each column of M will have sum pnT_B+rT_A. The rectangle M is thus a 2-sum rectangle which we will denote by M = A⊕B, the *dilation of A by B.* In Figure 4.10 an example of this construction is given in which A is a 3 by 7 2-sum rectangle and B is a 5 by 3 2-sum rectangle.

A

20	2	8	7	16	3	14
0	19	4	17	13	12	5
10	9	18	6	1	15	11

B

14	0	7
2	6	13
5	12	4
11	9	1
3	8	10

M

294+A	A	147+A
42+A	126+A	273+A
105+A	252+A	84+A
231+A	189+A	21+A
63+A	168+A	210+A

Figure 4.10

The same rectangle can also be *factored* C⊕D where C is a 3 by 3 magic square and D is a 5 by 7 2-sum rectangle as in Figure 4.11.

C

1	8	3
6	4	2
5	0	7

D

34	1	25	13	20	5	21
0	33	11	14	30	19	12
17	9	15	24	4	22	28
7	26	32	3	23	10	18
27	16	2	31	8	29	6

M

306+C	9+C	225+C	117+C	180+C	45+C	189+C
C	297+C	99+C	126+C	270+C	171+C	108+C
153+C	81+C	135+C	216+C	36+C	198+C	252+C
63+C	234+C	288+C	27+C	207+C	90+C	162+C
243+C	144+C	18+C	279+C	72+C	261+C	54+C

Figure 4.11

From the example in Figure 4.11 we can see that if R is a p by q rectangular polysquare and d = gcd{p,q} is the greatest common divisor of {p,q} and p/d >1, q/d>1, then R can be factored R = M⊕A where M is a magic square of order d and A is a 2-sum rectangle with relatively prime dimensions p/d, q/d. As a consequence the construction of 2-sum rectangles, (with odd dimensions), reduces to the construction of 2-sum rectangles whose dimensions are relatively prime, and to 2-sum rectangles one of whose dimensions is a multiple of the other.

4.5 Near Magic Partitions

If M(S) is a p by p square whose cells have been assigned values from a set S, and whose rows and columns, but not necessarily corner diagonals, have the same sum, we will call M(S) a *near magic square over S.* Let $\{S_1, \ldots, S_k\}$ be a partition of $Z(kp^2)$ into k sets each with p^2 elements, and such that $M(S_i)$ is a near magic square over S_i for each $i \in \{1, \ldots, k\}$. Require also that the all the rows and columns of every $M(S_i)$ have the same natural sum. Then we call $\{S_1, \ldots, S_k\}$ a *near magic partition of $Z(kp^2)$.* The sequence of squares $[M(S_1), \ldots, M(S_k)]$ form a p by kp 2-sum rectangle. In Figure 4.12 two examples of 2-sum rectangles obtained via near magic factorizations are illustrated. Figures 4.12(I,III) represent 2-sum rectangles obtained by means of equipartitions as in Section 4.2. Interchanges are then made within various 2 by 2 subsquares, to obtain the desired near magic partition. Note the flexibility of these constructions: in 4.12(II) the two column pairs {AD, BC} as well as {AB, CD} yield a near magic partition, and in 4.12(IV) the triples of column pairs {AB, CD, EF}, {AD, BC, EF}, {AD, BE, CF}, {AB, CF, DE} all yield near magic partitions.

I

31	1	29	3	4	26	6	24
0	30	2	28	27	5	25	7
23	9	21	11	12	18	14	16
8	22	10	20	19	13	17	15

III

47	1	45	3	43	5	6	40	8	38	10	36
0	46	2	44	4	42	41	7	39	9	37	11
35	13	33	15	31	17	18	28	20	26	22	24
12	34	14	32	16	30	29	19	27	21	25	23

II

31	1	2	28	27	5	6	24
0	30	29	3	4	26	25	7
23	9	10	20	19	13	14	16
8	22	21	11	12	18	17	15
A		B		C		D	

IV

0	46	45	3	4	42	41	7	8	38	37	11
47	1	2	44	43	5	6	40	39	9	10	36
12	34	33	15	16	30	29	19	20	26	25	23
35	13	14	32	31	17	18	28	27	21	22	24
A		B		C		D		E		F	

Figure 4.12

Define a *domino* to be a 2 by 4 2-sum rectangle. We will keep *domino* in italics to distinguish it from the usual meaning of domino as a polysquare consisting of two adjacent cells. If the first row of a *domino* is given by (a, b, c, d) and the second row by (e, f, g, h), then the equations a+e = b+f = c+g = d+h and a+b+c+d = e+f+g+h imply that the row sum is exactly twice the column sum. Our primary concern will be with the type of *dominoes* which constitute a dissection of both 2-sum rectangles in Figure 4.12. These involve two sequences of consecutive values each of length 4, threaded together so that both rows have the same sum.

Suppose p, q are dimensions of a rectangular polysquare and pq is a multiple of 8. Then Z(pq) has a partition $\Pi = \{S_0, S_4, S_8, \ldots, S_{pq-4}\}$ where $S_t = \{t, t+1, t+2, t+3\}$ for t = 0, 4,8, ..., pq-4, and the sets in Π can be put into pairs of the form $\{S_i, S_j\}$ where j = pq-4-i. Each such pair can be used to create a *domino* with column sum M = pq-1 and row sum 2M. Any rectangle with even area can be tiled using (standard) dominoes. If the dimensions p, q of such a rectangle are doubled then the resulting rectangle R has a similar tiling using our *dominoes,* as illustrated in Figure 4.13. Each column of R will have sum pM and each row will have sum qM. We have proved the following.

28	50	49	31	4	74	73	7
51	29	30	48	75	5	6	72
36	43	12	66	65	15	32	47
42	37	67	13	14	64	46	33
41	38	0	79	16	63	45	34
39	40	78	1	62	17	35	44
20	59	77	2	61	18	8	71
58	21	3	76	19	60	70	9
57	22	24	54	53	27	69	10
23	56	55	25	26	52	11	68

Figure 4.13♦

Theorem 4.6 *If p, q are even and pq is a multiple of 8, then there exists a 2-sum p by q rectangle which has a dissection into dominoes.* □

Note that if p, q are both multiples of 4 then a p by q rectangle can be tiled with 4 by 4 near magic squares since a *domino* tiling may be chosen with the *dominoes* paired to form 4 by 4 squares. If, in addition, p = q, then a near magic square is obtained which has a dissection into 4 by 4 near magic squares. Dissections into near magic squares of order p where p is even but not a multiple of 4 are alsfo possible.

Theorem 4.7 *Let p be an even integer which is not a multiple of 4 and let q = mp where m ≥2 is even. Then a p by q rectangle R over Z(pq) has a partition into m near magic squares of order p.*

Proof. Populate the cells of R with the elements of Z(pq) in the usual zig-zag fashion as in the example in Figure 4.14. Divide R into m p by p consecutive blocks $\{B_1, B_2, ..., B_m\}$, and consider these blocks in m/2 pairs $\{B_1, B_2\}$, $\{B_3, B_4\}$, ..., $\{B_{m-1}, B_m\}$. In each even subscripted block interchange consecutive pairs of rows. This will convert R into a 2-sum rectangle. In fact, each block pair $\{B_i, B_{i+1}\}$ will be a 2-sum rectangle, each row with sum p(pq-1) and each column with sum (p/2)(pq-1). The odd rows in B_i have sum $(1/2)(p^2q)$ as do the even rows in B_{i+1}. The even rows in B_i have sum (p/2)(pq-2) as do the odd rows in B_{i+1}. Thus the difference between sums of these rows is p.

For each odd subscript i, let C_i denote the second to last column of B_i and let D_i denote the column in B_{i+1} which is exactly p/2 units to the right of C_i. If $C_i = (a_1, a_2, ..., a_p)$ then $D_i = (a_1 - (p/2), a_2 + (p/2), a_3 - (p/2), a_4 + (p/2), ..., a_p + (p/2))$.

Consequently the effect of interchanging columns C_i, D_i is to reduce the odd rows of B_i by p/2, and increase the even rows of B_i by p/2. Then every row of B_i will have sum (p/2)(pq-1). The same interchange has the effect of increasing the sum of the odd rows of B_{i+1} by p/2, and decreasing the sum of the even rows of B_{i+1} by p/2. Thus every row of B_{i+1} will have sum (p/2)(pq-1). All the blocks B_1, ..., B_m have now been converted to near magic squares, completing the proof. □

143	1	141	3	139	5	6	136	8	134	10	132	131	13	129	15	127	17	18	124	20	122	22	120
0	142	2	140	4	138	137	7	135	9	133	11	12	130	14	128	16	126	125	19	123	21	121	23
119	25	117	27	115	29	30	112	32	110	34	108	107	37	105	39	103	41	42	100	44	98	46	96
24	118	26	116	28	114	113	31	111	33	109	35	36	106	38	104	40	102	101	43	99	45	97	47
95	49	93	51	91	53	54	88	56	86	58	84	83	61	81	63	79	65	66	76	68	74	70	72
48	94	50	92	52	90	89	55	87	57	85	59	60	82	62	80	64	78	77	67	75	69	73	71

143	1	141	3	136	5	6	139	8	134	10	132	131	13	129	15	124	17	18	127	20	122	22	120
0	142	2	140	7	138	137	4	135	9	133	11	12	130	14	128	19	126	125	16	123	21	121	23
119	25	117	27	112	29	30	115	32	110	34	108	107	37	105	39	100	41	42	103	44	98	46	96
24	118	26	116	31	114	113	28	111	33	109	35	36	106	38	104	43	102	101	40	99	45	97	47
95	49	93	51	88	53	54	91	56	86	58	84	83	61	81	63	76	65	66	79	68	74	70	72
48	94	50	92	55	90	89	52	87	57	85	59	60	82	62	80	67	78	77	64	75	69	73	71

Figure 4.14

Let M be a square matrix of order n with entries from an abelian group G. A *transversal* T in M is a collection of n entries which lie one in each row and column of M. A *complete set of transversals* in M is a collection of n mutually disjoint transversals of M. Thus every entry of M lies in exactly one transversal from a complete set. A pair of complete transversal sets are *complementary* in case each transversal in one set shares exactly one cell with each transversal in the other set. Each *slope (1)* diagonal, (including broken diagonals) is a transversal of M, and the set of all slope (1) diagonals constitutes a complete set of transversals in M. The same can be said of the collection of slope(-1) diagonals in M. Moreover these two transversal sets are complementary in case n is odd.

Let p be an odd integer (p≥3) and let $\{x_k : k \in Z(p)$ be a set of nonnegative integers. For each $k \in Z(p)$ let R_k denote the sequence

$$R_k = (x_k, 1+x_k, 2+x_k, \ldots, p-1+x_k),$$

and suppose these sequences are mutually disjoint. Let $R = \cup(\{R_k\} : k \in Z(p)\}$, and let M(R) denote the square matrix over R of order p with rows R_0, ..., R_{p-1}. We call M(R) a *seed square over R.*

Lemma 4.8 *The values in any seed square over R can be rearranged to form a near magic square over R.*

Proof. Let M(R) be defined as in the previous paragraph and let C= $x_0 + x_1 + \dots + x_{p-1}$ denote the sum of the elements in the first column of M(R). Let S denote the set of elements in the middle column of M(R), so ΣS is given by

$$\Sigma S = (x_0 + (1/2)(p-1)) + (x_1 + (1/2)(p-1))) + \dots + (x_{p-1} + (1/2)(p-1))) = C + (p/2)(p-1).$$

Let T = $\{a_0, a_1, \dots, a_{p-1}\}$ denote any transversal in M(R). The elements of T occur one in each column of M(R), and thus there is a permutation ϕ of Z(p) such that $a_i = x_i + \phi(i)$ for each $i \in Z(p)$. Then ΣT is given by

$$\Sigma T=(x_0 + \phi(0)) +(x_1 + \phi(1)) + \dots + (x_{p-1} + \phi(p-1)) = C + \Sigma\{\phi(i): i \in Z(p)\} = C+(p/2)(p-1).$$

Consequently ΣT = ΣS for every transversal T of M(R). In particular, the sum of the elements in any slope(-1) diagonal is ΣS. Construct a new square M′(R) by using the complete set of slope(-1) diagonals as columns. Thus every column of M′(R) has sum ΣS. Finally permute the values in the columns of M′(R) to obtain the square M″(R) in which each row contains the values belonging to a slope(1) diagonal of M(R). This is possible since the two sets of diagonals are complementary. The columns of M″(R) continue to have sum ΣS, and now the rows of M″(R) also have this sum. Thus M″(R) is a near magic square over R. □

An example of this construction is given for an order 5 seed square in Figure 4.15

0	1	2	3	4
15	16	17	18	19
35	36	37	38	39
55	56	57	58	59
70	71	72	73	74

M(R)

0	1	2	3	4
16	17	18	19	15
37	38	39	35	36
58	59	55	56	57
74	70	71	72	73

M'(R)

37	70	18	56	4
0	38	71	19	57
16	59	2	35	73
74	17	55	3	36
58	1	39	72	15

M"(R)

Figure 4.15

Theorem 4.6 and the following result will prove very useful in the construction of magic polygons with more than 4 sides.

Theorem 4.9 *Let m, p be odd integers with p≥3 and m≥3. Then $Z(mp^2)$ can be partitioned into m near magic squares of order p, all of which have the same row and column sums.*

Proof. Let A denote a p by q matrix with q = mp and with rows formed by the sequences

$$S(t) = (pt, pt+1, pt+2, \dots, pt+p-1),\ t = 0, \dots, q-1.$$

Let m = (1/2)(p-1), then the middle column of A is given by

$$S =(m, m+p, m+2p, \dots, m+(q-1)p).$$

Create a new set S′ from S by replacing each element according to the bijection $\phi(m+pt) = t$. S′ has an odd number pm of elements with both p, m odd. By Lemma 3.2 there exists an equipartition of S′ into m sets each with p elements. By retracting the mapping ϕ an equipartition of S into m sets each with p elements is obtained. The equipartition of S induces a partition of the rows of A into m sets each consisting of p rows of A; the rows in any one such set consisting of rows in A whose middle values lie in the same set determined by the equipartition of S.

The result is the creation of m p by p seed squares each of whose middle columns have the same sum, namely p(pq-1)/2. By Lemma 4.8 the elements in each of these seed squares can be rearranged to form near magic squares with all rows and columns having sum p(pq-1)/2. □

In Figure 4.16 the various components of the foregoing construction are illustrated for p = 5 and m = 3.

4.6 Magic Boxes

When pqr unit cubes are placed together to form a rectangular parallelopiped with dimensions p, q, r, a structure we call a *box* is obtained. Through any one cell of a box there are three mutually orthogonal linear polysquares, one in the X direction, one in the Y direction and one in the Z direction. If the elements of Z(pqr) are assigned one to each cell of a box so that all linear polysquares in the X direction have the same sum; all linear polysquares in the Y direction have the same sum and all linear polysquares in the Z direction have the same sum, we have a *magic box over Z(pqr)*. In case p=q=r, the well studied magic cube is obtained. According to [7] magic cubes of any odd order p≥3 can be constructed, as well as all magic cubes whose order is a multiple of 4. In Figure 4.16 magic cubes of order 3 and 4 are illustrated. To simplify the illustration all cells have been collapsed to points and linear polysquares are represented by lines.

A =

0	1	2	3	4
5	6	7	8	9
10	11	12	13	14
15	16	17	18	19
20	21	22	23	24
25	26	27	28	29
30	31	32	33	34
35	36	37	38	39
40	41	42	43	44
45	46	47	48	49
50	51	52	53	54
55	56	57	58	59
60	61	62	63	64
65	66	67	68	69
70	71	72	73	74

70	71	72	73	74
30	31	32	33	34
20	21	22	23	24
5	6	7	8	9
50	51	52	53	54
35	36	37	38	39
10	11	12	13	14
60	61	62	63	64
55	56	57	58	59
15	16	17	18	19
0	1	2	3	4
65	66	67	68	69
25	26	27	28	29
45	46	47	48	49
40	41	42	43	44

70	71	72	73	74
31	32	33	34	30
22	23	24	20	21
8	9	5	6	7
54	50	51	52	53
35	36	37	38	39
11	12	13	14	10
62	63	64	60	61
58	59	55	56	57
19	15	16	17	18
0	1	2	3	4
66	67	68	69	65
27	28	29	25	26
48	49	45	46	47
44	40	41	42	43

22	50	33	6	74
54	32	5	73	21
31	9	72	20	53
8	71	24	52	30
70	23	51	34	7
62	15	13	56	39
19	12	55	38	61
11	59	37	60	18
58	36	64	17	10
35	63	16	14	57
27	40	68	46	4
44	67	45	3	26
66	49	2	25	43
48	1	29	42	65
0	28	41	69	47

Figure 4.16

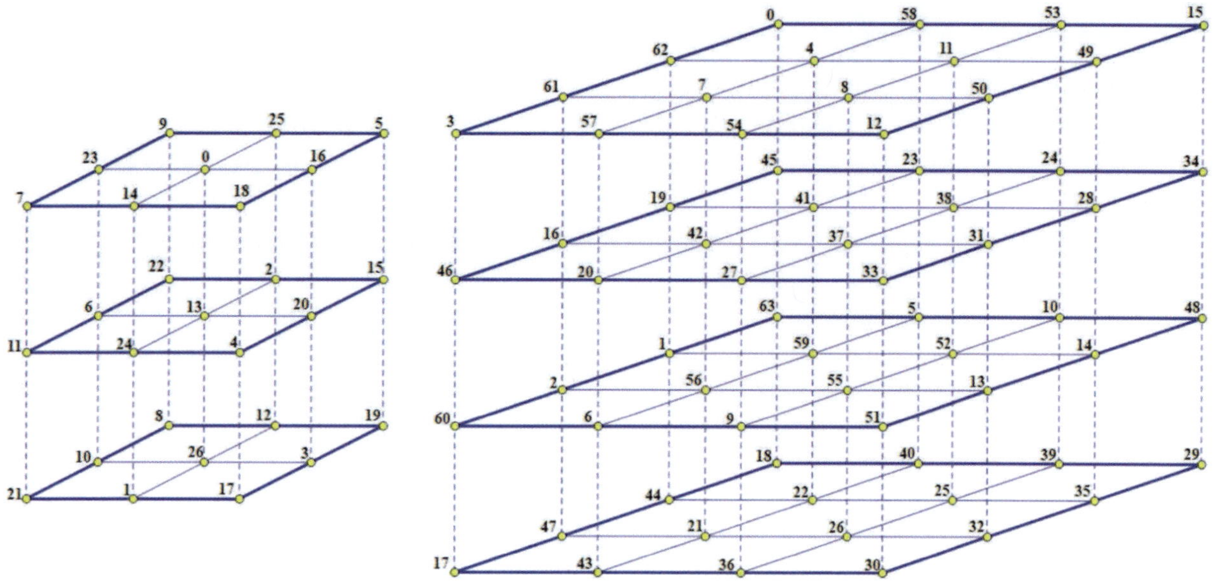

Figure 4.17

Obviously the collections of parallel planes in a magic box provide constructions of 2-sum rectangles.

Just as a 2-sum rectangle extends the notion of a magic square, a magic box extends the idea of a magic cube. The construction of magic boxes is an interesting subject in its own right, but not directly relevant to our main purpose of constructing magic polygons. So we will be content here just to exhibit a few of our constructions. Just as the dimensions of a 2-sum rectangle must have the same parity by Theorem 4.1, the three dimensions of a magic box must all have the same parity.

Theorem 4.10 *The dimensions p,q,r of a magic box over Z(pqr) must have the same parity. The three sums are (1/2)(p)(pqr-1), (1/2)(q)(pqr-1) and (1/2)(r)(pqr-1).*

Proof. Let R be a magic box with dimensions p,q,r. Suppose p is odd and q,r are even. Then R consists of r p by q parallel near magic squares. These can be reassembled to form a p by qr 2-sum rectangle. But since p, qr have opposite parity such a 2-sum rectangle cannot exist by Theorem 4.1. The same result occurs in case p is even and q,r are odd. R is composed of qr linear polysquares of length p. Since ΣZ(pqr) = (1/2)(pqr)(pqr-1), it must be that each linear polysquare of length p has sum (1/qr)(1/2)(pqr)(pqr-1) = (1/2)(p)(pqr-1). The sums in the other two directions follow similarly. □

A *platform* is a box in which one of the dimensions is 2. If P is a platform with dimensions 2, p, q and the cells of P are assigned elements from a subset S of an abelian group G such that all constituent linear polysquares in the same direction have the same sum, we say that P is *magic over S.* We will only consider the case in which G = Z(2pq) and distinct elements from S are assigned to the cells of P. In Figure 4.18 two examples of magic platforms are illustrated.

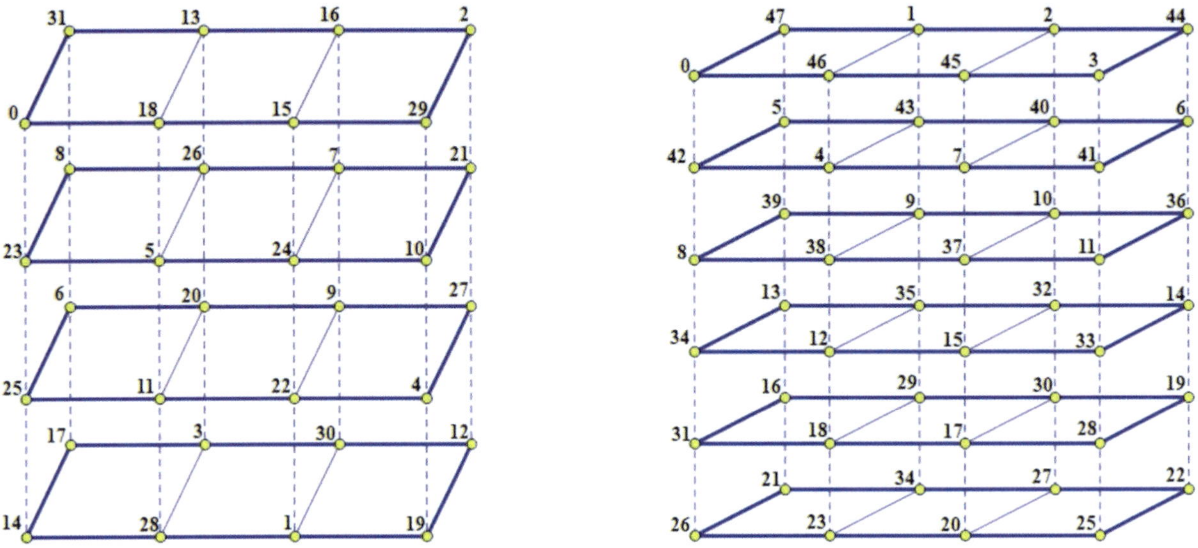

Figure 4.18

In Figure 4.18 both platforms have been represented conveniently using a linear representation. Linear polysquares in a box will just be referred to as *lines.* If M = 2pq-1, then a pair of points x, y with x+y = M will be referred to as *complements.* In any platform the lines of length 2 will consist of complementary pairs. The other two classes of mutually parallel lines will have sums (p/2)M and (q/2)M. We will construct magic boxes of various dimensions by combining *small, (i.e.,* 2 by 4 by 4) *platforms* like the one on the left in Figure 4.18.

Theorem 4.11 *Let m be a positive integer. Then Z(32m) can be partitioned into m platforms.*

Proof. Let M = 32m-1, and for each t ∈ {0,1, …, 4m-1)} let

$$D(t) = \{4t, 4t+1, 4t+2, 4t+3\} \cup \{M-4t, M-4t-1, M-4t-2, M-4t-3\},$$

and let

$$P(t) = D(t) \cup D(t+1)\ D(t+2) \cup D(t+3) \text{ for } t \in \{0,4,8, \ldots, 4(m-1)\}\ .$$

The collection of sets {P(t): t ∈ {0,4,8, …, 4(m-1)}} form a partition of Z(32m) into m sets each with 32 elements. For each t ∈ {0,4,8, …, 4(m-1)} the elements of P(t) can be organized into two 4 by 4 near magic squares as in Figure 4.19.

The rows and columns of each square in Figure 4.19 have sum 2M, and when the two squares are superimposed a small platform is obtained in which all 2 element lines consist of complementary pairs with sum M. □

4t	M-4t-5	M-4t-10	4t+15
M-4t-1	4t+4	4t+11	M-4t-14
M-4t-2	4t+7	4t+8	M-4t-13
4t+3	M-4t-6	M-4t-9	4t+12

M-4t	4t+5	4t+10	M-4t-15
4t+1	M-4t-4	M-4t-11	4t+14
4t+2	M-4t-7	M-4t-8	4t+13
M-4t-3	4t+6	4t+9	M-4t-12

Figure 4.19

Corollary 4.12 *A magic box can be constructed with even dimensions p, q, r if q, r are multiples of 4.*

Proof. Let q=4s and r=4t and form a p by q by r box B by taking pst/2 small platforms from the above partition of Z(pqr) = Z(32(pst/2)) and arranging them in a stack of p/2 platforms each of which is formed with s small platforms on one side and t small platforms on the other side. If M = pqr-1, then B will have sums (p/2)M, 2sM and 2tM in the three directions. □

Non-cube magic boxes with three odd dimensions are difficult to construct; we have not even been able to construct one with dimensions 3 by 3 by 5. There is, however, a way to construct non-cube magic boxes with odd dimensions by using the known constructions of magic cubes. Let U, V, W denote three parallel squares of the magic cube of order 3 as in Figure 4.17, and let A, B, C denote the three squares over the elements {0,27,54} given in Figure 4.20. If squares A,B,C are placed on top of each other, they form a *Latin cube* over *S* = {0,27,54}, i.e., a cube in which each element of S occurs exactly once in each of the X, Y and Z directions. The squares U+A, U+B, U+C, V+A, V+B, V+C, W+A,W+B, W+C are then formed and placed parallel to each other as in Figure 4.21. The result is a 3 by 3 by 9 magic box with sums 120, 120 and 360. Note that the set of three blue, (red, green) squares form a magic cube, so our box represents a partition of Z(81) into three magic cubes.

U

9	25	5
23	0	16
7	14	18

V

22	2	15
6	13	20
11	24	4

W

8	12	19
10	26	3
21	1	17

A

0	27	54
54	0	27
27	54	0

B

27	54	0
0	27	54
54	0	27

C

54	0	27
27	54	0
0	27	54

Figure 4.20

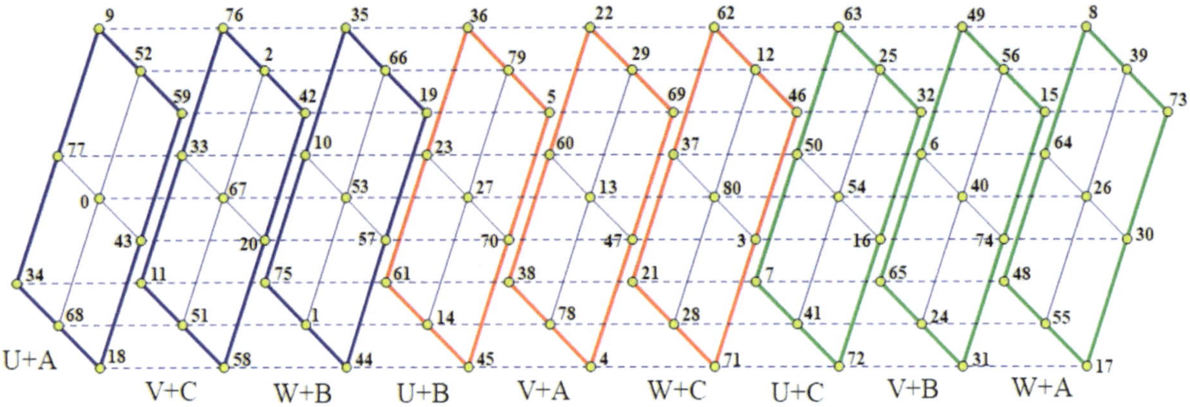

Figure 4.21

A Latin cube of order p is easily constructed for any positive integer greater than or equal to 3, and if also a magic cube of order p is obtained, then a p by p by p^2 magic box over $Z(p^4)$ can be constructed in similar fashion.

4.7 Integral Magic Rectangles

Perhaps the closest construction to a magic square which involves nonsquare rectangles is obtained using the integral base Z*(n) when n is odd. Let R be a p by q rectangle p, q odd and q>p≥3, and assign the set of elements

$$Z^*(pq) = \{0,1,-1, 2,-2, \ldots, (1/2)(pq-1), -(1/2)(pq-1)\}$$

one each to the cells of R so that all columns and rows of R have the same sum S. Then pS = qS = ΣZ*(pq) = 0 forces S = 0. If we can assign values to R, using Z*(pq) as base, so that each column and each row of R has sum 0, then we call R an *integral magic rectangle.*

Our method for constructing integral magic squares starts with an equipartition of Z(n), n = pq, into q subsets each with p elements as employed in Section 3.3. These subsets are used to form columns in a p by q rectangle R. The next step involves converting each element to its integral representation, i.e., if x ≤(n-1)/2 there is no change, and if x > (n-1)/2 then x is replaced by x-n. At this point we have a rectangle R* over Z*(n) in which neither rows nor columns have sum 0. In the third step (q+1)/2 is added to each element in every cell of R*. This will have the effect of equating the sum of the first row as well as all column sums to 0. Subsequent steps involve interchanging values within the same column until finally all rows also have sum 0. The resulting rectangle R** is an *integral mate of R,* and we say that R is *convertible.* Except in the case p=3, q = 9 where a different construction needs to be made, this method seems to be successful. in Figures 4.22. 4.23, 4.24 some examples are illustrated.

In Figure 4.23 the more complicated series of steps from I to VI illustrate the conversion of a 5 by 7 equipartition of Z(35) into an integral magic square, and in Figure 4.24 is illustrated via steps I through VIII, the conversion of a 7 by 9 equipartition of Z(63) into an integral magic square with 0 sum. The complex series of interchanges within columns makes it difficult to generalize this procedure.

If R is a p by q rectangular polysquare with q = mp, m≥3, p≥3, odd, then the magic square partition of Theorem 4.9 can be used to obtain an integral magic solution for R. In Figure 4.25 row I is a 3 by 21 2-sum rectangle R derived from a partition of Z(63) into 7 3 by 3 near magic squares. In row II the corresponding rectangle R* over Z*(63) is illustrated. In row III the rectangle 11+R*, which is a 0-sum magic rectangle, is obtained. In Figure 4.26 the same procedure is used to obtain a 0-sum

14	13	12	11	10
0	2	4	1	3
7	6	5	9	8

20	19	18	17	16	15	14
0	2	4	6	1	3	5
10	9	8	7	13	12	11

38	37	36	35	34	33	32	31	30	29	28	27	26
0	2	4	6	8	10	12	1	3	5	7	9	11
19	18	17	16	15	14	13	25	24	23	22	21	20

-1	-2	-3	-4	-5
0	2	4	1	3
7	6	5	-6	-7

-1	-2	-3	-4	-5	-6	-7
0	2	4	6	1	3	5
10	9	8	7	-8	-9	-10

-1	-2	-3	-4	-5	-6	-7	-8	-9	-10	-11	-12	-13
0	2	4	6	8	10	12	1	3	5	7	9	11
19	18	17	16	15	14	13	-14	-15	-16	-17	-18	-19

2	1	0	-1	-2
3	5	7	4	6
-5	-6	-7	-3	-4

3	2	1	0	-1	-2	-3
4	6	8	10	5	7	9
-7	-8	-9	-10	-4	-5	-6

6	5	4	3	2	1	0	-1	-2	-3	-4	-5	-6
7	9	11	13	15	17	19	8	10	12	14	16	18
-13	-14	-15	-16	-17	-18	-19	-7	-8	-9	-10	-11	-12

2	1	0	-1	-2
3	-6	-7	4	6
-5	5	7	-3	-4

3	2	1	0	-1	-2	-3
4	-8	8	-10	5	7	-6
-7	6	-9	10	-4	-5	9

6	5	4	3	2	1	0	-1	-2	-3	-4	-5	-6
-13	9	11	-16	-17	-18	-19	-7	10	12	14	16	18
7	-14	-15	13	15	17	19	8	-8	-9	-10	-11	-12

Figure 4.22

34	33	32	31	30	29	28
0	1	2	3	4	5	6
27	26	25	24	23	22	21
7	9	11	13	8	10	12
17	16	15	14	20	19	18

I

-1	-2	-3	-4	-5	-6	-7
0	1	2	3	4	5	6
-8	-9	-10	-11	-12	-13	-14
7	9	11	13	8	10	12
17	16	15	14	-15	-16	-17

II

3	2	1	0	-1	-2	-3
4	5	6	7	8	9	10
-4	-5	-6	-7	-8	-9	-10
11	13	15	17	12	14	16
-14	-15	-16	-17	-11	-12	-13

III

3	2	1	0	-1	-2	-3
4	5	6	-17	8	-12	10
-4	-5	-6	-7	-8	-9	-10
-14	-15	15	17	12	14	16
11	13	-16	7	-11	9	-13

IV

3	2	1	0	-1	-2	-3
4	5	6	-17	8	-12	10
-4	-5	15	17	-8	-9	-10
-14	-15	-6	-7	12	14	16
11	13	-16	7	-11	9	-13

V

3	2	1	0	-1	-2	-3
4	5	15	-17	-8	-9	10
-4	-5	6	17	8	-12	-10
-14	-15	-6	-7	12	14	16
11	13	-16	7	-11	9	-13

VI

Figure 4.23

62	61	60	59	58	57	56	55	54
0	1	2	3	4	5	6	7	8
53	52	51	50	49	48	47	46	45
9	10	11	12	13	14	15	16	17
44	43	42	41	40	39	38	37	36
18	20	22	24	26	19	21	23	25
31	30	29	28	27	35	34	33	32

I

-1	-2	-3	-4	-5	-6	-7	-8	-9
0	1	2	3	4	5	6	7	8
-10	-11	-12	-13	-14	-15	-16	-17	-18
9	10	11	12	13	14	15	16	17
-19	-20	-21	-22	-23	-24	-25	-26	-27
18	20	22	24	26	19	21	23	25
31	30	29	28	27	-28	-29	-30	-31

II

4	3	2	1	0	-1	-2	-3	-4	0
5	6	7	8	9	10	11	12	13	
-5	-6	-7	-8	-9	-10	-11	-12	-13	
14	15	16	17	18	19	20	21	22	
-14	-15	-16	-17	-18	-19	-20	-21	-22	
23	25	27	29	31	24	26	28	30	
-27	-28	-29	-30	-31	-23	-24	-25	-26	

III

IV

4	3	2	1	0	-1	-2	-3	-4	0
-27	6	7	8	9	-23	11	-25	13	
-5	-6	-7	-8	-9	-10	-11	-12	-13	
14	15	-29	-30	-31	19	20	21	22	
-14	-15	-16	-17	-18	-19	-20	-21	-22	
23	25	27	29	31	24	26	28	30	
5	-28	16	17	18	10	-24	12	-26	0

Figure 4.24

4	3	2	1	0	-1	-2	-3	-4	0
-27	15	-29	8	-31	19	11	21	13	0
-5	-6	-7	-8	-9	-10	-11	-12	-13	-81
14	6	7	-30	9	-23	20	-25	22	0
-14	-15	-16	-17	-18	-19	-20	-21	-22	-162
23	25	27	29	31	24	26	28	30	243
5	-28	16	17	18	10	-24	12	-26	0

V

4	3	2	1	0	-1	-2	-3	-4	0
-27	15	-29	8	-31	19	11	21	13	0
-5	-6	-7	-8	-9	-10	-11	-12	-13	-81
14	6	7	-30	9	-23	20	-25	22	0
23	25	27	-17	-18	24	-20	-21	-22	1
-14	-15	-16	29	31	-19	26	28	30	80
5	-28	16	17	18	10	-24	12	-26	0

VI

4	3	2	1	0	-1	-2	-3	-4	0
-27	15	-29	8	-31	19	11	21	13	0
-5	-6	-7	-8	31	-10	-11	28	-13	-1
14	6	7	-30	9	-23	20	-25	22	0
23	25	27	-17	-18	24	-20	-21	-22	1
-14	-15	-16	29	-9	-19	26	-12	30	0
5	-28	16	17	18	10	-24	12	-26	0

VII

VIII

4	3	2	1	0	-1	-2	-3	-4	0
-27	15	-29	8	-31	19	11	21	13	0
-5	-6	27	-17	-18	24	-20	28	-13	0
14	6	7	-30	9	-23	20	-25	22	0
23	25	-7	-8	31	-10	-11	-21	-22	0
-14	-15	-16	29	-9	-19	26	-12	30	0
5	-28	16	17	18	10	-24	12	-26	0

Figure 4.24

5 by 15 magic integral rectangle. The constant chosen to add to R* is always (1/2)(q+1). For Figure 4.26 the partition of Z(75) into three 5 by 5 near magic squares given in Figure 4.16 was used as the original 2-sum rectangle R, then 4.26(I) is the integral version R*, and 4.26(II) is the 0-sum magic integral rectangle obtained as 8 + R*.

	60	2	31	57	8	28	26	55	12	51	20	22	48	5	40	45	11	37	42	17	34
I	32	61	0	29	58	6	13	24	56	23	52	18	41	49	3	38	46	9	35	43	15
	1	30	62	7	27	59	54	14	25	19	21	53	4	39	50	10	36	47	16	33	44

	-3	2	31	-6	8	28	26	-8	12	-12	20	22	-15	5	-23	-18	11	-26	-21	17	-29
II	-31	-2	0	29	-5	6	13	24	-7	23	-11	18	-22	-14	3	-25	-17	9	-28	-20	15
	1	30	-1	7	27	-4	-9	14	25	19	21	-10	4	-24	-13	10	-27	-16	16	-30	-19

	8	13	-21	5	19	-24	-26	3	23	-1	31	-30	-4	16	-12	-7	22	-15	-10	28	-18
III	-20	9	11	-23	6	17	24	-28	4	-29	0	29	-11	-3	14	-14	-6	20	-17	-9	26
	12	-22	10	18	-25	7	2	25	-27	30	-31	1	15	-13	-2	21	-16	-5	27	-19	-8

Figure 4.25

	22	-25	33	6	-1	-13	15	13	-19	-36	27	-35	-7	-29	4
	-21	32	5	-2	21	19	12	-20	-37	-14	-31	-8	-30	3	26
I	31	9	-3	20	-22	11	-16	-38	-15	18	-9	-26	2	25	-32
	8	-4	24	-23	30	-17	-39	-11	17	10	-27	1	29	-33	-10
	-5	23	-24	34	7	35	-12	16	14	-18	0	28	-34	-6	-28

	30	-17	-34	14	7	-5	23	21	-11	-28	35	-27	1	-21	12
	-13	-35	13	6	29	27	20	-12	-29	-6	-23	0	-22	11	34
II	-36	17	5	28	-14	19	-8	-30	-7	26	-1	-18	10	33	-24
	16	4	32	-15	-37	-9	-31	-3	25	18	-19	9	37	-25	-2
	3	31	-16	-33	15	-32	-4	24	22	-10	8	36	-26	2	-20

Figure 4.26

Considering these various constructions, we feel that the following Conjecture is likely true.

Conjecture 4.13 *If p, q are odd integers both larger than or equal to 3, then a p by q 0-sum integral magic rectangle can be constructed.*

4.8 I-extensions

The notion of primary and composite integral magic squares was introduced in Section 3.8. The same ideas apply also to integral magic rectangles, but the determination of primary and composite appears to be much more difficult. In Figure 4.27 is illustrated a composite magic integral rectangle. It is a 5 by 9 rectangle obtained as an extension of a 3 by 9 kernel. The 3 by 9 magic integral rectangle was obtained by the same procedure used to obtain the constructions in Figures 4.25 and 4.26.

14	15	16	17	19	-18	-20	-21	-22
-14	-15	-16	-17	-19	18	20	21	22
5	-8	3	11	-11	0	8	-5	-3
4	6	-10	1	12	-13	-2	9	-7
-9	2	7	-12	-1	13	-6	-4	10

Figure 4.27

The following Lemmas are useful in the construction of composite integral magic rectangles.

Lemma 4.14 *Let S be a sequence of consecutive positive integers with |S| = m. If S has a partition into two nonempty subsets A, B such that ΣA = ΣB, then S is "splitable". If 4|m then S is splitable, but if m = 4k+2 for some nonnegative integer k, then S is not splitable.*

Proof. Let S = (a, ..., b), then ΣS = (1/2)(a+b)(b-a+1), where m = |S| = b-a+1. Also b = a+m-1, so a+b = 2a + m -1. If m = 4k+2 then both a+b and (1/2)(b-a+1) are odd. Thus ΣS is odd and cannot be split.

Now suppose m = 4k. Then S can be partitioned into m/2 pairs each with sum a+b. Since m/2 is even, half of these pairs can be put together to form a set A, and the other half can be used to form a set B. Then ΣA = ΣB = (m/4)(a+b) = (1/2)ΣS, and S is split. □

If S is an increasing sequence of positive integers with odd length m≥3, let α(S) denote the subset of {S} consisting of the first (1/2)(m+1) of S and let β(S) denote the remaining elements of {S}.

Lemma 4.15 *Let S be a sequence of strictly increasing positive integers with |S| = m, m≥3 odd. If Σα(S) > (1/2)ΣS then S is not splitable.*

Proof. Suppose Σα(S) > (1/2)ΣS. If Q is any subset of S with |Q| ≥ (1/2)(m+1), then we must have ΣQ ≥ Σα(S) > (1/2)ΣS. If Q is any subset of S with |Q|≤|β(S)| then ΣQ ≤ Σβ(S) < (1/2)ΣS. A subset Q of S must satisfy either |Q|≥(1/2)(m+1) or |Q|≤ (1/2)(m-1), and we have shown that in either case ΣQ≠(1/2)ΣS. The conclusion is that S is not splitable. □

Let S be a sequence of consecutive positive integers of length m and let A denote an initial segment of S, and B the remaining section of S. Suppose |A| = p and |B| = q with p+q = m. Let R be a p by q matrix with the sequence A as row headers and the sequence B as column headers. For each a ∈A and b∈B insert the integer b-a in the cell of R at the intersection of row a and column b. We call R a *difference matrix over S*, and denote it by R(S). In Figure 4.28(I) a difference matrix R(15, 16, …, 30) is illustrated.

R

	24	25	26	27	28	29	30
15	9	10	11	12	13	14	15
16	8	9	10	11	12	13	14
17	7	8	9	10	11	12	13
18	6	7	8	9	10	11	12
19	5	6	7	8	9	10	11
20	4	5	6	7	8	9	10
21	3	4	5	6	7	8	9
22	2	3	4	5	6	7	8
23	1	2	3	4	5	6	7

I

0	1	2	3	4
9	10	11	12	13
8	9	10	11	12
7	8	9	10	11
6	7	8	9	10
5	6	7	8	9

M

II

Figure 4.28

Note that the contents and structure of the difference matrix R remains unchanged if a different sequence S′ of consecutive integers is used for row and column headers, as long as |S′| = |S|: the lower left corner cell will contain 1, and the slope -1 diagonals will have the constant values 2, 3, 4, … as they move up from the corner. In Figure 4.28 we have extracted a square submatrix M of R. We will show that ΣT is the same for any transversal T in M.

Lemma 4.16 *Let M be any square submatrix of a difference matrix R. Then ΣT is the same for any transversal T in M.*

Proof. Suppose M has order t and let $(m_1, m_2, \ldots, m_t)$ denote the consecutive values which form the first column of M. These will have sum $(1/2)(m_1 + m_t)(m_t - m_1 + 1)$. Label the columns of M with the sequence S = (0, 1, …, t-1) as in the example in Figure 4.28(II). Let $T = (b_1, b_2, \ldots, b_t)$ denote the contents of a transversal in M, with b_i occupying the i'th of M for i ∈ {1, …, t}. Then $b_i = m_i + s_i$ for each i ∈ {1, …, t} and for some s_i ∈S. We then have

$$\Sigma T = (m_1 + s_1) + \ldots + (m_t + s_t) = (1/2)(m_1 + m_t)(m_t - m_1 + 1) + (s_1 + \ldots + s_t).$$

Since T is a transversal, its elements occur each at a different distance from the first column of M. Thus $\{s_1, \ldots, s_t\} = \{S\}$, and consequently $s_1 + \ldots + s_t = \Sigma S = (1/2)t(t-1)$. Then

$$\Sigma T = (m_1 + s_1) + \ldots + (m_t + s_t) = (1/2)(m_1 + m_t)(m_t - m_1 + 1) + (1/2)t(t-1).$$

This expression is dependant only on parameters associated with the square M, so we can conclude that it represents the common sum of every transversal in M. □

If R is a difference matrix then a *difference set* in R is a nonempty subset of the entries in R with the property that no two entries are taken from the same column or row.

Corollary 4.17 *Let R be a difference matrix and let D* denote the diagonal of R which terminates at the upper right corner cell of R. Then the maximum value $\Delta(R) = \{\Sigma D : D$ is a difference set in $R\}$ occurs when $D = D^*$.*

Proof. Let M denote the largest subsquare of R which contains the upper right corner of R. It is clear that the maximum value of

$$\{\Sigma T : T \text{ is a transversal in a subsquare of } R\}$$

occurs when the chosen subsquare is M. Since a difference set in R cannot contain more than one value from any row or column, it must be that Δ is maximized by a difference set contained in M. Among the difference sets in M, certainly a transversal will yield the greatest sum. By Lemma 4.16 all transversals in M have the same sum, so we can take D* as the maximizing transversal. □

Lemma 4.18 *Let R be a p by q difference matrix and let $\Delta(R) = \{\Sigma D : D$ is a difference set in $R\}$. Then $\Delta(R) = \{1, 2, \ldots, pq\}$.*

Proof. We will proceed by induction on the area of R. The Lemma is clearly true for difference matrices with area 1,4,6 or 9. Assume inductively that the Lemma holds true for any difference matrix with area less than pq. Note that every subrectangle of R which contains the lower left corner cell of R is also a difference matrix. Let R_1 denote the p by q-1 subrectangle of R consisting of all but the last column of R, and let R_2 denote the p-1 by q subrectangle of R consisting of every row of R except the first one. Let $\Delta(R_1) = \{\Sigma D : D$ is a difference set in R_1 and define $\Delta(R_2)$ similarly. The rectangles R_1, R_2 are difference matrices with area less than R, so by the inductive hypothesis, $\Delta(R_1) = \{1, \ldots, p(q-1)\}$ and $\Delta(R_2) = \{1, \ldots, (p-1)q\}$. Assume without loss of generality that $p \geq q$ and thus $\Delta(R_2) \geq \Delta(R_1)$. Of course $\Delta(R_2) \subseteq \Delta(R)$, which leaves only the q sums $\{pq - q + 1, \ldots, pq\}$ to account for. (Note that by Corollary 4.17, pq is the maximum sum of any difference set in R.)

Let M(t) denote the subsquare of order t which contains the upper right corner of R. Let D denote the portion of the corner diagonal of R which lies in M(t) and terminates at the upper right corner cell of R. By lemma 4.16 the maximum value of $\Delta(M(t))$ is given by

$$\Sigma D = (p+q-1) + (p+q-2) + \ldots + (p+q-(2t-1) = (p+q)t - t^2.$$

Let R_3 denote the p-t by q-t subrectangle of R which contains the lower left cell of R. R_3 is a difference matrix with area smaller than the area of R, so by the inductive hypothesis $\Delta(R_3) = \{1, 2, \ldots, (p-t)(q-t)\}$.

Suppose q is odd and let $t = (1/2)(q+1)$. Then

$$\Sigma D = (1/4)(q+1)(2p+q-1) \text{ and } \Delta(R_3) = \{1, 2, \ldots, (1/4)(2p-q+1)(q-1).$$

Then $\Delta(M \cup R_3)$ contains all the values in $\Sigma D + \Delta(R_3)$. (See Figure 4.29.) This includes pq which is the maximum sum over all difference sets in R as well as the (p-t)(q-t) values which lie directly below pq. For t = (1/2)(q+1) we would like to have (p-t)(q-t) ≥ q so that all the missing sums are accounted for. Consider:

$$(p-t)(q-t) \geq q \leftrightarrow (p - (1/2)(q+1))(q - (1/2)(q+1)) \geq q \leftrightarrow pq - (1/2)(q+1)(p-1) \geq q \leftrightarrow p(q-1) + 1 \geq 0,$$

which is clearly true. Now suppose q is even. It is easy to check that the Lemma holds if q = 2, so we assume q≥4. Let t = q/2, and consider:

$$(p-t)(q-t) \geq q \leftrightarrow (p - (q/2))(q - (q/2)) \geq q \leftrightarrow (1/4)(2p-q)q \geq q \leftrightarrow 2p-q \geq 4.$$

But we have p≥q so 2p≥2q, and thus 2p-q ≥ q ≥ 4. So in this case also, all the missing sums are accounted for, and we conclude that Δ(R) = {1, 2, …, pq}. □

8	9	10	11	12	13	14
7	8	9	10	11	12	13
6	7	8	9	10	11	12
5	6	7	8	9	10	11
4	5	6	7	8	9	10
3	4	5	6	7	8	9
2	3	4	5	6	7	8
1	2	3	4	5	6	7

M(4)

R_3

R

Figure 4.29

Theorem 4.19 *Let S =(a, …, b) be a sequence of consecutive positive integers with |S| = m, and suppose m≥3 is odd. Let K = (1/4)(m-1)^2. Then S is splitable iff 1≤a≤K and a has the same parity as K.*

Proof. If a = K then the sequence S is given by

$$S = ((K, K+1, \ldots, K+m-1), \text{ and } 2\Sigma S = ((1/2)(m-1)^2+m-1)(m)$$

which yields ΣS = (1/4)m(m^2-1). Now let T denote the sequence consisting of the first (1/2)(m+1) elements of S. Then T is given by

$$T = ((K, K+1, \ldots, K+ (1/2)(m-1)), \text{ and } 2\Sigma T = ((K+(1/2)(m-1))((1/2)(m-1)+1)$$

which yields ΣT = (1/8)m(m^2-1) = (1/2)ΣS. Thus S is splitable. If a >K, then

$$U = (a, a+1, \ldots, a+(1/2)(m-1))$$

denotes the subsequence of S consisting of the first (1/2)(m+1) elements of S, and

$$\Sigma U = (1/2)(2a + (1/2)(m-1))((1/2)(m+1) = (a + (1/4)(m-1))((1/2)(m+1).$$

Thus ΣU > (K + (1/4)(m-1))((1/2)(m+1) = (1/8)m(m^2-1) = (1/2)ΣS, and by Lemma 4.15 S is not splitable.

It remains to show that if $1 \leq a < K$, and a, K have the same parity then S is splitable. Suppose K is even. Since m is odd, either m-1 = 4t or m-1 = 4t+2 for some positive integer t. If m-1 = 4t+2 then $K = (1/4)(m-1)^2 = (2t+1)^2$ which is odd. So we must have m-1 = 4t. Now $(1/2)\Sigma S = (1/4)m(2a+m-1)$ must be an integer if S is splitable, so it must be that 4|(m(2a+m-1), and since m is odd it must be that 4|(2a+m-1). Then since 4|(m-1) we must have 4|2a and thus a must be even.

Now suppose $K= (1/4)(m-1)^2$ is odd. Since m is odd we can set m = 2t+1. Consider: $(1/2)\Sigma S = (1/4) m(2a+m-1) = (1/2)m(a+t)$ must be an integer, so 2|m(a+t), and since m is odd we must have 2|(a+t). Since K is odd $(1/4)4t^2 = t^2$ is odd and thus t is odd. Then 2|(a+t) implies that a is odd. We have shown that if S is splitable then a, K must have the same parity. It remains to show that if this condition is satisfied and $a \leq K$, then S is splitable.

With |S| = m = 2p+1, let S_1 denote the initial segment of S with $|S_1|$= p+1, and let S_2 denote the remainder of S, so $|S_2|$ = p. Construct a difference matrix R(S) with the sequence S_1 as row headers and S_2 as column headers as in the example in Figure 4.30(I). We seek to interchange values in S_1 with values in S_2 in order to create new sequences S_1' and S_2' with $\Sigma S_1' = \Sigma S_2'$, and thus show that S is splitable. Each pair of values x,y to be interchanged with $x \in S_1$, $y \in S_2$ corresponds to the entry y-x in R which represents the increase in ΣS_1 due to the interchange. The total number of interchanges required to create S_1' and S_2' corresponds to a difference set D in R with $\Sigma D = (1/2)(\Sigma S_2 - \Sigma S_1)$. By Lemma 4.18, $\Delta(R) = \{1, \ldots, p(p+1)\}$, so we must show that $(1/2)(\Sigma S_2 - \Sigma S_1) \in \{1, \ldots, p(p+1)\}$.

Consider:

$S_1 = (a, a+1, \ldots, a+p)$ so $\Sigma S_1 = (1/2)(2a+p)(p+1)$, and

$S_2 = (a+p+1, a+p+2, \ldots, a+2p)$, so $\Sigma S_2 = (1/2)(2a+3p+1)p$, and thus

$\Sigma S_2 - \Sigma S_1 = (1/2)(2a+3p+1)p - (1/2)(2a+p)(p+1) = p^2 - a$, and then

$(1/2)(\Sigma S_2 - \Sigma S_1) = (1/2)(p^2 - a)$.

Since $a < K = p^2$, and since clearly $(1/2)(p^2 - a) < p(p+1)$, we can conclude that $(1/2)(\Sigma S_2 - \Sigma S_1) \in \{1, \ldots, p(p+1)\}$, and thus that S is splitable. □

In Figure 4.30(II) the exchanges indicated in 4.30(I) have been made to obtain the sequences S_1'= {20,29,22,23,24,25,26} and S_2'={7,28,21,30,31,32} which split S, i.e., $\Sigma S_1' = \Sigma S_2'$= 169. A 2 by 13 matrix I is then created with the bottom row consisting of the values in S_1' followed by the negatives of the values in S_2', and the other top row equal precisely to the negative of the bottom row. Thus the rows and columns of I all have sum 0, and it can be adjoined to the 3 by 13 0-sum integral rectangle to create a 5 by 13 0-sum integral rectangle as in Figure 4.30(II). We call I an *I-extension,* and denote it by I(20, …, 32) to indicate the range of positive values in the extension.

It is never possible to make a repeated I-extension of a given 0-sum integral rectangle since the first value (a) in the extension changes parity, while the parity of K remains constant. On the other hand, it is easy to show that if $q \geq 9$ is odd and R is a 3 by q 0-sum integral rectangle, then R has an I-extension.

$\sum S_1 = 161;\ \sum S_2 = 177;\ (1/2)(\sum S_2 - \sum S_1) = 8$

	27	28	29	30	31	32	S_2
20	7	8	9	10	11	12	
21	6	7	8	9	10	11	
22	5	6	7	8	9	10	
23	4	5	6	7	8	9	
24	3	4	5	6	7	8	
25	2	3	4	5	6	7	
26	1	2	3	4	5	6	
S_1							

I

-20	-29	-22	-23	-24	-25	-26	27	28	21	30	31	32
20	29	22	23	24	25	26	-27	-28	-21	-30	-31	-32
6	5	4	3	2	1	0	-1	-2	-3	-4	-5	-6
-13	9	11	-16	-17	-18	-19	-7	10	12	14	16	18
7	-14	-15	13	15	17	19	8	-8	-9	-10	-11	-12

I(20, ... , 32)

II

Figure 4.30

4.9 L-extensions

Another method for extending integral magic rectangles involves a polysquare formed by appending at right angles two 0-sum rectangles each with two rows. We refer to this structure as an *L-extension.* We have already effectively seen examples of L-extensions of integral magic squares in Section 3.8. In Figure 3.9, if the bottom row in either square is detached and placed above the first row, and then the leftmost column is detached and reattached to the right side of the square, we obtain an L-extension as in Figures 4.31(I,II). Note, however, that the resulting squares are only semimagic. In Figure 4.3(III) an L-extension of a 3 by 7 magic integral rectangle is illustrated. We give the full designation of "magic" since nonsquare rectangles have no corner to corner diagonals to bother with.

-14	-18	-13	21	22	19	-17
14	18	13	-21	-22	17	-19
-5	-9	12	8	-6	15	-15
5	9	-12	6	-8	16	-16
-1	4	-3	7	-7	-20	20
-2	0	2	-10	10	-23	23
3	-4	1	-11	11	-24	24

I

31	-26	28	-24	26	-21	-29	19
-30	27	-27	25	-25	22	-18	30
17	-12	14	-10	-15	9	32	-31
-16	13	-13	11	-8	16	24	-23
1	-2	7	-4	18	-17	23	-22
8	-5	2	-3	12	-11	20	-19
-6	5	0	3	10	-9	-20	21
-1	4	-7	6	-14	15	-28	29

II

-11	-12	-15	-18	17	19	20	-16	16
11	12	15	18	-17	-19	-20	-14	14
3	2	1	0	-1	-2	-3	-13	13
4	-8	8	-10	5	7	-6	22	-22
-7	6	-9	10	-4	-5	9	21	-21

III

Figure 4.31

We will start by considering L-extensions of p by q integral rectangles with p, q odd and q > p ≥ 3. Let S denote the set of positive values in the extension, then |S| = p+q+2 which is even. The L-extension is either composed of 2 by p and 2 by q+2 0-sum rectangles, or by 2 by p+2 and 2 by q 0-sum rectangles. In either case the corresponding values in S must form splitable sets. As a consequence S itself must be splitable, so by Lemma 4.14 it must be that |S| is a multiple of 4.

Thus p+q = 4k+2 for some positive integer k. Unlike I-extensions, L-extensions can be repeated ad infinitum as in the following constructions.

Theorem 4.20 *Let p ≥ 3 be odd and let q = p+8. Then a p by q integral magic rectangle has an L-extension.*

Proof. Instead of an algebraic proof, a sequence of examples will be provided which reveal the general pattern of the construction. Starting with a 3 by 11 integral magic rectangle R as in Figure 4.33, Let S = {17, ...,32) denote the positive elements in an L-extension. The sequence S is partitioned into 5 groups, A,B,C,D,E as indicated in Figure 4.32 and a 3-cycle of interchanges is indicated. After the interchanges are made new groups A′=A, B′, C′, D′, E′=E are obtained with ΣB′=ΣE′ and ΣA′+ΣC′ = ΣD′. A 2 by 5 0-sum rectangle composed of the elements in B′∪E′ is appended to the short side of R, and a 2 by 11 0-sum rectangle composed of the elements in A′∪C′∪ D′ is placed against the top side of R completing the L-extension. In Figure 4.32 the consecutive sequences of positive integers are given along with the interchanges used to create the next two L-extensions. The 3-cycle of interchanges occupies the same position in each sequence; the lengths of subsequences A, B, D, E increase by one and the length of C remains at 4, so the pattern for further L-extensions is clear. □

In Figure 4.33 the three L-extensions corresponding to the sequences in Figure 4.32 are illustrated.

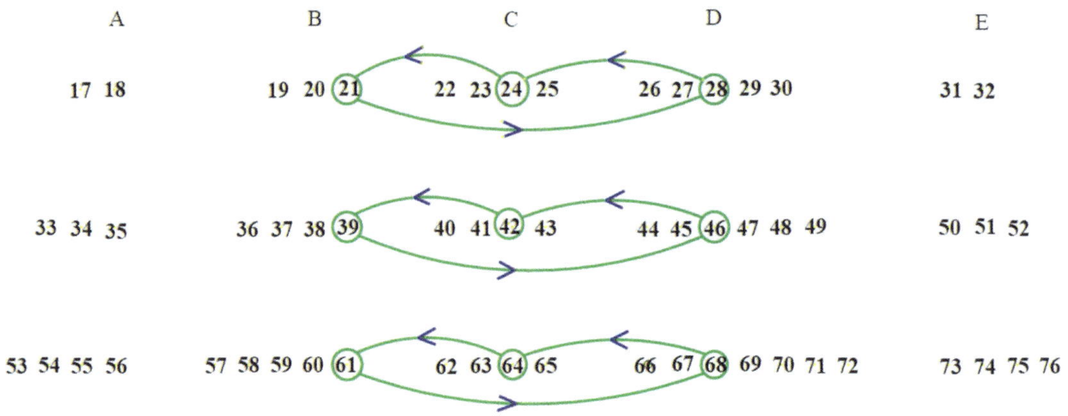

Figure 4.32

Theorem 4.21 *Let p ≥ 3 be odd and let q = p+4. Then a p by q integral magic rectangle has an L-extension.*

Proof. We will proceed along the same lines as in the proof of Theorem 4.20 by providing a series of L- extensions, from which the procedure for making further L-extensions will be evident. The extending consecutive sequence of positive integers will be again divided into five groups A,B,C,D,E as in Figure 4.34, and the interchanges, this time involving a 3-cycle and a transposition, are indicated. □

-53	-54	-55	-56	-62	-63	-68	-65	66	67	61	69	70	71	72	-57	57
53	54	55	56	62	63	68	65	-66	-67	-61	-69	-70	-71	-72	-58	58
-33	-34	-35	-40	-41	-46	-43	44	45	39	47	48	49	-36	36	-59	59
33	34	35	40	41	46	43	-44	-45	-39	-47	-48	-49	-37	37	-60	60
-17	-18	-22	-23	-28	-25	26	27	21	29	30	-19	19	-38	38	-64	64
17	18	22	23	28	25	-26	-27	-21	-29	-30	-20	20	-42	42	73	-73
5	4	3	2	1	0	-1	-2	-3	-4	-5	-24	24	50	-50	74	-74
-11	-12	-13	12	-15	-16	7	9	11	13	15	31	-31	51	-51	75	-75
6	8	10	-14	14	16	-6	-7	-8	-9	-10	32	-32	52	-52	76	-76

Figure 4.33

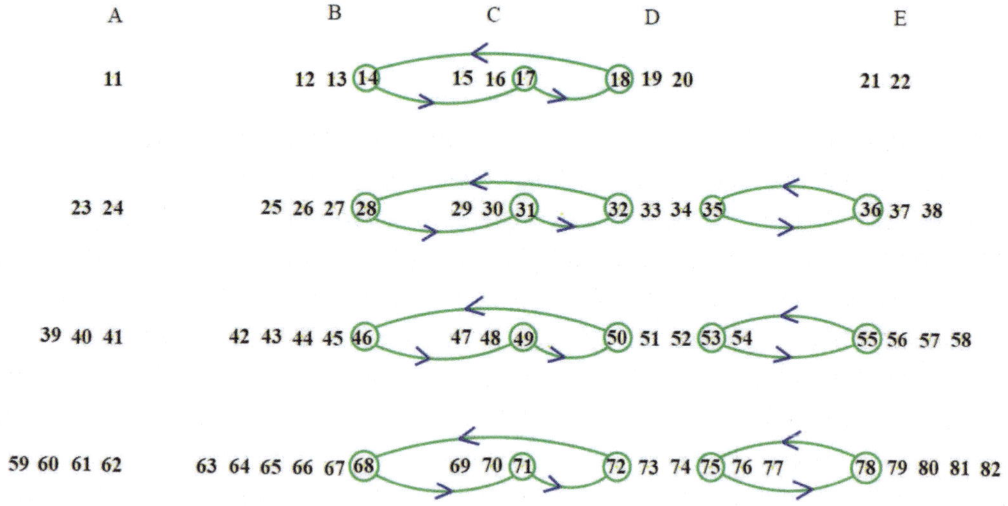

Figure 4.34

In Figure 4.35 the four L-extensions corresponding to the sequences in Figure 4.34 are illustrated.

-59	-60	-61	-62	-68	-69	-70	71	73	74	76	77	78	-63	63
59	60	61	62	68	69	70	-71	-73	-74	-76	-77	-78	-64	64
-39	-40	-41	-46	-47	-48	49	51	52	54	55	-42	42	-65	65
39	40	41	46	47	48	-49	-51	-52	-54	-55	-43	43	-66	66
-23	-24	-28	-29	-30	31	33	34	36	-25	25	-44	44	-67	67
23	24	28	29	30	-31	-33	-34	-36	-26	26	-45	45	-72	72
-11	-14	-15	-16	17	19	20	-12	12	-27	27	-50	50	75	-75
11	14	15	16	-17	-19	-20	-13	13	-32	32	53	-53	79	-79
3	2	1	0	-1	-2	-3	-18	18	35	-35	56	-56	80	-80
4	-8	8	-10	5	7	-6	21	-21	37	-37	57	-57	81	-81
-7	6	-9	10	-4	-5	9	22	-22	38	-38	58	-58	82	-82

Figure 4.35

L-extensions and I-extensions are not the only types of extensions that can be constructed. For example, the 3 by 5 integral magic rectangle has neither an I-extension nor an L-extension but it does have a *double L-extension,* or *L^2-extension* illustrated in Figure 4.36(I). Note that an L^2-extension must be distinguished from repeated L-extensions. In Figure 4.36(II) a *quadruple I-extension,* or *I^4-extension* of the same 3 by 5 rectangle is illustrated. It is clear that the subject of extensions of integral rectangles is a rich and interesting subject in its own right.

-8	-9	-12	14	15	-21	21	-17	17
8	9	12	-14	-15	-22	22	-19	19
-10	-11	-13	16	18	-23	23	-20	20
10	11	13	-16	-18	-24	24	-25	25
2	1	0	-1	-2	29	-29	26	-26
3	-6	-7	4	6	30	-30	27	-27
-5	5	7	-3	-4	31	-31	28	-28

I

2	3	-5	-16	16	-12	12	-10	10	-8	8
1	-6	5	-18	18	-17	17	-14	14	-9	9
0	-7	7	-19	19	-20	20	-21	21	-11	11
-1	4	-3	26	-26	24	-24	22	-22	13	-13
-2	6	-4	27	-27	25	-25	23	-23	15	-15

II

Figure 4.36

4.10 Integral Magic Boxes

If a box B has three odd dimensions p,q,r then it may be possible to populate the cells of B with the full set of elements from Z*(pqr) so that a 0- sum is obtained in all three directions. This interesting topic is not particularly relevant to our main purpose of constructing magic polygons, so we will just provide a few examples.

Let B denote the magic cube over Z(27) illustrated in Figure 4.17, and let B* denote the corresponding cube over Z*(27). The reader may wish to check that the cube 5+B* is a 0-sum integral magic cube. If instead, B denotes the cube of order 4 over Z(64) in Figure 4.17, the reader might also wish to check that replacing 32 by -32 in B*, and then inverting the elements in every cell, an integral magic cube with sum 2 is obtained. Finally, the 3 by 3 by 9 box B in Figure 4.21 can be converted to a 0-sum integral magic square by adding 14 to each cell in B*. Can every magic cube over $Z(n^3)$ be converted so easily to an integral magic cube ? Can an integral magic 3 by 3 by 5 box or 3 by 5 by 5 box be constructed ?

CHAPTER 5

MAGIC HEXAGONS

5.1 Regular Magic Hexagons

A polyparallelogram with an hexagonal boundary is *regular* in case each side has the same number of basic edges. We let H(e) denote a regular hexagonal polyparallelogram with e basic edges per side. In Figure 5.1(I) is illustrated the hexagon H(2) with a magic solution over Z(12) with stream sum 22. In 5.1(II) the central dashed hexagon in 5.1(I) is inverted. Inversion of simple hexagons preserves stream sums and thus 5.1(II) is also a magic H(2). Three simple hexagons in 5.1(II) are simultaneously inverted to obtain the magic *flower* in 5.1(III). In accordance with Theorem 1.5, every parallelogram tiling of H(2) can be obtained by a sequence of simple hexagon inversions from the polar tiling, and thus a magic solution for the polar tiling implies the existence of magic solutions for every parallelogram tiling of H(2). This applies to any polyparallelogram with a semiregular boundary so we can confine our attention to polar tilings; *a magic solution for the polar tiling of a semiregular polyparallelogram P implies the existence of magic solutions for every parallelogram tiling of P.*

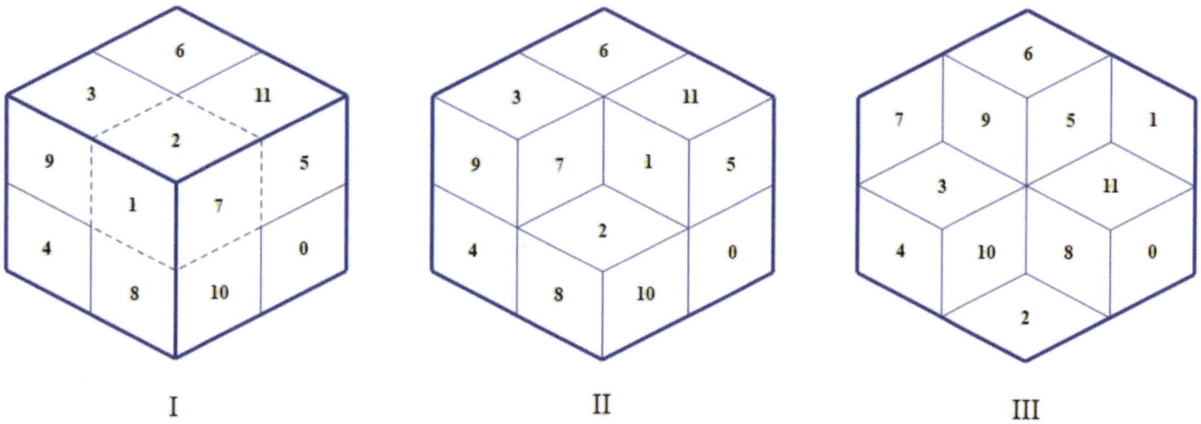

Figure 5.1

In Figure 5.2 the genesis of a magic H(2) over Z(2)×Z(6) is depicted. Figure 5.2(I) is magic over $(Z(2))^6$ and 5.2(II) is magic over $(Z(6))^2$. These two hexagons are orthogonal and their superimposition yields the magic H(2) over Z(2)×Z(6) in 5.2(III).

Although Z(12) and Z(2)×Z(6) are, up to isomorphism, the only abelian groups of order 12, we cannot say that H(2) is universally magic; no natural magic H(2) with base Z(3)×Z(4) can be constructed, and no natural magic H(2) over Z(2)×Z(2)×Z(3) can be found.

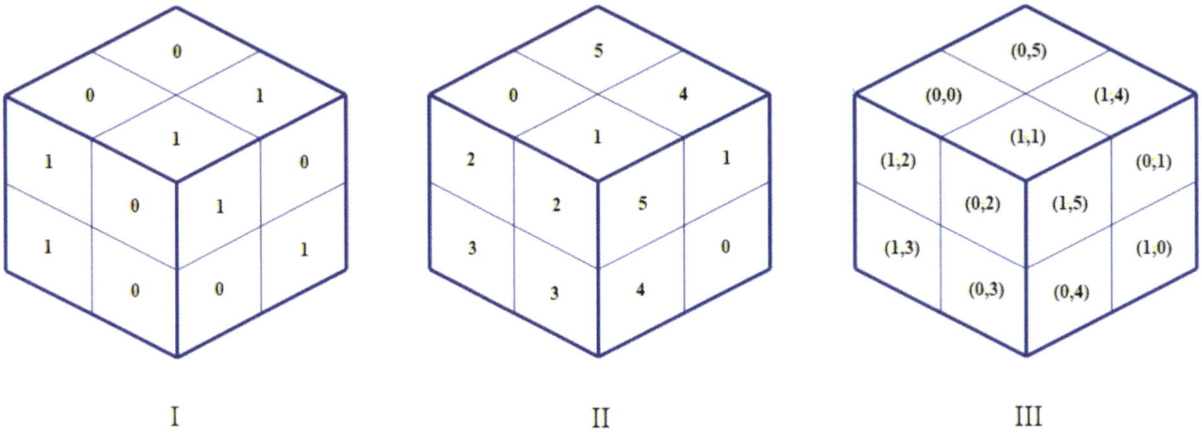

Figure 5.2

In Figure 5.3 the derivation of a magic integral H(2) is illustrated. Figure 5.3(I) is a standard magic H(2) over Z(12); Figure 5.3(II) is the direct conversion H*(2) of 5.3(I) over Z*(12), and the magic integral version in 5.3(III) is obtained as 1+H*(2). We will say simply that a polyparallelogram is *integral* in case there exists a magic integral version. We can thus say that H(2) is integral.

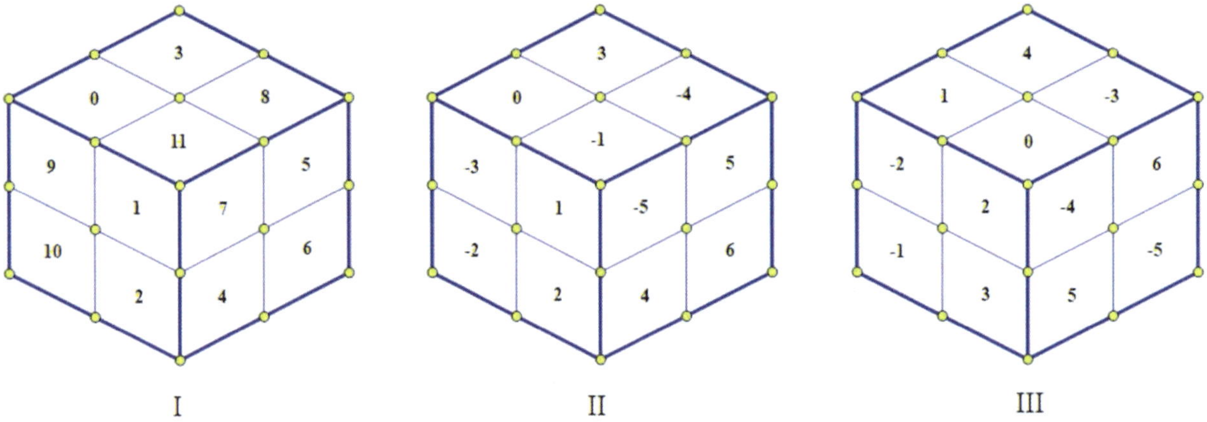

Figure 5.3

In the following result we display some basic facts about regular standard magic hexagons.

Theorem 5.1 *(i) The number of tiles in H(e) is given by* $n = 3e^2$ *; (ii) stream length is 2e; (iii) the semiperimeter s = 3e = the number of streams; (iv) stream sum is S = e(n-1); (v) In the polar tiling of H(e) the sum in each of the three major parallelograms is (1/6)n(n-1).*

Proof. (i), (ii) and (iii) are obvious. To prove (iv) the expression 3eS is the sum of all the stream sums. In this sum the value in each tile is counted twice and thus 3eS = 2ΣZ(n) = n(n-1), and thus S = (1/3e)n(n-1) = $(1/3e^2)$(ne)(n-1) = e(n-1). To prove (v) let A, B, C denote the three major parallelograms in the polar tiling of H(e). Each of the e streams joining tiles in A and B has sum S = e(n-1) and thus $\Sigma A + \Sigma B = e^2(n-1)$. Similarly $\Sigma B + \Sigma C = e^2(n-1)$ and $\Sigma A + \Sigma C = e^2(n-1)$. Thus $\Sigma A + \Sigma B = \Sigma B + \Sigma C = \Sigma A + \Sigma C = e^2(n-1)$ with the consequence that $\Sigma A = \Sigma B = \Sigma C$. Since also $\Sigma A + \Sigma B + \Sigma C = \Sigma Z(n) = (1/2)n(n-1)$, we obtain $\Sigma A = \Sigma B = \Sigma C = (1/6)n(n-1)$. □

In Figure 5.4 two magic H(3) hexagons over Z(27) are illustrated. Figure 5.4(I) is a rather difficult to find ad hoc solution and 5.4(II) is derived from the partition of Z(27) into three 3 by 3 near magic squares provided by Theorem 4.9. In Figure 5.5(I, II) orthogonal $(Z(3))^9$ and $(Z(9))^3$ are provided, which when superimposed yield a magic H(3) over Z(3)×Z(9). Note that in 5.5(II) three copies of the 3 by 3 Lo Shu square have been substituted for the major parallelograms. These are overlaid by the three Latin squares over Z(3) in 5.5(I) whose orientation can be chosen to create orthogonality.

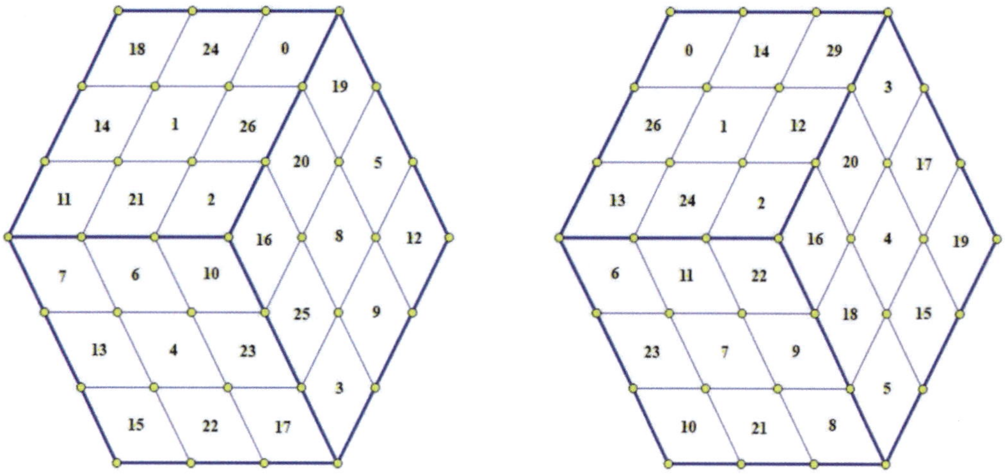

I II

Figure 5.4

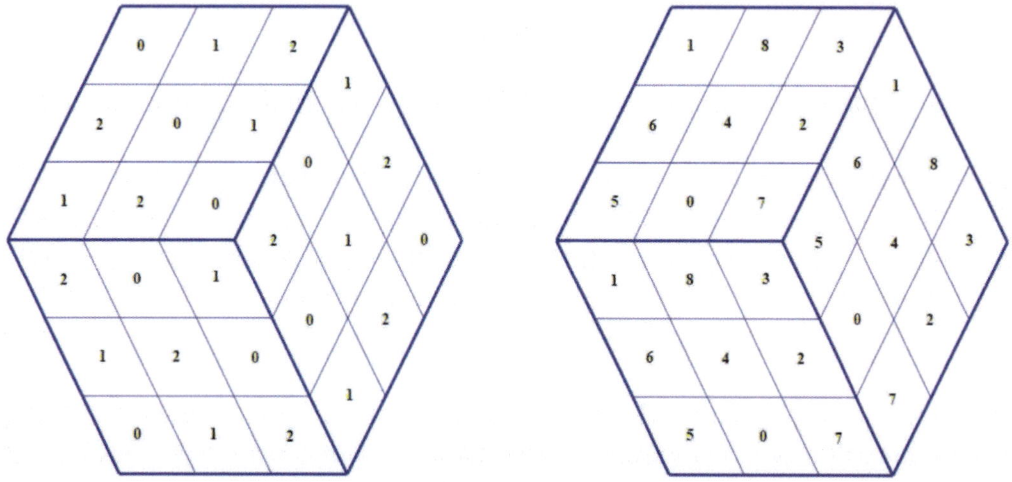

I II

Figure 5.5

The 3 by 3 magic square over Z(3)×Z(3) given in Figure 3.1(IV) can be substituted for each Lo

Shu square in Figure 5.5(II) to obtain an H(3) hexagon with base $(Z(3)\times Z(3))^3$. When this hexagon is superimposed over the $(Z(3))^9$ hexagon in 5.5(I), a natural magic H(3) over Z(3)×Z(3)×Z(3) is obtained. Since Z(27), Z(3)×Z(9) and Z(3)×Z(3)×Z(3) are the only direct product representations of an abelian group of order 27, we can conclude that H(3) is universally magic. In Figure 5.6(I,II) the conversion to integral of the hexagon in Figure 5.4(II) shows, in addition, that H(3) is integral.

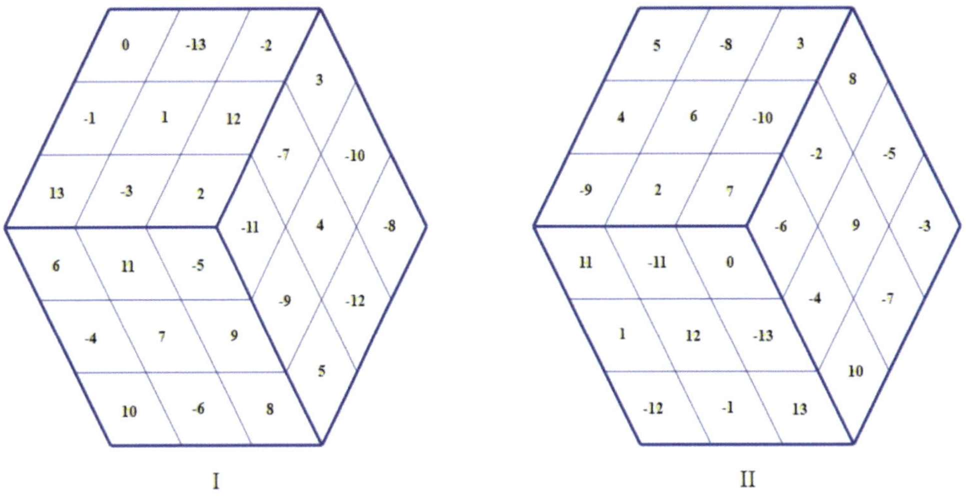

Figure 5.6

To study H(4) we develop a method for partitioning Z(48) into three 4 by 4 near magic squares. In Figure 5.7(I) the elements of Z(48) are listed in 12 groups of 4 in the matrix A, each group containing 4 consecutive elements. One element from each group is selected to form the set S and a one to one correspondence ϕ between S and Z(12) is established as in Figure 5.7(II). In 5.7(III) the natural equipartition of Z(12) and the bijection ϕ are used to determine a corresponding equipartition of S. This equipartition is indicated by the rows in the matrix R.

Each element in R selects the row of A which contains it, so the elements in each row of R select four corresponding rows of A. The three rows of R thus determine the three 4 by 4 squares S_1, S_2, S_3 indicated in 5.7(IV). Finally the elements in each of these squares are rearranged with the two larger groups of 4 elements placed consecutively in the two corner diagonals. The result is the three near magic squares M_1, M_2, M_3, also indicated in 5.7(IV), which form a near magic partition of Z(48). These squares can then be *compressed* to form the major parallelograms in the polar tiling of H(4) to obtain the magic H(4) hexagon over Z(48) in Figure 5.8(I). Figure 5.8(II) illustrates the conversion of 5.8(I) to base Z*(48). All streams except the two which contain 24 have sum -4. When 24 is replaced by -24 and then every value is replaced by its inverse, an integral magic H(4) hexagon is obtained with sum 4.

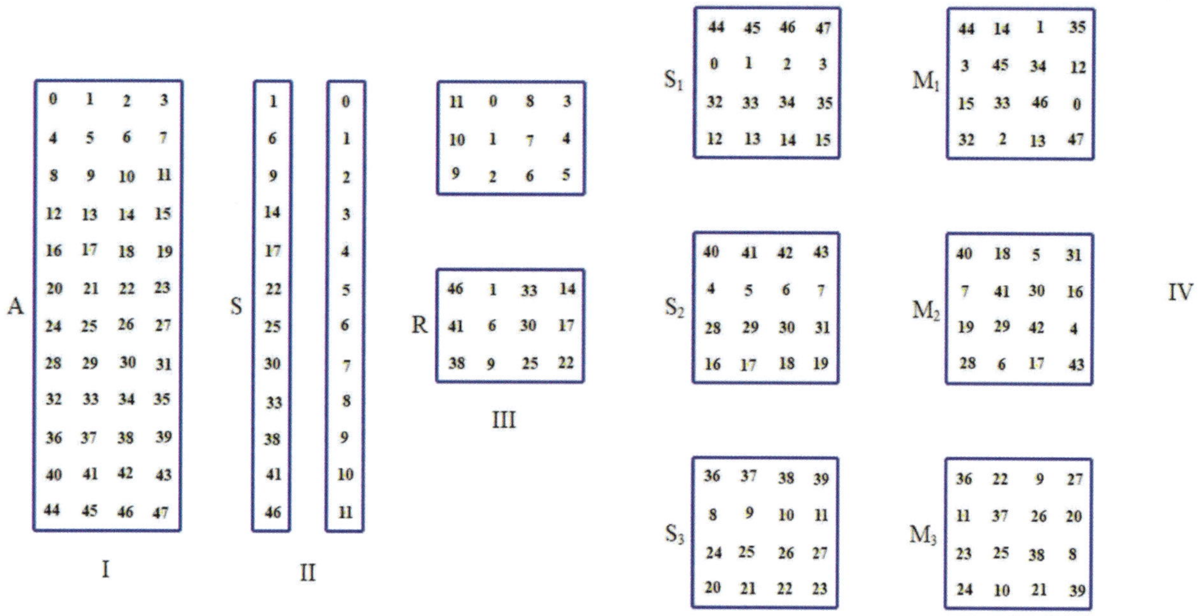

Figure 5.7

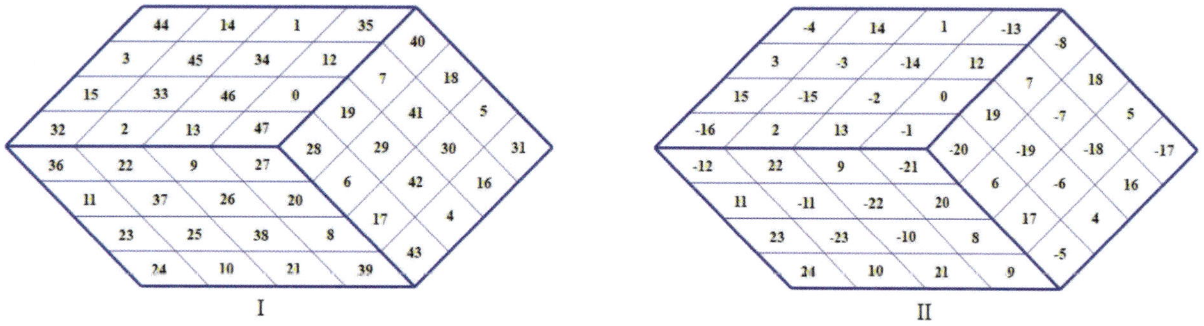

Figure 5.8

Before proceeding further with the analysis of H(4), a few general results are in order.

Lemma 5.2 *Let P be a magic polyparallelogram over Z(r) with a regular polygonal boundary where r denotes the number of tiles in P . Let M denote a magic square of order n over Z(n^2). Then the dilation of M by P yields a magic polyparallelogram M⊕P over Z(rn^2). M⊕P has the same number of sides as P but now has ne basic edges per side where e denotes the number of basic edges per side in P.*

Proof. Replace each value x in P by the square xn^2 + M. The resulting polyparallelogram will be magic over Z(rn^2) and will have a regular polygonal boundary with the same number of sides as P. If P[M] denotes the polyparallelogram obtained by replacing each element x in P by M, and n^2P denotes the polyparallelogram obtained by multiplying each element of P by n^2, then M⊕P = n^2P + P[M]. □

In Figure 5.9 the construction of M⊕P is illustrated with the Lo Shu square M and with the H(2) hexagon P of Figure 5.1(I).

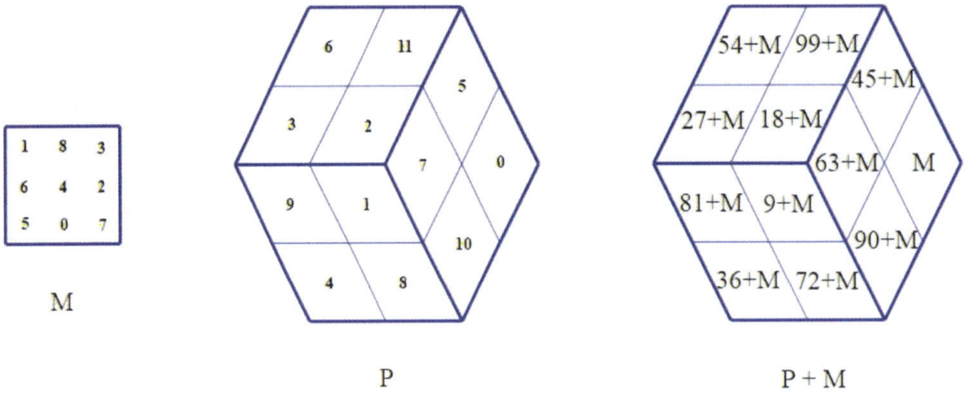

Figure 5.9

Theorem 5.3 *For each integer e with e≥2, the regular hexagon H(e) has a magic solution over Z($3e^2$).*

Proof. The constructions for e=2,3,4 are given in Figures 5.1, 5.4 and 5.8(I) respectively. In case e≥3 is odd, Theorem 4.9 provides a partition of Z($3e^2$) into three near magic squares of order e, each with the same row and column sums. These may be used to populate the cells of the three major parallelograms in the polar tiling of H(e). The result is a magic solution for H(e) over Z($3e^2$). Figure 5.4(II) is an example of this construction for e=3.

Now suppose e is even and e≥6. Let P denote the magic H(2) hexagon in Figure 5.1, and let M be a magic square of order e/2 over Z($e^2/4$). Then M⊕P is a magic H(e) hexagon over Z($3e^2$). □

Theorem 5.4 *Let e≥4 be an even integer and let M be a natural magic square over the abelian group G of order e^2. Then the regular hexagon H(e) has a natural magic solution over Z(3)×G.*

Proof. Let J=J(e,1) denote the all ones square of order e; let A = J(e/2,0) and B= J(e/2,2) denote respectively the all 0 and all 2 squares of order e/2. Let K, L denote the squares of order e as defined in Figure 5.10 and let H_1(e) be defined as in 5.10 by substituting K, L, J for the major parallelograms in the polar tiling of H(e). Let H_2(e) be constructed by placing three copies of M, one each, in the major parallelograms which constitute the polar tiling of H(e). Then H_1(e) is a natural magic square over $(Z(3))^{ee}$ and H_2(e) is a natural magic square over G^3. Moreover, H_1(e) is orthogonal to H_2(e), and thus H_1(e)×H_2(e) is a natural magic square over Z(3)×G. □

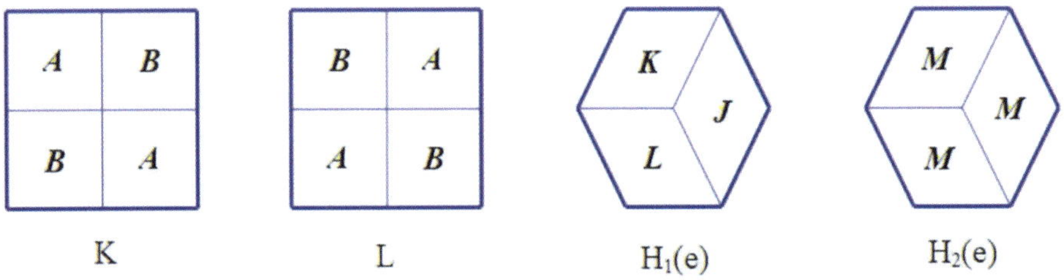

Figure 5.10

In Figure 5.11 the construction of Theorem 5.4 is illustrated with M equal to the order 4 natural magic square over Z(4)×Z(4) from Figure 3.2(IV).

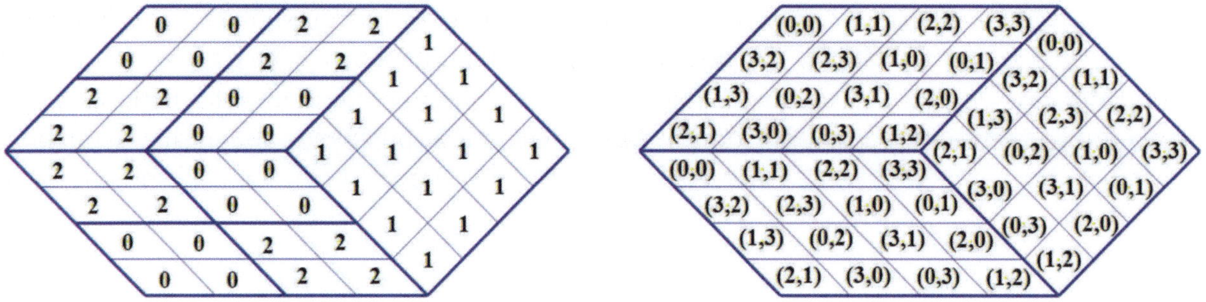

$H_1(4)$ $H_2(4)$

Figure 5.11

In Theorem 5.4, M can be taken to be any of the natural magic squares over abelian groups of order 16 as constructed in Figures 3.2(I), 3.2(IV), 3.2(VII), 3.3(III) and 3.26. This accounts for all abelian groups of order 48 which have Z(3) as a factor. To prove that H(4) is universally magic it remains to construct natural magic solutions over the following representations of abelian groups: Z(6)×Z(8), Z(6)×Z(2)×Z(4), Z(6)×Z(2)×Z(2)×Z(2), Z(2)×Z(24), Z(2)×Z(2)×Z(12), Z(4)×Z(12).

In Figure 5.12(I) a natural magic H(4) over $(Z(4))^{12}$ is illustrated, and in Figure 5.12(II) a natural magic H(4) hexagon over $(Z(12))^{4}$ is provided. These two hexagons are orthogonal and thus their superimposition yields a natural magic H(4) over Z(4)×Z(12). Note that the hexagon in 5.12(II) is obtained from the H(2) hexagon in Figure 5.12(III) via the substitution x →J(2,x). If each element in 5.12(I) is replaced by its base 2 equivalent, a natural magic H(4) over $(Z(2)\times Z(2))^{12}$ is obtained. This square remains orthogonal to the hexagon in 5.12(II), and consequently their superimposition yields a natural magic H(4) over Z(2)×Z(2)×Z(12).

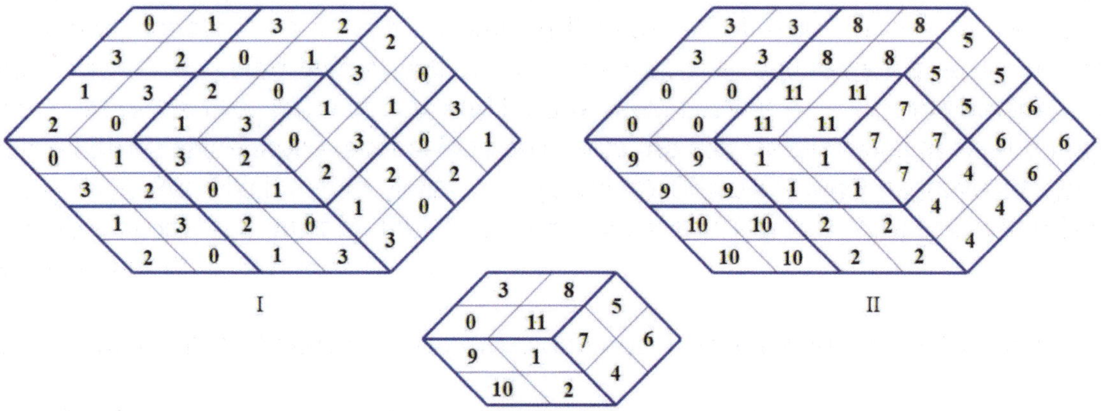

I II III

Figure 5.12

The natural magic H(4) over $(Z(6))^8$ in Figure 5.13(I) and the natural magic H(4) over $(Z(8))^6$ in 5.13(II) are orthogonal. When superimposed a natural magic H(4) over Z(6)×Z(8) is crated. When the substitution x →x base 2 is applied to each element in 5.13(II) a natural magic H(4) over $(Z(2)\times Z(2)\times Z(2))^6$ is obtained. When this hexagon is superimposed over 5.13(I), a natural magic H(4) over Z(2)×Z(2)×Z(2)×Z(6) is created. Finally, when the substitution x→x base 4 is applied to each element in 5.13(II) a natural magic H(4) over $(Z(2)\times Z(4))^6$ is obtained. When this hexagon is superimposed over 5.13(I), a natural magic H(4) over Z(2)×Z(4)×Z(6) is created.

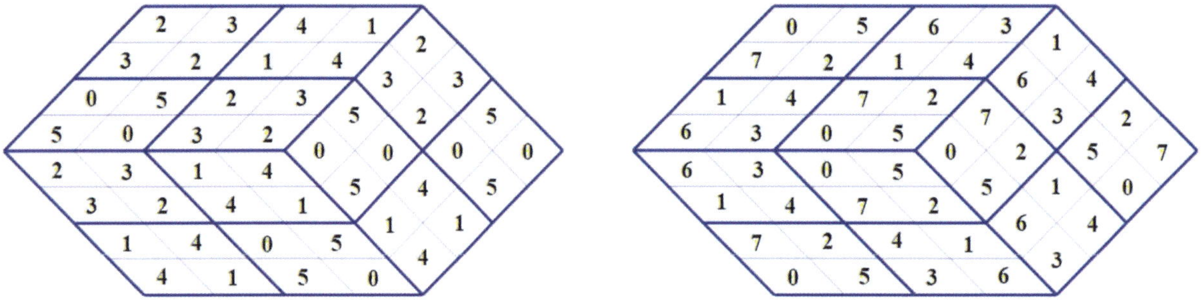

I II

Figure 5.13

It remains only to construct a natural magic H(4) over Z(2)×Z(24). All constructions involving direct product require the creation of a pair of orthogonal magic polygons. In case a natural construction is desired, the orthogonal polygons must themselves be natural. In Figure 5.14(I) a natural magic H(4) over $(Z(24))^2$ is illustrated. It was obtained by applying the homomorphism x→x (mod 24) to the H(4) over Z(48) given in Figure 5.8(I). In Figure 5.8(II) a natural magic H(4) over $(Z(2))^{24}$ is depicted. These two hexagons are orthogonal, and thus their superimposition yields a natural magic H(4) hexagon over Z(2)×Z(24).

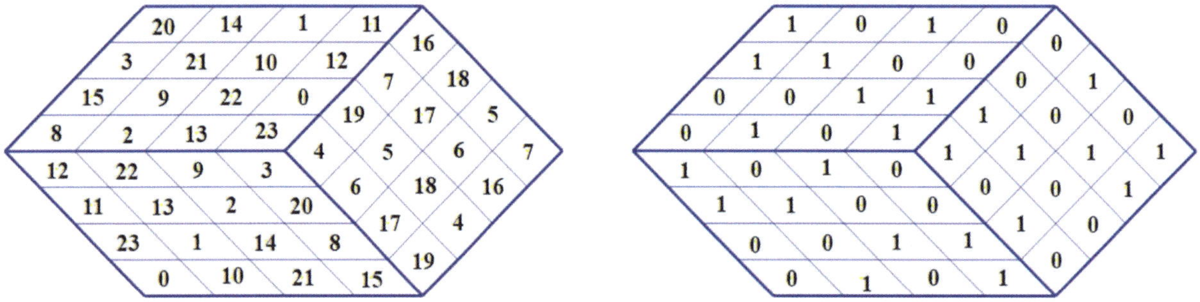

I II

Figure 5.14

We have shown that H(4) is universally magic and integral. No doubt the constructions in Figures 5.11 - 5.14 yield generalizations, but this is as far as we will take the subject. It is possible that H(e) is naturally universally magic and integral whenever e≥3.

5.2 Chevron Sequences

By *irregular hexagon* we mean a semiregular hexagon which is not regular. Our main focus will be integral irregular hexagons, since natural ones are hard to find and basic ones, although many can be constructed, are less proximitous to the concept of a magic polygon. The simplest type of irregular integral hexagons are *chevron sequences.* These consist of a linear sequence of chevrons with a *cap* consisting of a single parallelogram. The three smallest integral chevron sequences are illustrated in Figure 5.15.

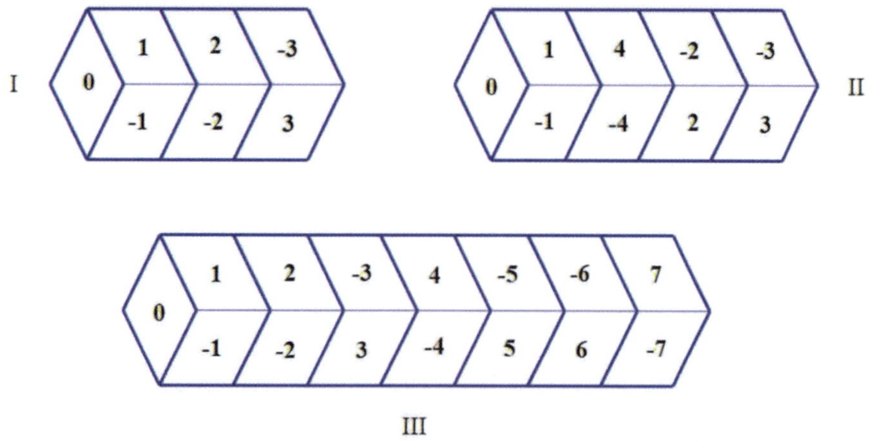

Figure 5.15

First we will quickly show that no natural magic solution is possible for any chevron sequence.

Theorem 5.5 *No natural solution is possible for any chevron sequence.*

Proof. Let x denote the value assigned to the diamond cap of a chevron sequence S consisting of cap and m chevrons, and let y denote the common sum of the pair of values assigned to each chevron. Let w_1 denote the sum of the values in the top half of each chevron, and let w_2 denote the sum of the values in the bottom half of each chevron. The requirement that all streams have the same sum yields $x+w_1 = x + w_2 = y$, and thus $w_1 = w_2$. Let $w = w_1 = w_2$ and note that $my + x = 2w + x = m(2m+1)$ since these expressions represent the sum of all values in S. Then $my = 2w$ implies $m(w+x) = 2w$, which in turn yields $x = w((2-m)/m)$. Since x is a nonnegative integer and w, m are positive integers, it must be that $2-m \geq 0$ and thus $m = 1$ or $m = 2$. But no natural chevron sequence exists with $m = 1$ or $m = 2$. □

On the other hand integral chevron sequences do exist. These are determined in the next result.

Theorem 5.6 *An integral chevron sequence exists over $Z^*(n)$ precisely when $n \equiv 1$ or $7 \pmod 8$.*

Proof. With x, w, y defined as in Theorem 5.5 we again have $x+w = y$. Since n is odd, $\Sigma Z^*(n) = 0$, and thus $2w+x = 0 = (1/2)(n-1)y + x$. From $2w+x = 0$ and $x+w = y$ we obtain $w+y = 0$. Along with $2w = (1/2)(n-1)y$ we then obtain $(2 + (1/2)(n-1))y = 0$ and thus $y = 0$. It then follows that $w = x = 0$.

Let S = {1, 2, ..., (1/2)(n-1)}, then ΣS = (1/8)(n^2 - 1). Since w = 0, it must be that S can be partitioned into two sets A, B each with the same sum, namely (1/16)(n^2 - 1). Thus $n^2 \equiv 1$ (mod 16). This congruence has solutions n ≡1, 7 (mod 8). In case n ≡ 7(mod 8), then S can be partitioned into (1/4)(n+1) sets each with sum (1/2)(n-1):

S = {{(1/2)(n-1)}, {1, (1/2)(n-3)}, {2, (1/2)(n-5), ..., {(1/4)(n-3), ((1/4)(n+1)}}.

The sets in this partition are even in number, so we can select exactly half of them for A and use these to determine half of the values in w_1. The other half are negated, (-B) and used to constitute the other half of the values in w_1. The result is $\Sigma w_1 = 0$ as desired. For w_2 we choose the values in -A ∪ B and situate these so that the sum of the two values in each chevron is 0. In case n ≡ 1 (mod 8), (n≠1) the set S is partitioned into (1/4)(n-1) sets each with sum (1/2)(n+1):

S = {{1, (1/2)(n-1)}, {2, (1/2)(n-3)}, {3, (1/2)(n-5), ..., {(1/4)(n-1), (1/4)(n-3)}}.

The sets in this partition of S are also even in number and so we can select exactly half of them for A and use these to populate half the cells belonging to w_1. The other half are negated (-B) and used to populate the remaining cells of w_1. In this way we obtain again $\Sigma w_1 = 0$. For w_2 we choose the values in -A ∪ B and situate these so that the sum of the two values in each chevron is 0. In either case a sum 0 magic integral chevron sequence is obtained. □

5.3 Platforms

We use the notation H(p,q,r) to denote a semiregular hexagon with consecutive sides of a semiperimeter of length p, q, r. With a few exceptions, our construction of nonregular magic hexagons will use the integral base Z*(n) with n odd. All streams will have 0-sum. One exception is the H(1,3,3) hexagon illustrated in Figures 5.16(I,II). In 5.16(I) an integral version is given, and the natural solution in 5.16(II) is obtained by a direct conversion from the integral solution. All streams in this hexagon have a natural sum of 30. Note that 5.16(I) is constructed as an extension of H(1,3,1). The extending piece, an *I-bar,* is essentially an I-extension as encountered in Section 4.8; only its shape is modified in order to attach it to the H(1,3,1). When two opposite sides of a semiregular hexagon consist of only one basic edge, we call the hexagon a *platform.* Thus all chevron sequences as well as H(1,3,3) are platforms.

In Figure 5.16(III) an H(1,2,3) platform is illustrated. This platform, as well as the chevron sequences in Figure 5.15(I,II) are examples of *primary* magic integral hexagons. This means that they cannot be obtained as extensions of smaller integral hexagons. Any integral hexagon which can be so obtained is called *composite,* and thus the H(1,3,3) in Figure 5.16(I) is composite with H(1,1,3) as kernel. We have seen that every chevron sequence can be obtained as a sequence of I-extensions from the primary H(1,3,1) and H(1,4,1) in Figure 5.15. These I-extensions will be called *0-dominoes* since they bear a close resemblance to the 2 by 4 2-sum *dominoes* introduced in Section 4.5.

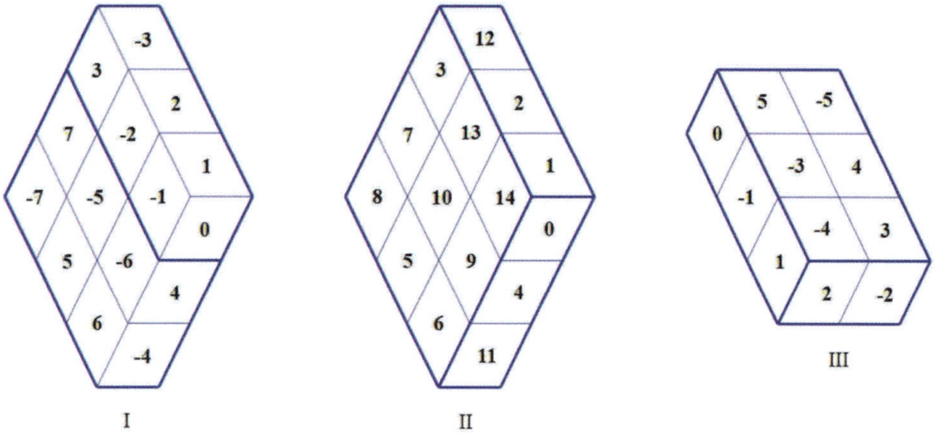

Figure 5.16

Repeated I-extensions of H(1,3,1) and H(1,2,3) as indicated in Figure 5.17 allow the creation of integral platforms H(1,q,3) for every positive integer q≥3. In Figure 5.18 an extension of H(1,4,1) is featured which involves the attachment of three 0-sum sections each isomorphic to a 2 by 5 rectangle. We will refer to such constructions generally as *0-isorectangles,* and if the isorectangle consists of streams of length 2 as in this case, we call it an *I-bar.* I-bars occur prominently in our constructions; note that the integral platform in Figure 5.18 consists of the 0- tile and 6 I-bars, two of which are 0-dominoes.

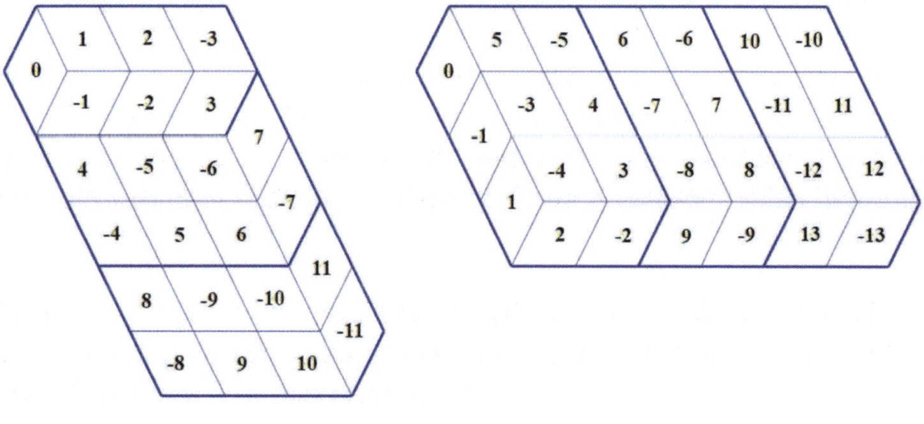

Figure 5.17

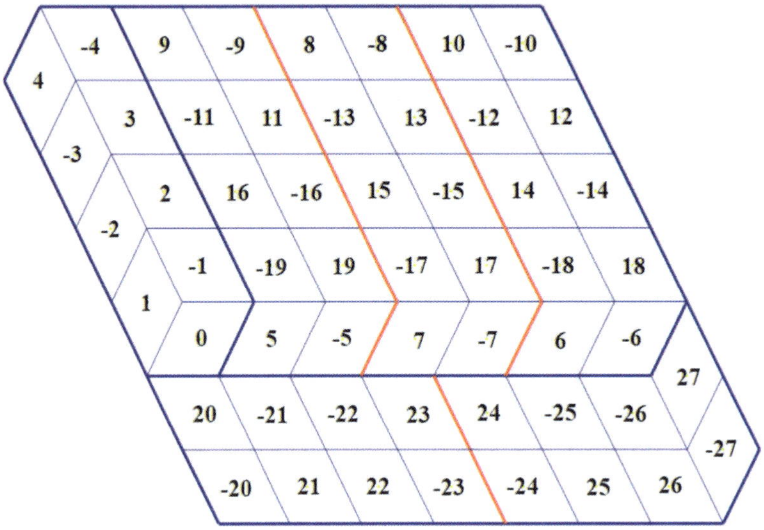

Figure 5.18

A second important type of extension, a *C-extension,* is illustrated in Figure 5.19. 5.19(I) contains H(1,2,3) as kernel, and 5.19(II) contains H(1,2,3) as kernel and H(1,3,4), H(1,3,6) as inclusions. All I-extensions attach to two sides of an integral hexagon, whereas C-extensions attach to 3 or 4 sides. In Figure 5.20 a C-extension of H(1,2,3) is illustrated which is composed of a 3 by 4 *0-isorectangle* and an I-bar.

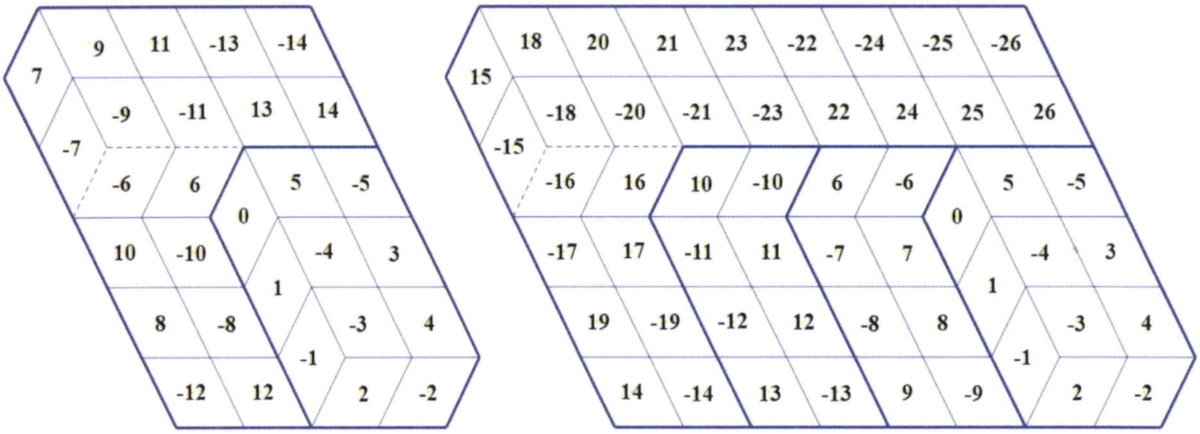

I II

Figure 5.19

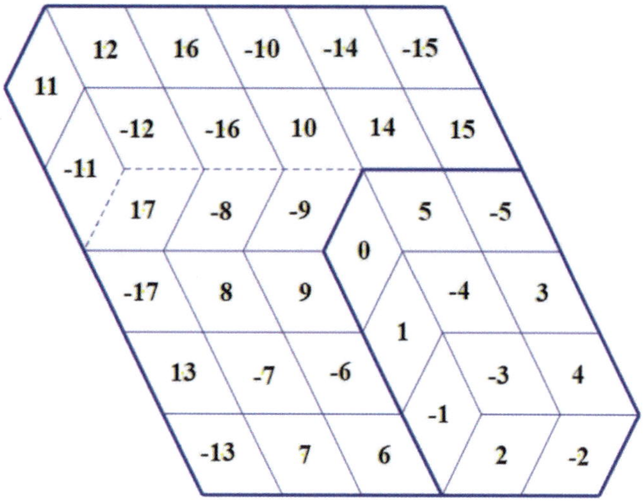

Figure 5.20

5.4 Odd Order Integral Hexagons: Class(1)

The order of a hexagon H = H(p,q,r) is given by n = pq+pr+qr. If n is odd then there are two possibilites: (1) p,q,r are all odd, or (2) exactly one from among p,q,r is even. In case (1) holds we will say H is *Class 1,* and if (2) holds H is *Class(2).* In this Section we will construct integral hexagons in Class(1). First some general considerations involving 2-sided extensions are given.

A *2-sided extension* of an integral hexagon is one which attaches to exactly two adjacent sides of the hexagon to be extended. The constructions in Figures 5.17, 5.18 feature series of two-sided extensions. If the sum of the lengths of the two sides to be extended is even, then we have the following simple result. A *0-domino extension* is an extension which has a dissection into 0-dominoes.

Theorem 5.7 *Let H(p,q,r) be an integral hexagon with p, q odd. If 4|(p+q) then H has a 0- domino extension H(p,q,r)→ H(p,q,r+2) consisting of a single I-bar, If p+q is twice an odd number then H has an H(p,q,r)→ H(p,q,r+4) 0-domino extension .*

Proof. Let S be a consecutive sequence of positive integers whose length m is a multiple of 4, say m = p+q =4t, then the elements in S ∪(-S) can be used to form t 0-dominoes. These dominoes can be joined end to end to create an I-extension for H which attaches to the sides of H with lengths p, q. If p+q = 2t where t is odd then the elements of S can again be partitioned into t 0-dominoes. This time these dominoes are stacked with their long sides, (sides of length 4), adjacent. This results in a 0-domino extension *H(p,q,r)→ H(p,q,r+4)* which is be attached to sides p, q of H. □

The extension in Figure 5.18 over S ∪(-S) with S = {20, ..., 27} is an example of a single I-bar extension with a dissection into 0-dominoes. In Figure 5.21 a 0-domino extension H(1,5,5)→H(5) is illustrated, and thus the regular hexagon H(5) is shown to be composite.

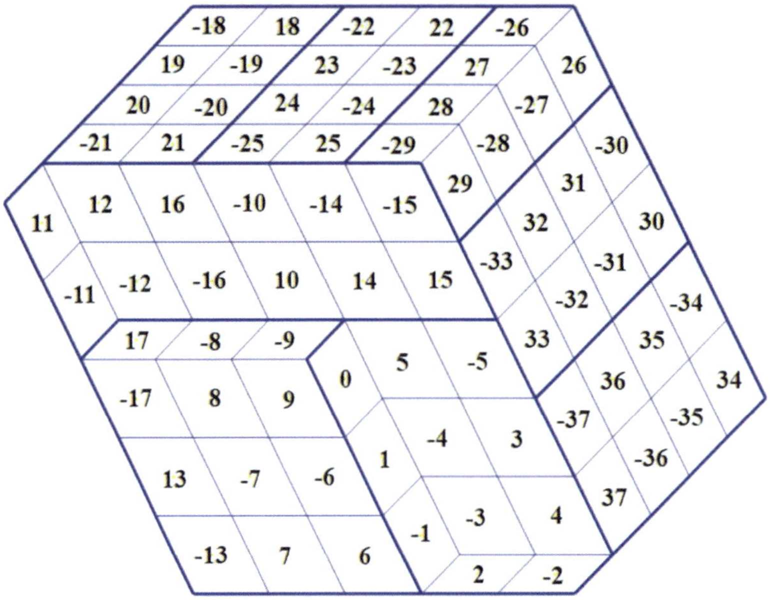

Figure 5.21

If two sides of an integral hexagon have an odd length sum, then it is much more difficult to find extensions which attach to the two sides. One example is already given as an H(1,7,4) inclusion in the platform in Figure 5.18. A second example is provided in Figure 5.22 which features an extension of H(1,2,3).

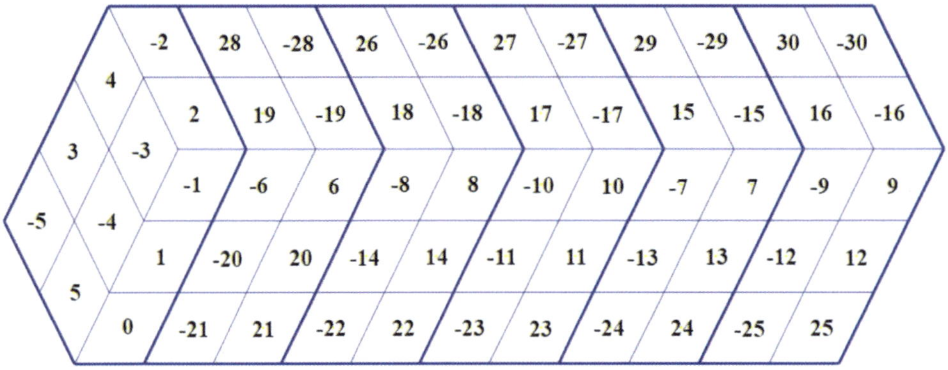

Figure 5.22

Extensions involving 0-dominoes can be used to construct regular integral hexagons as in the following observation.

Theorem 5.8 *Let p≥3 be odd. Then the integral regular hexagon H(p+4) can be obtained from H(p) by a 0-domino extension.*

Proof. By the construction given in Theorem 5.7, H(p) = H(p,p,p) has a 0-domino extension to H(p,p,p+4) which, in turn, has a 0-domino extension to H(p,p+4,p+4). Finally, since p + (p+4) is twice an odd number, H(p,p+4,p+4) has a 0-domino extension to H(p+4,p+4,p+4) = H(p+4). □

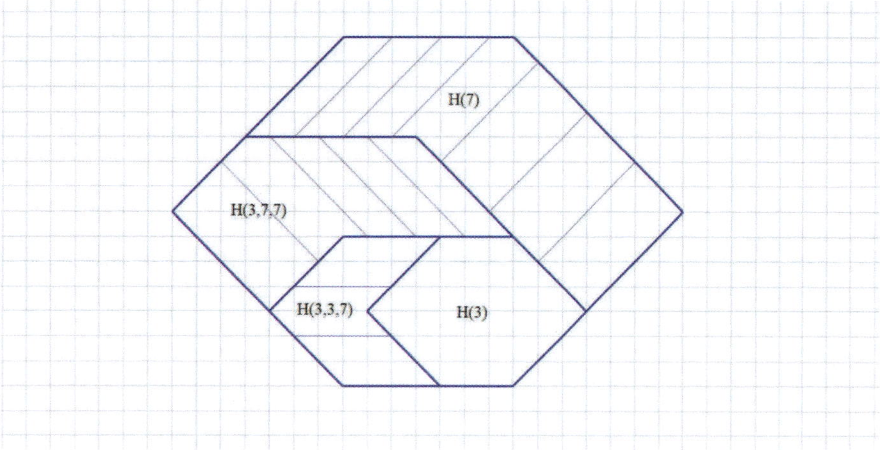

Figure 5.23

In Figure 5.23 a 0-domino extension from H(3) to H(7) is illustrated. The construction is essentially a tiling problem with the tiles representing 0-dominoes. The integral hexagon in Figure 5.6(II) can be substituted for H(3).

By means of 0-domino extensions all Class(1) integral hexagons can be obtained, as we show next.

Theorem 5.9 *Let p, q, r be odd positive integers with no two having a value of 1. Then an integral hexagon H(p,q,r) can be obtained by 0-domino extensions of H(1,5,5) or one of the primary integral hexagons H(1,1,3), H(3,3,3) = H(3).*

Proof. We proceed by induction on the semiperimeter s = p+q+r. If s = 7, we have only H(1,3,3) which is obtained by the 0-domino extension H(1,3,1)→ H(1,3,3) in accordance with Theorem 5.7. In Section 5.3 it was observed that repeated I-extensions of H(1,1,3) yield the integral hexagon H(1,3,q) for every odd positive integer q. These are 0-domino extensions, so we can consider this class of hexagons to be proved. If s = 9 we have only H(1,3,5) and H(3,3,3). As just stated, we have H(1,3,5), and H(3,3,3) = H(3) is one of the primaries, an integral solution for which can be found in Figure 5.6. For s = 11 we have H(1,3,7), H(1,5,5) and H(3,3,5). The integral H(1,5,5) we are taking as primary; an integral solution can be found as an inclusion in Figure 5.21. The extension H(3,3,1)→ H(3,3,5) is assured by Theorem 5,7 since p+q = 6 is twice an odd number.

Now let the semiperimeter s be larger than 11, and assume inductively that H(p,q,r) is integral whenever p, q, r are odd; p+q+r < s, and no two among {p,q,r} are equal to 1. We consider the following cases.

CASE 1: *Neither p nor q nor r is equal to 1 and p+q is a multiple of 4 .*

Then H(p,q,r-2)→ H(p,q,r), and this extension can be partitioned into (1/4)(p+q) 0-dominoes.

CASE 2: *Neither p nor q nor r is equal to 1 and p+q is twice an odd number.*

Then $H(p,q, r-4) \rightarrow H(p,q,r)$ provided $r \geq 5$. This extension can be accomplished with $(1/2)(p+q)$ 0-domino extensions. In case $r = 3$, we have either $H(3,3,3)$, $H(5,5,3)$ or $H(p, q,3)$ with at least one of p,q is greater than 5. $H(3,3,3)$ is one of our kernels, and $H(5,5,3) = H(3,5,5)$ comes under Case 2 so we have the extension $H(3,5,3) \rightarrow H(3,5,5)$. Finally in case we want $H(p, q,3)$ with at least one of p,q greater than 5, suppose without loss of generality that $q \geq 7$. Then the extension $H(p,3,q-4) \rightarrow H(p,3,q) = H(p,q,3)$ yields the desired result in case p+3 is a multiple of 4, and the extension $H(p,3, q-2) \rightarrow H(p,3,q) = H(p,q,3)$ works in case p+3 is twice an odd number. All these extensions can be accomplished with 0-domino extensions.

CASE 3: *Exactly one of p,q,r is equal to 1.*

Without loss of generality, let $r = 1$. If $p = 3$ or $q = 3$, we have all integral hexagns of the form $H(1,3,q)$ with q odd. So we can assume that $p \geq 5$ and $q \geq 5$. The integral $H(5,5,1)$ is included in the Theorem statement; it is constructed as a C-extension of $H(1,2,3)$ in Figure 5.21 but cannot be obtained by means of a 0-extension of a smaller integral hexagon. It then remains to consider $H(p,q,1)$ with $p \geq 5$ and $q \geq 7$. If p+1 is a multiple of 4 then $H(p,1,q-2) \rightarrow H(p,1,q) = H(p,q,1)$, and if p+1 is twice an odd number then the extension $H(p,1,q-4) \rightarrow H(p,1,q) = H(p,q,1)$ provides the desired result. Again these extensions can be accomplished by means of a series of 0-domino extensions.

The above 0-domino extensions are all made from a hexagon with semiperimeter $s < p+q+r$, so the induction is complete and the Theorem proved. □

5.5 Transformations

There are many ways to select a series of 0-domino extensions to obtain a given integral hexagon. There is a specific set of simple transformations which may connect any two such extensions. We will now describe such transformations in connection with 0-domino extensions of $H(1,1,3)$.

Let $H = H(p,q,r)$ be a 0-domino extension Ω of $H(1,1,3)$ and let L_1, L_2 be a pair of parallel broken line segments in Ω. The set of tiles which lie between L_1 and L_2 will be called a *chasm.* A *slide transformation* occurs when the chasm is removed by sliding together the blocks of Ω separated by the chasm. When this sliding is accomplished an untiled *emergent* region is left in its wake. The emergent region is then tiled, completing the transformation.

A *0-square* is a 0-sum polygon whose stream structure is the same as a square. There are two ways to tile a 0-square with two 0-dominoes; the conversion between these two tilings is a *0-square transformation.* The *platform* $H(1,3,3)$ can be partitioned into an $H(1,1,3)$ and a 0-domino in four ways by placing the long side of the $H(1,1,3)$ against each of the four long sides of $H(1,3,3)$. Converting from one of these tilings to another will be called a *small platform transformation.* These types of transformations are illustrated In Figure 5.24, (I-II) is a slide transformation of $H(1,1,3)$ in $H(1,3,5)$; (II-III) is a slide transformation of $H(1,3,3)$ inside $H(3,3,5)$; (III-IV) is a 0-square transformation, and (IV-V) illustrates both a 0-square transformation and a small platform transformation. This discussion of transformations is intended as a prelude to the following Conjecture.

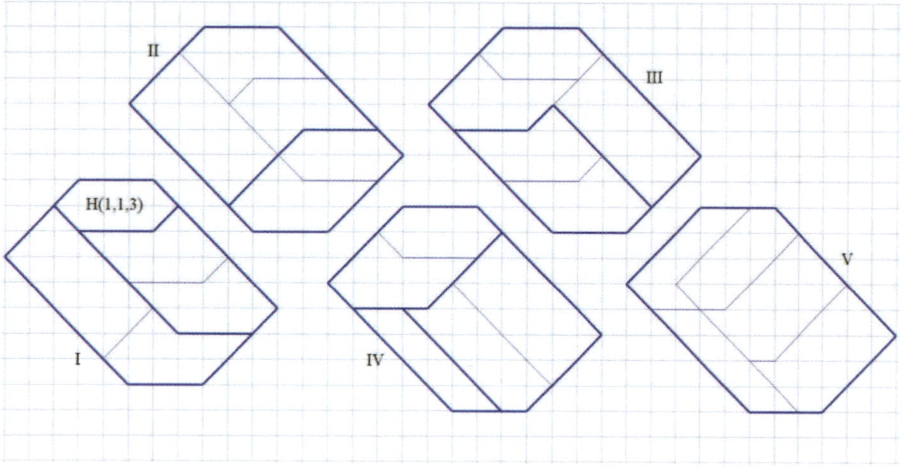

Figure 5.24

Conjecture 5.10 *Let H = H(p,q,r) with p,q,r odd be obtainable as a 0-domino extension of the primary integral hexagon H(1,1,3), and let Ω, Γ denote two 0-domino extensions from H(1,1,3) to H. Then there exists a series of slide, 0-square and small platform transformations which convert Ω to Γ.*

To test this Conjecture, the reader may try to find the series of transformations which carries the 0-domino extension of H(1,1,3) in Figure 5.25(I) to the one in 5.25(II). One solution is given in Appendix I.

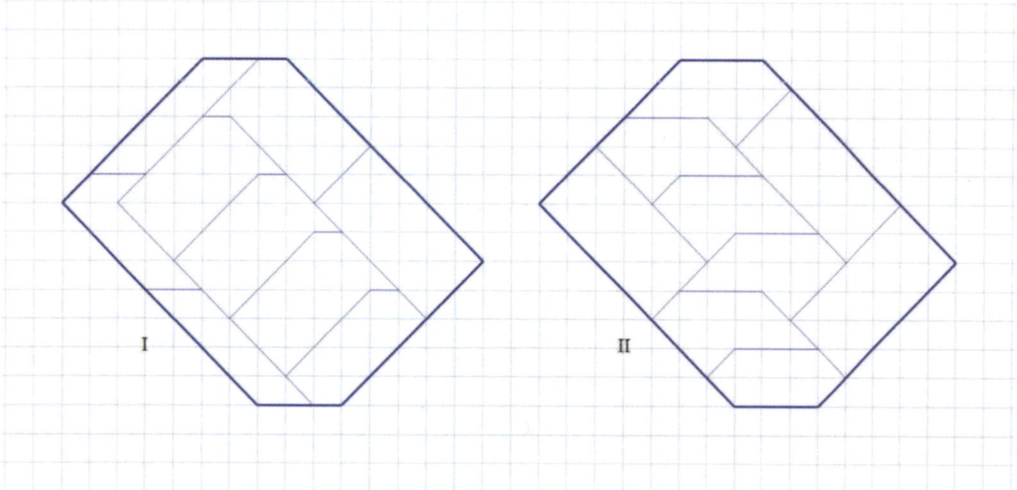

Figure 5.25

We provide one further example in Figure 5.26. The tiling in 5.26(I) is the starting point and 5.26(VI) is the goal. The parallel jagged line segments L_1, L_2 which indicate the boundaries of the chasm, are indicated in bold. The chasm consists only of 0-domino R. When tiles A, B are slid downward to eliminate the chasm, the tiling in 5.26(II) results. In this tiling the emergent region is indicated with a red boundary and tiled with two 0-dominoes R′, S′ which replace the original tiles R, S. Also a new chasm is formed consisting of tiles E, S′, A. In 5.26(II-III) the chasm is closed and the new emergent, indicated in red, is tiled with tiles E′, S″, A′. In III-IV a small platform transformation is performed in the indicated 0-square. In III - IV a slide transformation moves the H(1,3,3) consisting of tiles F′, H to the lower boundary and the emergent region is retiled. Only a few transformations are still required to obtain the goal in Figure 5.26(VI). These the reader can easily find.

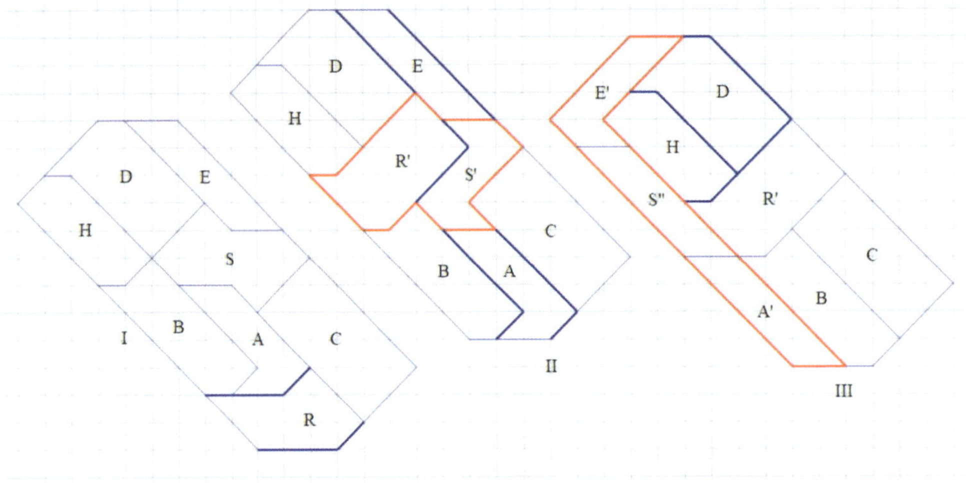

Figure 5.26

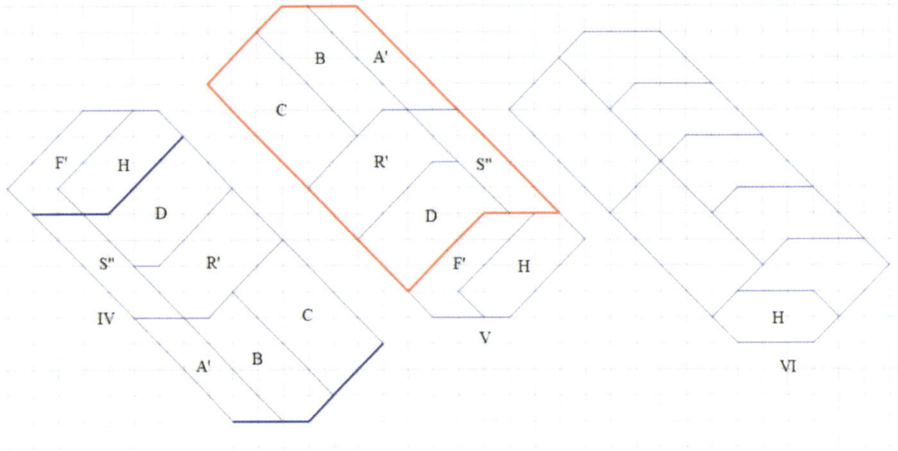

Figure 5.26

5.6 Nested Extension Sequence for regular Integral Hexagons of Odd Order

It is possible to create a nested sequence of regular integral hexagons of odd order analogous to the nesting of integral magic squares determined in Section 3.8. This will be accomplished by means of a series of 4-sided C-extensions. The following easy Lemma will be useful.

Lemma 5.11 *Let S be a consecutive sequence of nonnegative integers of length 12. Then S can be partitioned into two splitable sets each with 6 elements.*

Proof. Let S be given by

$$S = (a, a+1, a+2, a+3, a+4, a+5, a+6, a+7, a+8, a+9, a+10, a+11),$$

And let {A, B} be a partition of S obtained just by interchanging the two central values a+5, a+6. Thus

$$A = \{a, a+1, a+2, a+3, a+4, a+6\}; \quad B = \{a+5, a+7, a+8, a+9, a+10, a+11\}.$$

Then $A = A_1 \cup A_2$ with $A_1 = \{a, a+2, a+6\}$ and $A_2 = \{a+1, a+3, a+4\}$ and $B = B_1 \cup B_2$ with $B_1 = \{a+5, a+9, a+11\}$ and $B_2 = \{a+7, a+8, a+10\}$. Since $\Sigma A_1 = \Sigma A_2 = 3a+8$ and $\Sigma B_1 = \Sigma B_2 = 3a+25$, the sets A, B are splitable. □

With q≥3 odd, let H(q), as usual, denote a regular integral hexagon with q edges per side. The C-extension which extends H(q) to H(q+2) is partitioned into three I-bars, I_1, I_2, I_3 as indicated in Figure 5.27.

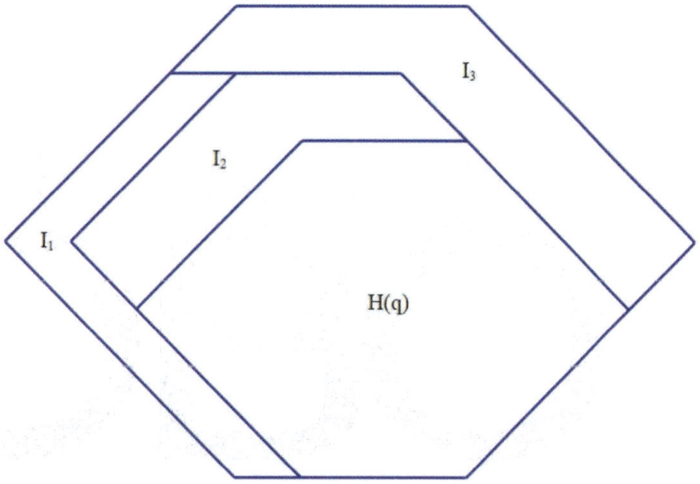

Figure 5.27

Theorem 5.12 *Let q be an odd integer with q≥3. Then there exists a C-extension H(q)→ H(q+2).*

Proof. Let q = 2t+1, and refer to Figure 5.27. The length of I_1 is 2q+2 = 4(t+1) so I_1 can be realized by appending t+1 0-dominoes. The length of I_2 is 2q = 4(t-1)+6 so it can be realized by appending t-1 0-dominoes and one I-bar of length 6. The length of I_3 is 2q+4 - 4t+6, so it can be realized by appending t 0-dominoes and one I-bar of length 6.

Let S denote the consecutive sequence of positive integers used in the C-extension comprised of the three I-bars. Take the first 12 elements of S for the two I-bars of length 6. By Lemma 5.11, this set of 12 elements is partitionable into two splitable subsets each with 6 elements. These two sets can be used to create the two I-bars of length 6. Thus all three I-bars are realized, and their combination forms the desired C-extension. □

In Figure 5.28 the extension H(3)→H(5)→H(7) is illustrated.

5.7 Odd Order Integral Hexagons of Class 2 : Platforms

The simplest odd order integral hexagons in Class 2 are H(1,1,4) illustrated in Figures 5.15(II) and H(1,3,2) illustrated in Figure 5.16(III). If H(1,1,q) is integral with q even, then in accordance with Theorem 5.6, q must be a multiple of 4. The next simplest Class 2 integral hexagons are the platforms H(1,3,q) with q even. Some of these are illustrated in Figure 5.29. The integral H(1,5,2) in 5.29(I) has a stand alone construction, but for q≥7 a standard procedure for construction will be provided.

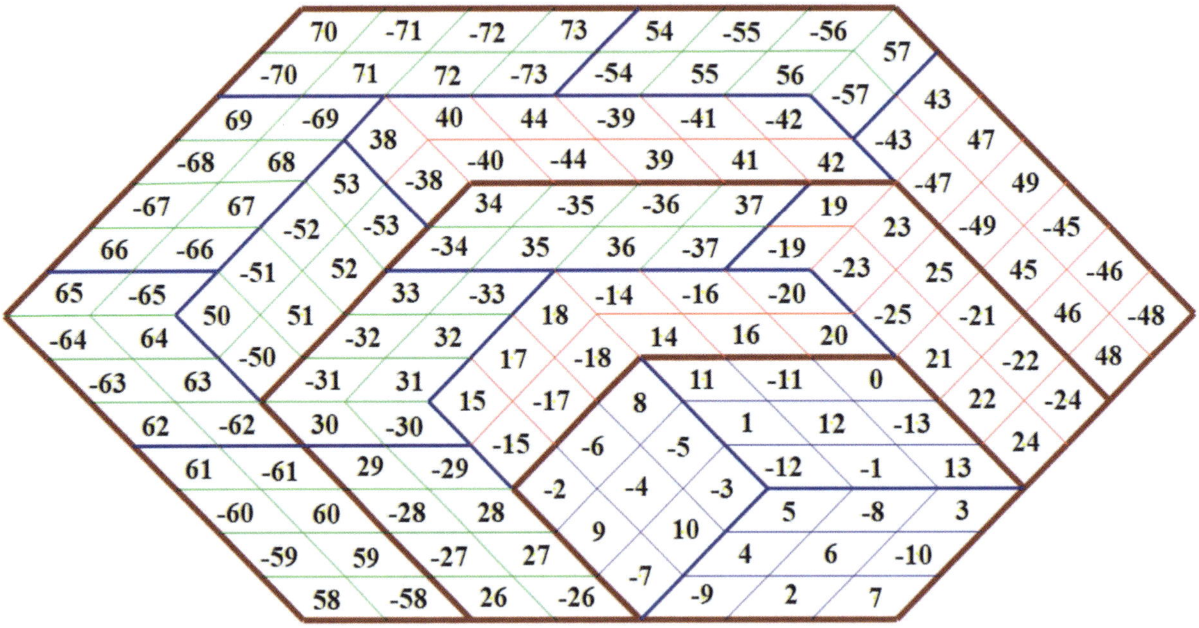

Figure 5.28

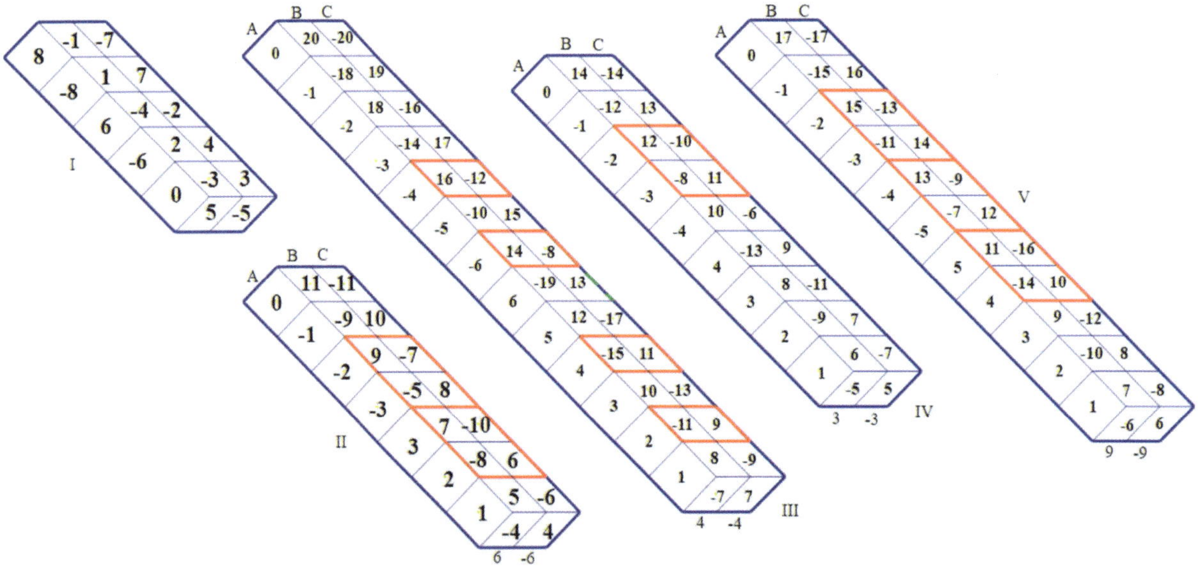

Figure 5.29

In Figures 5.29(II-V) the starting configuration for this standard procedure is illustrated. The interchanges required to complete the construction are indicated in red.

Theorem 5.13 *For each odd positive integer q ≥3, an integral H(1,q,2) hexagon can be constructed.*

Proof. The primary integral H(1,3,2) is illustrated in Figure 5.16(III), and the integral H(1,5,2), which is also primary, requires a special construction given in Figure 5.29(I) . For the remaining H(1,q,2) hexagons there is a standard starting configuration which insures that all the *short streams,* (i.e., streams of length 3), have 0 sum. The *long streams* are denoted by A, B,C as in Figure 5.29 . Stream A has the assignment

$$A = (0,-1,-2,-3, \ldots, -(1/2)(q-1), (1/2)(q-1), \ldots, 3, 2, 1)$$

with $\sum A = 0$. Interchanges within the short streams must be made in order to obtain $\sum B = 0$, and then $\sum C = 0$ will be automatically obtained. We set $\sum B = \varepsilon$ and $\sum C = -\varepsilon$ and thus interchanges must be found which shift a sum equal to ε from the B stream to the C stream.

The order n of H(1,q,2) is 3q+2 and the integral representation Z*(n) is

$$Z^*(n) = \{0, 1,-1, 2, -2, \ldots, (1/2)(3q+1), -(1/2)(3q+1)\}.$$

If m = (1/2)(3q+1), then our starting assignment for the B, C streams consists of a zig-zag with m in the top position of B and ending with (1/2)(q+1) at the bottom of C. The remaining values in the B and C streams are assigned so as to produce 0-sums in all short streams.

It is necessary to determine the ε value associated with each (odd) q. These correspondences are given in the table below.

q	7	9	11	13	15	17	19	21	23	...
ε	6	3	9	4	12	5	15	6	18	...

Note that the ε values follow a regular pattern: in particular, the ones that are multiples of 3 occur alternately in triples and singles. Interchanges which reduce ΣB by multiples of 3 are easily located: partition streams B,C into consecutive 2 by 2 blocks and note that, with the possible exception of the central block, interchanging the two values in a block reduces ΣB by 3. The ε values which remain have the form 4+3k or 5+3k.

First suppose ε = 5+3k, so we need a transfer value of -5-3k = -2 -3(k+1) in order to reduce ΣB to 0, and thus increase ΣC to 0. Since we can easily reduce ΣB by multiples of -3, it suffices just to find an additional reduction of -2. If we start with the top pair of values in B, C and then consider alternate pairs as we move from top to bottom in these streams, we see that interchanges of such pairs either increase or decrease ΣB by even values. These transfer values are indicated for H(1,q,2) with q = 29 in Figure 5.30 . Among these values can be found -2q-12, -2q, 2q+8 and 2q+2 which sum to -2. For the example in Figure 5.30, it is required to reduce ΣB only by an additional -6, which is accomplished as indicated by interchanging two 2 by 2 blocks. Again, all required interchanges are indicated in red.

Now suppose ε = 4+3k. If k = 0 we have the construction given in Figure 5.29(III). For k≥1 we have -ε = -7 + (k-1)(-3). Since k-1 2 by 2 blocks can be easily found each of which adds -3 to ΣB, it suffices to find interchanges which add -7 to ΣB. Interchanging values in the central 2 by 2 block adds q-3 to ΣB and interchanging the second to last values in streams B, C reduces ΣB by q+4. Combining these interchanges adds -7 to ΣB as desired. □

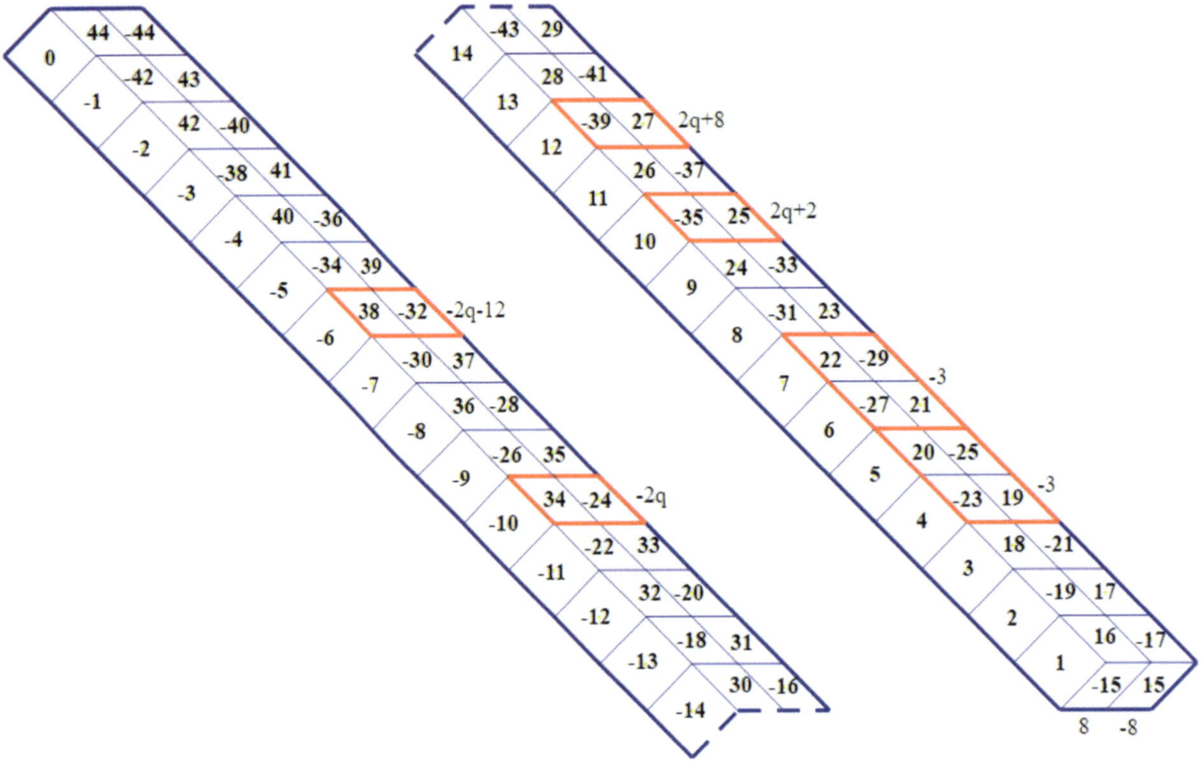

Figure 5.30

It is possible to realize some of the H(1,q,2) platforms as extensions using the following observations.

Lemma 5.14 *Let S =(a, ..., b) be an increasing sequence of consecutive positive integers with m = |S|. Let W be an initial segment of S with |W|= (2/3)m, and let L denote the remainder of S with |L|= m/3. If ΣW = ΣL then b = 7a-4.*

Proof. Let s denote the high value in the sequence W. Then ΣW = (1/2)(a+s)|W| = (1/3)m(a+s), and ΣL = (1/2)(b+s+1)|L| = (1/6)m(b+s+1). Equating ΣW and ΣL yields

(1) s= (b+1)-2a.

From |L|= (1/3)m =b-s we obtain s = b - (1/3)(b-a+1 which yields

(2) 3s = 2b+a-1.

Multiplying (1) by 3 gives

(3) 3s = 3b+3-6a.

Eliminating 3s from (1), (3) and solving for b yields b = 7a-4. □

An extension of an integral polygon by means of a series of equal odd length I-bars will be called an *odd I-bar extension.* We have already seen examples of odd I-bar extensions in Figures 5.18. 5.22. The I-bars in these examples have length 5. With the help of Lemma 5.14 we can construct odd I-bar extensions in which each extending I-bar has length 3.

Theorem 5.15 *Let q≥3 be odd. Then there exists an integral extension H(1,q,2)→ H(1, 7q+4, 2) by means of an odd I-bar extension.*

Proof. Let S = (a, …, b) be a consecutive sequence of positive integers with b = 7a-4, and dissect S into the following three segments each with length 2a-1.

S_1 = (a, a+1, a+2, …, 3a-3, 3a-2)

S_2 = (3a-1, 3a, 3a+1, …, 5a-4, 5a-3)

S_3 = (5a-2, 5a-1, 5a, …, 7a-3, 7a-4).

Now form a 3 by 2a-1 matrix A with rows R_1, R_2, R_3 are rearrangements of the sequences S_1, S_2, $-S_3$ respectively. The matrix A is illustrated in Figure 5.31.

A =

4a-3	4a-4	4a-5	…	3a-1	5a-3	5a-4	5a-5	…	4a-2
a+1	a+3	a+5	…	3a-3	a	a+2	a+4	…	3a-2
-(5a-2)	-(5a-1)	-5a	…	-(6a-4)	-(6a-3)	-(6a-2)	-(6a-1)	…	-(7a-4)

Figure 5.31

Note that each column of A has sum 0. If the matrix -A is annexed to A, a 0-sum 3 by 4a-2 rectangle R over S∪ -S is created. The columns in R can be rearranged to form 2a-1 I-bars each with length 3. The integral hexagon H(1,q,2) has base Z*(3q+2) = {0, 1,-1,2,-2, …, (1/2)(3q+1),-(1/2)(3q+1)}, so the first value in an extension sequence S is a = (1/2)(3q+3). The 0-rectangle R with a = (1/2)(3q+3) is precisely what is needed to extend H(1,q,2). The resulting integral hexagon is H(1, q+4a-2, 2) = H(1, 7q+4, 2). □

In Figure 5.32 the odd I-bar extension H(1,3,2)→H(1,25,2) is illustrated.

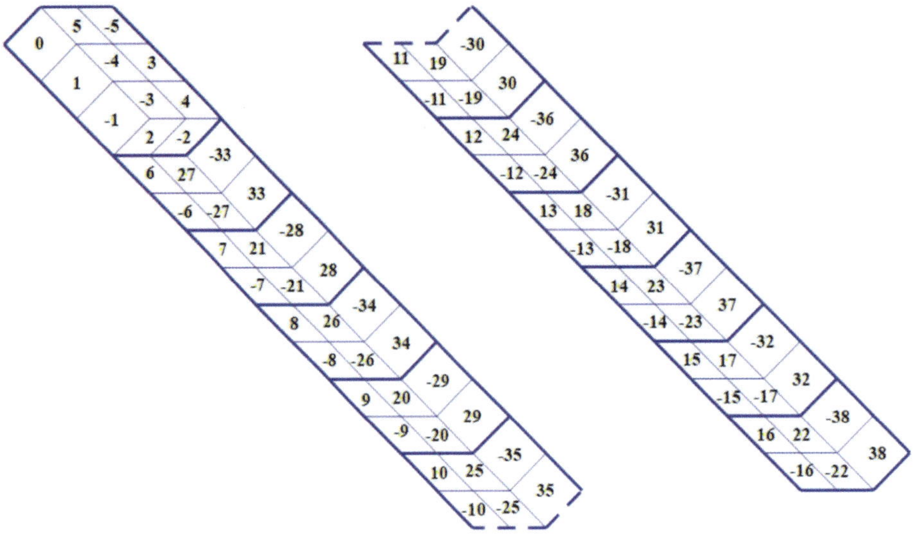

Figure 5.32

The construction in Theorem 5.15 can also be used to extend odd order integral rectangles with 3 rows. Let R(3,q) denote such a rectangle with q≥3 odd. Then using the technique used in Theorem 5.15 the extension R(3,q)→R(3, 6q+3) can be obtained. In Figure 5.33 the integral magic square of order 3 is extended to R(3,21).

-1	4	-3	5	-5	6	-6	7	-7	8	-8	9	-9	10	-10	11	-11	12	-12	13	-13
-2	0	2	22	-22	17	-17	21	-21	16	-16	20	-20	15	-15	19	-19	14	-14	18	-18
3	-4	1	-27	27	-23	23	-28	28	-24	24	-29	29	-25	25	-30	30	-26	26	-31	31

Figure 5.33

Corollary 5.16 *An integral extension H(1,q,2)→ H(3,q,2) exists for every odd integer q≥5.*

Proof. The extension sequence for H(1,q,2)→ H(3,q,2) is given by S = (a, … b) with a = (1/2)(3q+3) and b = a + (q+1). Then m = |S| = b-a+1 = q+2 which is odd. Since q≥5 by hypothesis, 6(q+1) ≤ $(q+1)^2$ and thus a = (1/2)(3)(q+1) ≤ (1/4)$(q+1)^2$ = K. (Here K is defined by K = (1/4)$(m-1)^2$ as in Theorem 4.19.) Since q is odd, either (i) q = 4k+1 or (ii) q = 4k-1. If (ii) then a = (1/2)(3)(4k) is even, and K = (1/4)$(q+1)^2$ = (1/4)$(4k)^2$ is also even. If (i) then a = (1/2)(3)(4k+2) is odd, and K = (1/4)$(q+1)^2$ = (1/4)$(4k+2)^2$ is also odd. Thus a, K always have the same parity, and by Theorem 4.9, S is splitable. The splitability of S makes the stipulated extension possible via a single I-bar. □

Two examples of the construction described in Corollary 5.16 are illustrated in Figure 5.34.

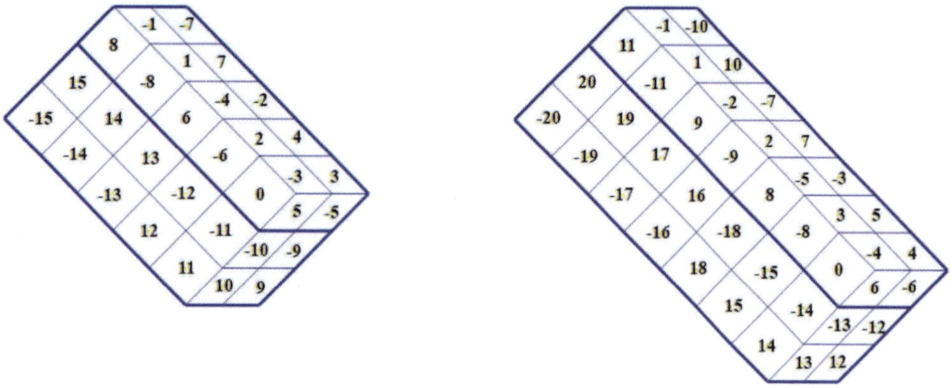

Figure 5.34

An increasing sequence S of positive integers has a *domino dissection* of the form S = {S_1, S_2, ... S_t} if each subsequence S_i has length 4 and is splitable. Each S_i has the form (a, b, c, d) where a<b<c<d and a+d = b+c, and consequently the elements in S_i along with their inverses can be used to form a *domino.* If a C-extension is to be composed of two I-bars one of which has length a multiple of 4, then the creation of a sequence with a domino dissection can be used to complete the extension. As an example consider the extension H(5,7,2)→H(5,9,4) illustrated in Figures 5.35, 5.36.

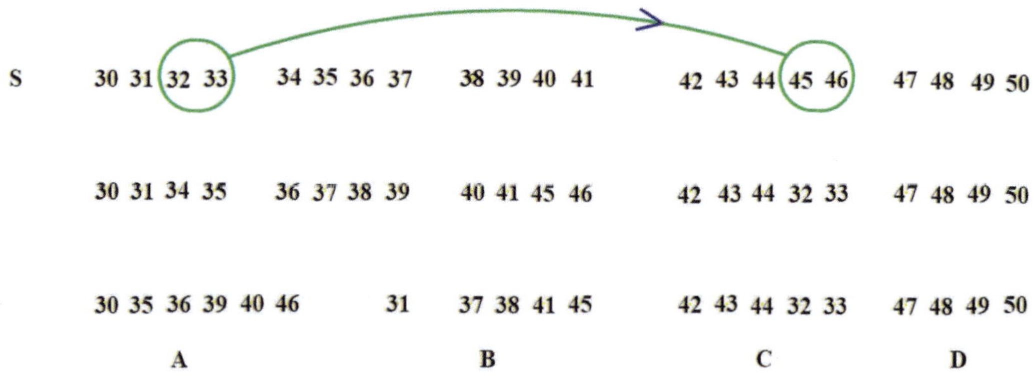

Figure 5.35

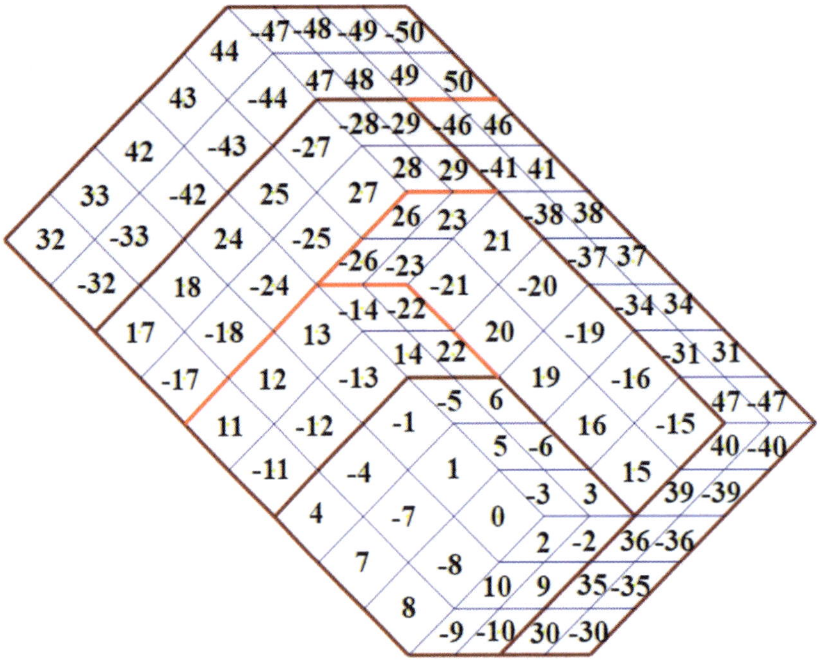

Figure 5.36

Domino dissections can also be used to determine platforms as in the following construction.

Theorem 5.17 *If q+1 is a multiple of 4, then the extension H(1,q,2)→H(1,q+2,4) can be constructed.*

Proof. Let S = (a, ..., b) denote the intended H(1,q,2)→H(1,q+2,4) extension sequence. Then a = (1/2)(3q+3) and b = (1/2)(5q+13). For q<15 the desired extensions do not follow the general pattern; for q=3 and q=7 constructions are given in Figure 5.37. Suppose q≥15 and let S_1 denote an initial segment of length q+1, and S_2 the remaining terminal segment of S with length 5. Let $\beta(S_2)$ denote the highest two elements b, b-1 of S_2 and let $\alpha(S_2)$ denote the set of three remaining elements of S_2. Then $\Sigma\beta(S_2)$ = b + (b-1) = 5q+12.

We will carefully select three elements {x,y,z} from S_1 such that x+y+z = 5q+12. These three elements will be removed from S_1 and substituted for the three elements in $\alpha(S_2)$. The elements of S_1 are then compressed together to fill the empty spaces left by the removal of x, y, z, and the elements in $\alpha(S_2)$ are appended at the end of S_1. The resulting sequence S_1' will be increasing and will remain with q+1 elements, and the new sequence S_2' will now be splitable into two segments each with sum 5q+12.

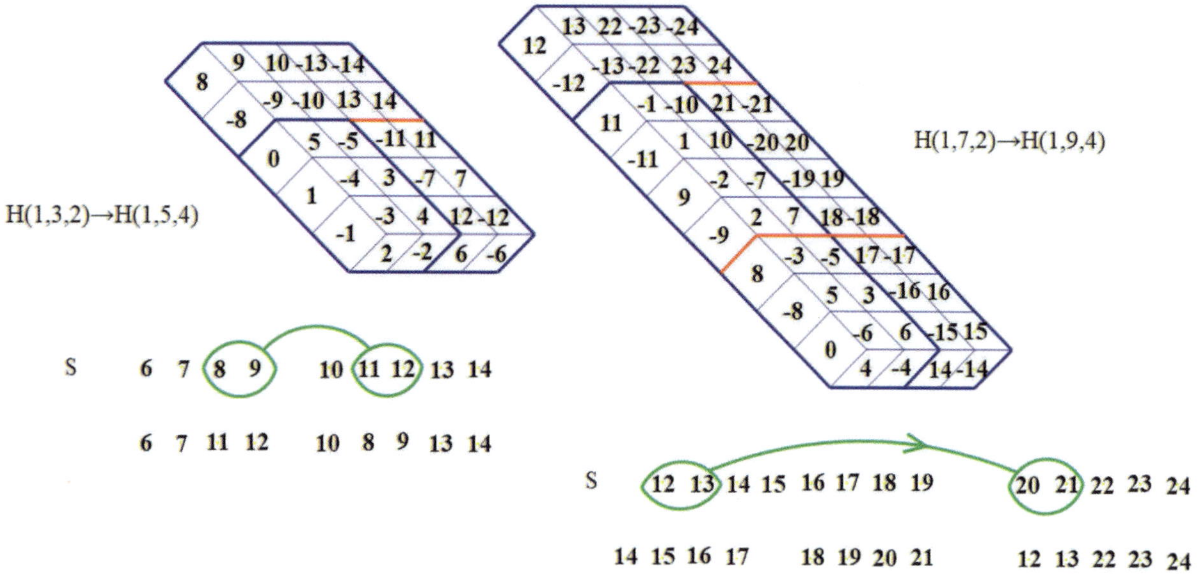

Figure 5.37

Since q+1 is a multiple of 4 by hypothesis, we have q = 3+2t where t is an even integer. We will take x = a+4, y = a+5 and z = a+t. Then

$$x+y+z = 3a+9+t = (3/2)(3q+3) + 9 + t = (3/2)(3q+3) + 18 + (1/2)(q-3) =$$
$$(1/2)\{9q+9 + 18 + q - 3\} = 5q+12 \text{ as desired.}$$

The elements x, y, z have also been chosen so that the newly constructed sequence S_1' has a *domino* dissection. This is due to the fact that as x = a+4, there are exactly four consecutive elements of S, (or S′), which precede it, and these constitute the positive elements of a *domino.* In addition, there are an even number of elements in S which lie between y and z and this makes possible the desired *domino* dissection in S_1' when y, z are removed and the subsequent compression is instituted. S_1' is then split into halves by taking for each constituent *domino (a,b,c,d)* the elements a, d for one half, and b,c for the other. The splitable sequences S_1' and S_2' are then used to construct I-bars of length q+1, 5 respectively, and these in turn, are combined to create the C-extension H(1,q,2)→H(1,q+2,4). □

The extension sequence S for H(1,19,2)→H(1,21,4) is illustrated in Figure 5.38 along with its reconfiguration into the two splitable subsequences S_1', S_2' .

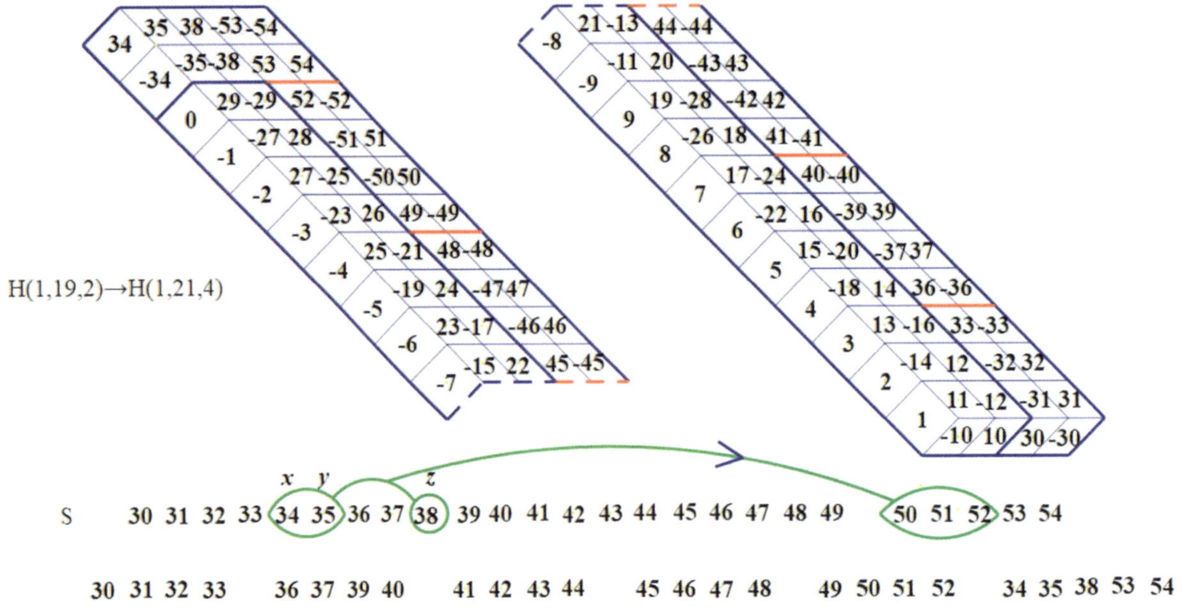

Figure 5.38

Corollary 5.18 *Every integral Class2 platform H(1,q,r), with r even and q≥3 odd, can be obtained by a series of domino extensions from H(1, 3+4k, 2) or from H(1, 5+4k,2).*

Proof. By Corollary 5.18 an integral H(1,q,2) can be constructed for every odd q≥3. By Theorem 5.7 we have the extension H(1, 3+4k,2)→H(1,3+4k,4) for every nonnegative integer k, and by Theorem 5.17 we have the extension H(1,3+4k,2)→ H(1, 5+4k, 4) for every nonnegative integer k. By Theorem 5.7 H(1, 3+4k, r) can be obtained by repeated domino extensions from H(1,3+4k,4) for every even r≥6. Also by Theorem 5.7 H(1,5+4k,r) can be obtained by repeated domino extensions whenever r≥8 is a multiple of 4. If instead, r is twice an odd number, then H(1,5+4k,r) can be obtained by repeated domino extensions of H(1,5+4k,2). □

In Appendix II another simple type of domino extension is illustrated.

5.8 Near Regular Hexagons

A semiregular hexagon H(q,q,r), r≠q, with exactly 4 sides of equal length will be called *near regular.* We will show that this type of integral hexagon, with q odd and r even, can be obtained by a series of extensions. The following observation will be useful.

Lemma 5.19 *Let S be an increasing sequence of consecutive positive integers with |S| a multiple of 4, say |S|=4t. Let F be an initial segment of S, and L a terminal segment of S, each with length t. Let M denote the middle section of S with length 2t between F and L. Then ΣM = ΣF + ΣL.*

Proof. Let l be a line which divides S into two equal length halves. If x ∈ S, let x′ denote the element in the opposite half of S which is equidistant from l. If S = (a, …, b) then x+x′ = a+b for every x ∈

S. Since $|M| = 2t$, it contains t pairs (x,x'), and thus $\Sigma M = t(x+x') = t(a+b)$. Similarly, there are t pairs (x,x') in $F \cup L$ with $x \in F$, and thus $\Sigma F + \Sigma L = t(a+b)$ as well. □

Theorem 5.20 *For each odd integer $q \geq 3$ there exists an integral extension $H(q,q,2) \to H(q+2,q+2,2)$.*

Proof. Let S be an increasing sequence of consecutive positive integers of length $2q+6$ where $q \geq 3$ is odd, and partition S into consecutive subsequences (S_1, S_2, S_3, S_4) with $|S_1| = (1/2)(q+3)$; $|S_2| = (1/2)(q+1)$; $|S_3| = (1/2)(q+5)$ and $|S_4| = (1/2)(q+3)$. If $q = 2t+1$, then $|S| = 4t+8$ is a multiple of 4. In the terminology of Lemma 5.19, $F = S_1$, $L = S_4$ and $M = S_2 \cup S_3$, and thus $\Sigma S_1 + \Sigma S_4 = \Sigma S_2 + \Sigma S_3$.

Let $S_1 = (a, \ldots, b)$; $S_2 = (b+1, \ldots, c)$; $S_3 = (c+1, \ldots, d)$ and $S_4 = (d+1, \ldots, e)$. Then

$\Sigma S_1 = (1/4)(a+b)(q+3)$; $\Sigma S_2 = (1/4)(b+1+c)(q+1)$;

$\Sigma S_3 = (1/4)(c+1+d)(q+5)$; $\Sigma S_4 = (1/4)(d+1+e)(q+3)$.

We will show that there exists a subset E of $\{S_3\}$ with $(1/2)(q+1)$ elements, such that $\Sigma E = \Sigma S_1$. The $(1/2)(q+1)$ elements in S_2 will then be interchanged with the elements of E. This will result in a partition of S into two splitable subsets, one with $q+2$ elements and the other with $q+4$ elements. The splitable sets will be used to form two I-bars which combine to form a C-extension $H(q,q,2) \to H(q+2,q+2,2)$.

Consider: $|S_1 \cup S_2| = q+2 = c - a+1$, so the first element of S_3 is $c+1 = q+a+2$ and the third element of S_3 is $q+a+4$. Let E consist of every element of S_3 except these two. Then

$$\Sigma E = \Sigma S_3 - (2q+2a+6) = (1/4)(c+1+d)(q+5) - (2q+2a+6) =$$
$$(1/4)(q+a+2+d)(q+5) - (2q+2a+6),$$

and

$$\Sigma S_1 = (1/4)(a+b)(q+3) = (1/2)(a+b)(1/2)(q+3)=$$
$$(1/4)(4a+q+1)(1/2)(q+3) = (1/8)(q+3)(4a+q+1).$$

Thus $\Sigma E = \Sigma S_1$ iff

$$(1/4)(q+a+2+d)(q+5) - (2q+2a+6) = (1/8)(q+3)(4a+q+1) \text{ iff}$$
$$(q+a+2+d)(q+5) - (8q +8a +24) = (1/2)(q+3)(4a+q+1) \text{ iff}$$
$$aq + q^2 +2q +dq +5a +10 + 5d - 8q -8a -24 = (1/2)(q+3)(4a+q+1) \text{ iff}$$
$$2(aq +q^2 + dq + 5d -q -3a -14) = (q+3)(4a+q+1) = q^2 + 4aq + 4q + 12a + 3 \text{ iff}$$
$$q^2 + 2dq + 10d = 2aq +6q +18a + 31.$$

Now using $d - (a+q+2) +1 = |S_3| = (1/2)(q+5)$ we obtain $2dq = 3q^2 +2aq + 7q$, and $10d = 15q + 10a + 35$. Then continuing the above sequence of equations we obtain

$$\Sigma E = \Sigma S_1 \text{ iff } 4q^2 +2aq + 22q + 1a +35 = 2aq +6q +18a + 31 \text{ iff}$$
$$4q^2 + 16q + 4 = 8a.$$

Consider: a = (1/2)(n+1) where $n = q^2 + 4q$ is the order of H(q,q,2). Thus

$$8a = 4(q^2 +4q +1) = 4q^2 + 16q + 4,$$

and we have shown $\Sigma E = \Sigma S_1$. Now interchange the (1/2)(q+1) elements in E with the (1/2)(q+1) elements in S_2 so that the sequence S is reconstituted by $S = (S_1, E, S_3', S_4)$ where $\{S_3'\} = \{S_3 - E + S_2\}$. Then since $\Sigma E = \Sigma S_1$, the q+2 element set $\{S_1 \cup E\}$ is splitable, and since $\Sigma S_3' = \Sigma S_3 - \Sigma E + \Sigma S_2 = \Sigma S_3 - \Sigma S_1 + \Sigma S_2 = \Sigma S_4$, the q+4 element set $\{S_3' \cup S_4\}$ is also splitable. As mentioned above, these two splitable subsets of {S} can be used to construct two I-bars which complete the H(q,q,2)→H(q+2,q+2,2) extension. □

In Figure 5.39 the technique described in Theorem 5.20 is used to construct the extension series H(3,3,2)→H(5,5,2)→H(7,7,2). Next we establish a similar result with even edge length 4.

Theorem 5.21 *For each odd integer q≥3 there exists an integral extension H(q,q,4)→H(q+2,q+2,4).*

Proof. Let the extension sequence S be given by $S = (S_1, S_2, S_3, S_4)$ with S_1 = (a, …,b); S_2 = (b+1, …, c); S_3 = (c+1, …, d) and S_4 = (d+1, …, e) and with

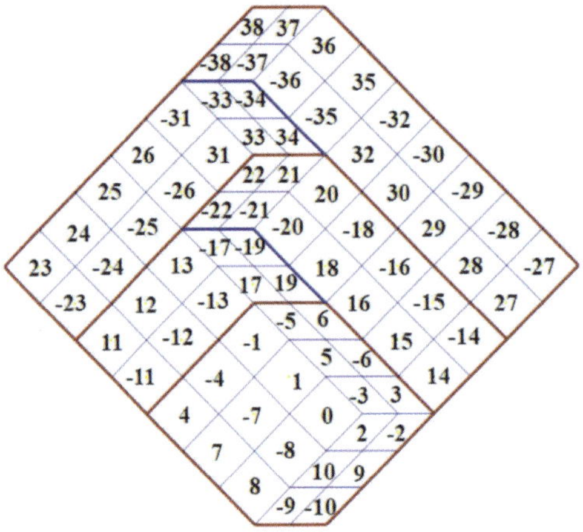

Figure 5.39

$$|S_1| = (1/2)(q+7), \quad |S_2| = (1/2)(q+5), \quad |S_3| = (1/2)(q+5), \quad |S_4| = (1/2)(q+3).$$

Then $\Sigma S_1 = (1/4)(a+b)(q+7)$, $\Sigma S_2 = (1/4)(c+b+1)(q+5)$, $\Sigma S_3 = (1/4)(d+c+1)(q+5)$ and $\Sigma S_4 = (1/4)(e+d+1)(q+3)$. Our first objective is to show $\Sigma S_2 - \Sigma S_4 = q-2$. Consider

(1) $\Sigma S_2 - \Sigma S_4 = (1/4)\{(c+b+1)(q+5) - (e+d+1)(q+3)\} =$
$(1/4)\{2 + q((c+b) - (e + d)) + 5(b+c) - 3(e+d)\}$.

Now we use

$(c+b) - (e+d) = -2q-9$; $b+c = (3/2)(q+5)+2a = (3/2)(q+5)+q^2+8q+1$, and

$(e+d) = 2q + 9 + (b+c)$ in (1) to obtain

$\Sigma S_2 - \Sigma S_4 = (1/4)\{2 - 2q^2 -9q + 5(b+c) - 3(2q+9+(b+c))\} =$

$(1/4)\{2 - 2q^2 - 9q +2(b+c) - 6q - 27\} =$

$(1/4)\{2 - 2q^2 - 15q - 27 + 3q+ 15 +2q^2 + 16q + 2\} = (1/4)(4q-8) = q-2.$

We will now select elements x, y, z with $x \in S_1$, $y \in S_2$ and $z \in S_3$ so that $y = x+q - 2$ and $z = y + 7$. For q = 3, 5, 7 this is easily done. Suppose $q \geq 9$ and let z be the smallest element c+1 in S_3. Since $|S_2| = (1/2)(q+5) \geq 7$, the element c-6 lies in S_2. Let y = c-6, and thus x = c - (q+4). To check that $x \in S_1$, note that $a = c+1 - |S_1 \cup S_2| = c - (q+5)$, and thus a<x. Consider: $q+4 > (1/2)(q+5)$, and thus $x = c-(q+4) < c - (1/2)(q+5) = b$. We have shown that a < x < b, and thus $x \in S_1$.

Next the elements x, y, z are permuted in accordance with the 3-cycle (x y z) as indicated below for q = 5, x = 38, y = 41, z = 48.

33,34,35,36,37,<u>38</u>	39,40,<u>41</u>,42,43	44,45,46,47,<u>48</u>	49,50,51,52
33,34,35,36,37,<u>48</u>	39,40,<u>38</u>,42,43	44,45,46,47,<u>41</u>	49,50,51,52

This results in a new partition of S given by $\{S\} = \{S_1', S_2', S_3', S_4\}$ with $S_1' = \{S_1 - x + z\}$, $S_2' = \{S_2\} - \{y\} + \{x\}$ and $S_3' = \{S_3\} - z + y$. Now we have $\Sigma S_2' = \Sigma S_2 -(q-2) = \Sigma S_4$ in accordance with the above argument. As a consequence, the set $\{S_2' \cup S_4\}$ is splitable and can be used to construct an I-bar of length q+4.

It remains to show that the set $\{S_1' \cup S_3'\}$ is also splitable. To this end we must show $\Sigma S_1' = \Sigma S_3'$. Consider: $\Sigma S_1' = \Sigma S_3'$ iff $\Sigma S_1 - x + z = \Sigma S_3 - z + x$ iff

$(1/4)(a+b)(q+7) - c + q + 4 + c + 1 = (1/4)(d+c+1)(q+5) - 7$ iff

$(a+b)(q+7) + 4q + 20 = (d+c+1)(q+5) - 28$ iff

$(a+b)(q+7) + 4q + 20 = (q+5)(c+d) + q - 23$

$(a+b)(q+7) + 3q + 43 = (q+5)(c+d).$

Now using $a = (1/2)(q^2 + 8q + 1)$ and $b = a + (1/2)(q+5)$ we obtain

(2) $a+b = (q^2 + 8q + 1) + (1/2)(q+5)$,

and using $c = (1/2)(q^2 + 10q + 11)$ and $d = (1/2)(q^2 + 11q + 16)$, we obtain

(3) $c+d = (1/2)(2q^2 + 21q + 27)$.

Using (2), (3) to substitute for a+b and c+d we obtain

$\Sigma S_1' = \Sigma S_3'$ iff $\{(q^2 + 8q + 1) + (1/2)(q+5)\}(q+7) + 3q + 43 =$
$\{(1/2)(2q^2 + 21q + 27)\}(q+5)$ iff
$q^3 + 15q^2 + 60q + 50 + (1/2)(q+5)\}(q+7) = (1/2)(q+5)(2q^2 + 21q + 27)$ iff
$2q^3 + 31q^2 + 132q + 135 = 2q^3 + 31q^2 + 132q + 135.$

We may now conclude that $\Sigma S_1' = \Sigma S_3'$ and thus that the set $S_1' \cup S_3'$ is splitable. From this splitable subset of {S} an I-bar of length q+6 can be constructed. The two I-bars of length q+4 and q+6 combine to form a C-extension H(q,q,4)→H(q+2,q+2,4) as desired. □

In Figure 5.40 the series of extensions H(1,1,4)→H(3,3,4)→H(5,5,4)→H(7,7,4) using the method described in Theorem 5.21 is illustrated.

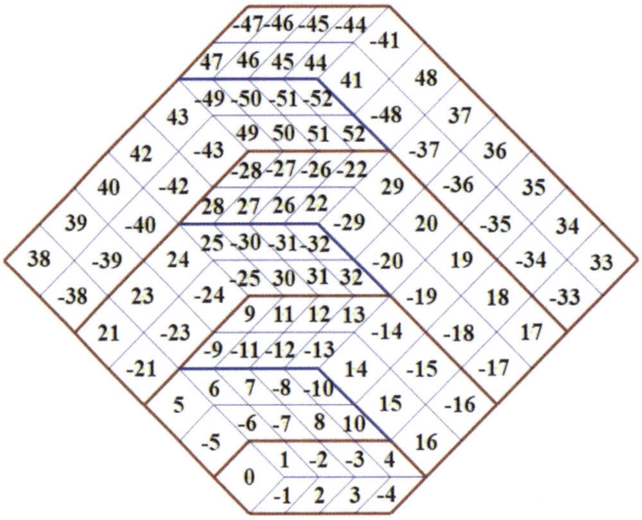

Figure 5.40

With the constructions of Theorems 5.20 and 5.21 in hand, domino extensions can then be used to obtain an integral H(q,q,r) for any even positive integer r.

5.9 Various Class 2 Extensions

The following Conjecture provides the basis for extensions of Class 2 integral hexagons by means of a series of odd length I-bars.

Conjecture 5.22 *Let S be a splitable sequence of consecutive positive integers with* $|S| = m = m_1 + m_2 + \ldots + m_k$ *where each* m_i *is odd and larger than or equal to 3. Let* S_1 *be the initial segment of S with* $|S_1| = \Sigma\{(1/2)(m_i+1), i = 1, \ldots, k\}$ *and let* S_2 *denote the sequence formed by the remaining elements of S. If* $\Sigma S_1 \leq \Sigma S_2$ *Then {S} an be partitioned into k splitable subsets* $\{T_1, T_2, \ldots, T_k\}$ *with* $|T_i| = m_i$ *for i = 1, …, k.*

By Lemma 4.15, $\Sigma\alpha(T_i)\leq\Sigma\beta(T_i)$ for i = 1, …, k. Consider: $\Sigma\{|\alpha(T_i)|$, i = 1, …, k} = $\Sigma\{(1/2)(m_i+1)$, i = 1, …, k} = $|S_1|$, and consequently $\Sigma S_1 \leq \Sigma\{\alpha(T_i)$: i = 1, …, k} $\leq\Sigma\{\beta(T_i)$, i = 1, …, k}. Since $|S_2|$ = m - $|S_1|$ = (1/2)(m-k) = $\Sigma\{|\beta(T_i)|$, i = 1, …, k}, and since among all subsets of S with length (1/2)(m-k), S_2 achieves the maximum sum, we must have $\Sigma\{\beta(T_i)\leq\Sigma S_2$. Then we also have $\Sigma S_1\leq\Sigma S_2$. So the condition $\Sigma S_1\leq\Sigma S_2$ is necessary; the sufficiency however, may be quite difficult to prove. We provide a selection of examples in the following Figures as support for the Conjecture.

Note that the two examples in Figure 5.41 feature extensions of integral isorectangles. The 3 by 13 isorectangle in 5.41(I) is taken from the integral rectangle in Figure 4.22.

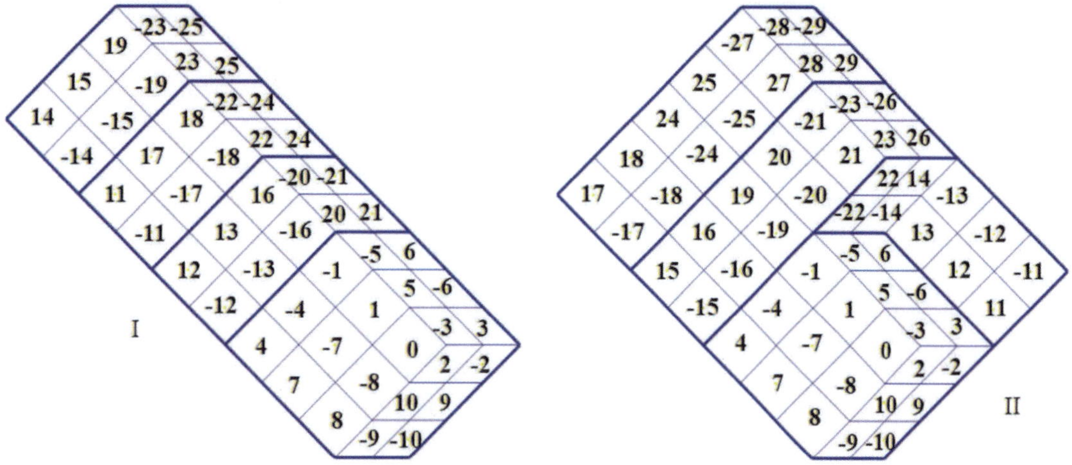

Figure 5.39

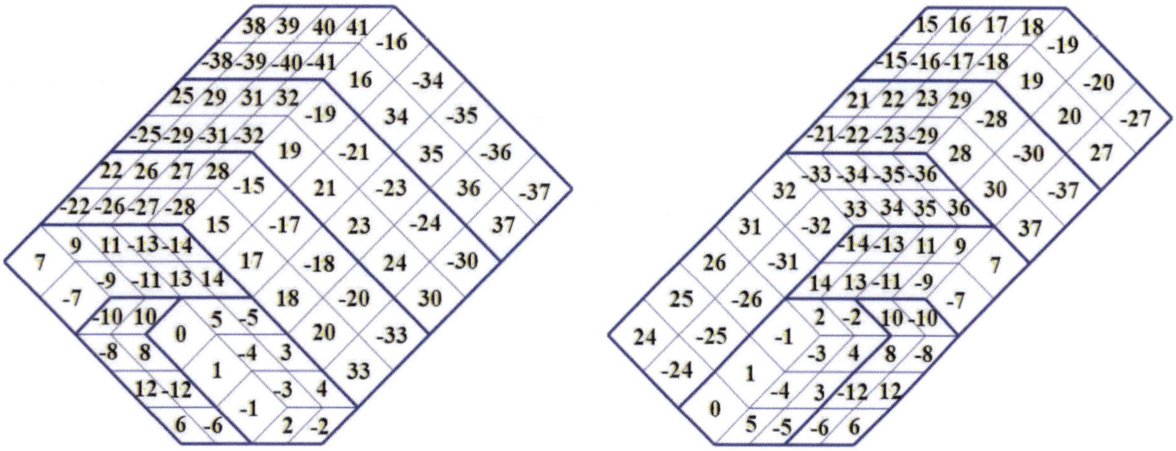

Figure 5.40

Figure 5.42 illustrates an extension of H(1,3,2) by a series of 12 I-bars of length 3. The initial segment S_1 of the extension sequence has ΣS_1 = 420, and the terminal segment S_2 has ΣS_2 = 426 so the condition of Conjecture 5.22 is satisfied. If the extension sequence is reduced by just 3 elements, then $\Sigma S_1 = \Sigma S_2$ = 363 and the Conjecture is also satisfied with the construction illustrated in Figure 5.32 resulting.

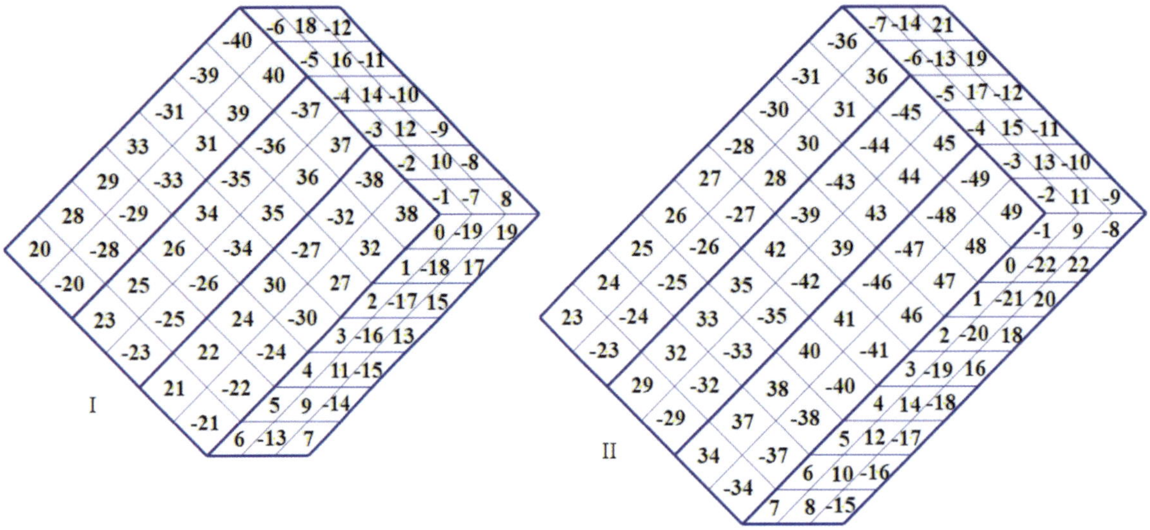

Figure 5.41

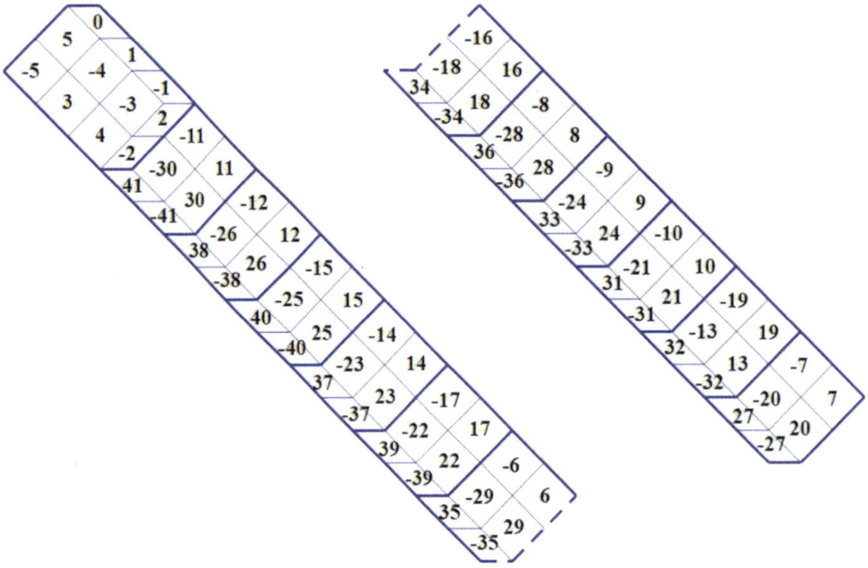

Figure 5.42

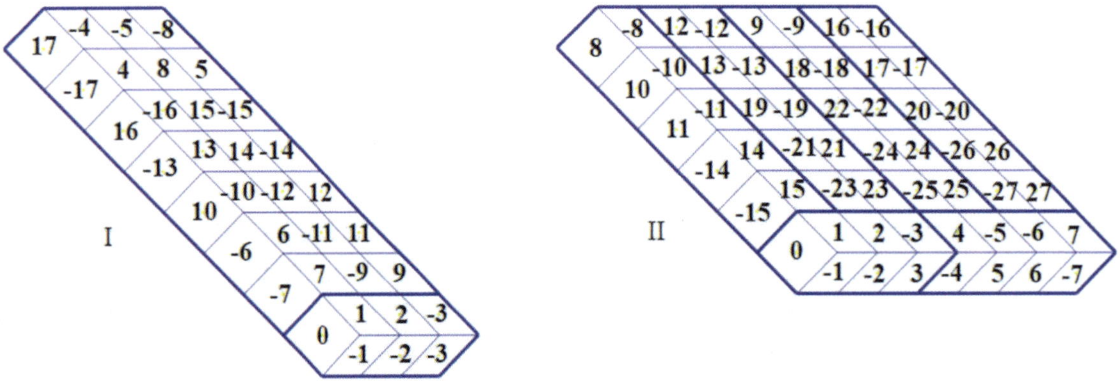

Figure 5.43

A *hybrid extension* occurs whenever a Class 1 integral hexagon is extended to a Class 2 integral hexagon, or vice versa. The H(1,3,2)→H(1,5,5) extension in Figure 5.20 is our first example and the constructions in Figure 5.43 provide two more examples. Note that Conjecture 5.22 requires that the extension sequence be splitable, but in 5.43(I) the extension sequence is *not* splitable, and yet a construction is achieved. The extension sequence for H(1,1,7)→H(1,7,6) in Figure 5.43(II) is splitable and does support Conjecture 5.22, with an extension consisting of four I-bars.

In Figure 5.44 two simple domino extensions of integral parallelograms are illustrated. The parallelograms are derived from the integral rectangles in Figure 4.22. These examples as well as the ones in Figure 5,41, although not extensions of semiregular hexagons, nevertheless provide a simple method for constructing integral semiregular hexagons. In the next Section we will provide further examples involving extensions of nonsemiregular integral polyparallelograms.

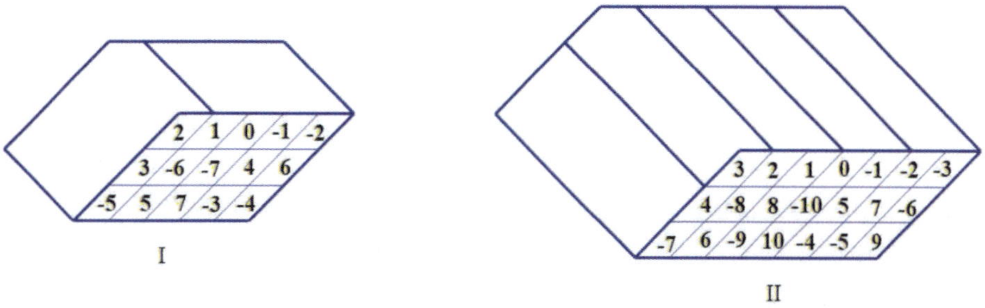

Figure 5.44

5.10 Extensions of Nonsemiregular Integral Polyparallelograms

In order to obtain constructions with some degree of similarity to magic squares, we have confined our attention to polyparallelograms with semiregular boundary. It is not then unnatural to require inclusions to also have semiregular boundaries; it follows the spirit of general algebra in which subalgeras share the properties of the algebra which contains them. But in the case of magic polygons it seems appropriate to only require that an inclusion in an odd order integral polygon have the property that all its streams have 0-sum. This opens the possibility for the construction of odd shaped polyparallelogram inclusions as we illustrate in the following Figures. We continue to require that an extension to a semiregular integral polygon can be constructed. In our examples this is achieved by means of domino extensions.

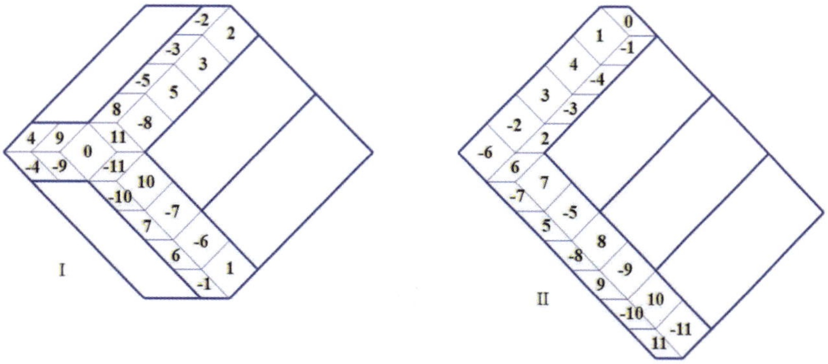

Figure 5.45

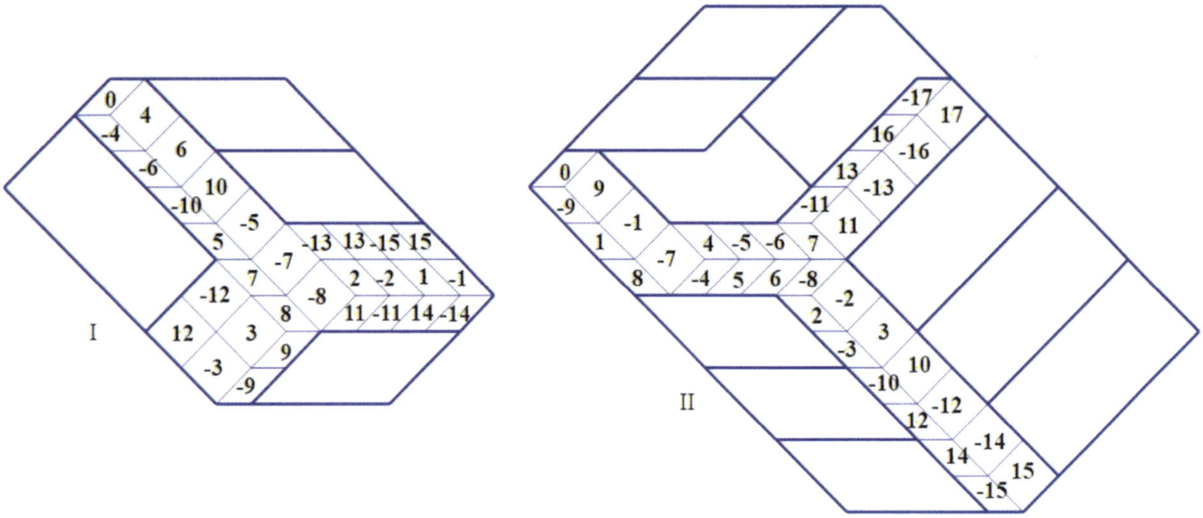

Figure 5.46

5.11 Integral Hexagons of Even Order

Semiregular hexagons over Z*(n) have an essentially different character depending on whether n is even or odd. This is due to the fact that ΣZ*(n) = 0 if n is odd and ΣZ*(n) = n/2 if n is even. The order of H(m,p,q) is given by n = mp+mq+pq and the number of streams is given by m+p+q. If S denotes the common sum of all streams then (m+p+q)S = 2ΣZ*(n) so if n is odd we must have S = 0 as we have seen in so many examples. But if n is even then we obtain S = (mp+mq+pq)/(m+p+q) which must be an integer. This observation is recorded in the following Theorem.

Theorem 5.23 *Let H(m,p,q) be an even order integral hexagon. Then the common sum of streams is given by S = (mp+mq+pq)/(m+p+q) which must be an integer.* □

Regarding mp+mq+pq and m+p+q as polynomials in p, long division yields

$$S = (mp+mq+pq)/(m+p+q) = (m+q) - (m^2+mq+q^2)/(m+p+q).$$

Thus S is an integer iff $W = (m^2+mq+q^2)/(m+p+q)$ is an integer. If we now let

$p = mq-q+q^2 + m^2-m = (q+m)(m-1) + q^2$ then W = 1, and S = m+q-1. We have thus obtained at least one set of parameters for which S is a positive integer. This leads to the following Conjecture.

Conjecture 5.24 *Let H = H(m,p,q) be an even order semiregular hexagon with p, q even. If* $p = (q+m)(m-1) + q^2$, *then H has an integral solution with common stream sum S = m+q-1.*

As evidence for this Conjecture consider the examples in Figure 5.47(I,II,IV).

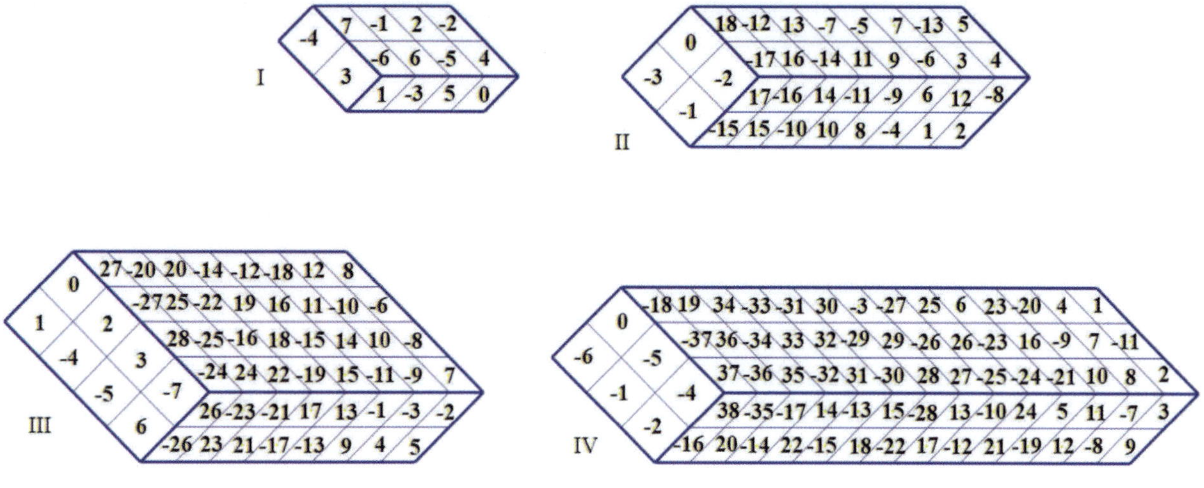

Figure 5.47

In 5.47(I) the parameters are m=1, q=2, p=4; in 5.47(II), m=q=2 and p = 8, and in 5.47(IV) the parameters are m=3, q=2 and p=14. For these constructions the stream sums are 2, 3, 4 respectively in agreement with the formula given in Conjecture 5.24.

In 5.47(III) a construction *not* stipulated by the Conjecture is obtained. If we take m=8, p=2, q=4, then S = 4 is obtained. This suggests a second Conjecture.

Conjecture 5.25 *If* $m = p^3$ *and* $q = p^2$, *(p even), then H(m,p,q) has an integral solution with common stream sum* $S = p^2$.

CHAPTER 6

MAGIC OCTAGONS

6.1 Streamorphisms

Similar to the notation used for magic hexagons, the notation O(m,p,q,r) will be used to represent a semiregular octagon with semiperimeter consisting of 4 sides with lengths m, p, q, r. The sides will occur in the order indicated by the notation. For a semiregular hexagon H(p,q,r) the order in which the sides are listed makes no difference in the structure of the hexagon; for example H(p,r,q) when read counterclockwise is H(q,r,p) which is just a different representation of the cycle (p,q,r). However, for semiregular octagons, or any semiregular polygon with more than 6 sides, different permutations of the sides can yield very different structures. For octagons there are just two permutations of the sides, namely O(m,p,q,r) and O(m,q,p,r) which can produce essentially different semiregular octagons. These will be referred to as *dual semiregular octagons.* In Figure 6.1 dual octagons O(1,1,2,2) and O(1,2,1,2) are illustrated, each with a magic solution over Z(12).

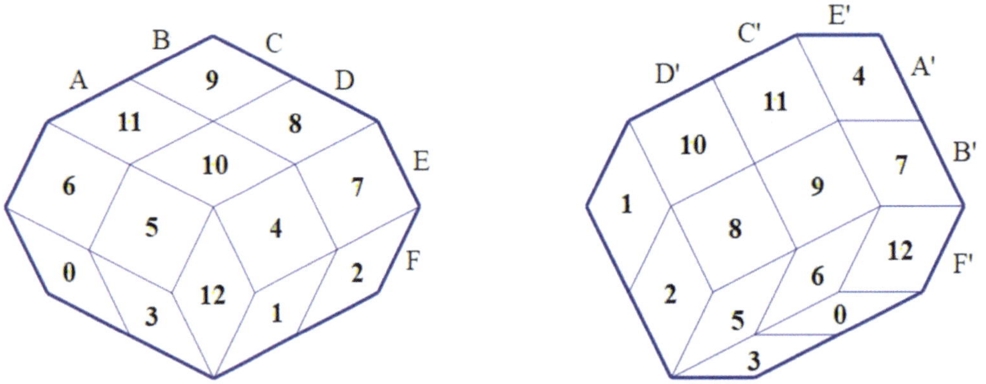

Figure 6.1

Once a magic solution is found for a semiregular octagon, then there is an easy procedure for obtaining a magic solution for the dual. This is a consequence of the following general result.

Theorem 6.1 *Let P be a semiregular polygon with a parallelogram tiling and with semiperimeter formed by the cyclic sequence of side vectors (*$\mathbf{s}_1, \ldots, \mathbf{s}_t$*). Let* ϕ *denote any permutation of the set* $\{\mathbf{s}_1, \ldots, \mathbf{s}_t\}$ *and let P' denote the semiregular polygon with any parallelogram tiling and with* *semiperimeter formed by the sequence* $(\phi(\mathbf{s}_1), \ldots, \boldsymbol{\phi}(\mathbf{s}_t))$. *If P has a magic solution over an abelian group G, then P′ also has a magic solution over G.*

Proof. As a consequence of having the same set of side vectors, there is a one-one correspondence X→X′ between streams in P, P′ such that if X, Y lie in the same parallel class in P, then X′, Y′ lie in the same parallel class in P′. Suppose P has a magic solution over G and let v(X) denote the set of elements from G which occupy the tiles belonging to stream X in P. If X is any stream in P, let N(X) denote the collection of streams in P which are *not* parallel to X. Then v(X) = {v(X) ∩ v(Y) : Y∈ N(X)}. If X, Y are two intersecting streams in P, then X′, Y′ are intersecting streams in P′.

Assign elements from G to the tiles of P′ by setting v(X′) ∩ v(Y′) = v(X) ∩ v(Y) for each pair X, Y of intersecting streams in P. If S is the common stream sum for P and X is any stream in P, then

$$S = \Sigma v(X) = \Sigma\{v(X) \cap v(Y) : Y \in N(X)\} = \Sigma\{v(X') \cap v(Y') : Y' \in N(X')\} = \Sigma v(X').$$

It follows that every stream in P′ has sum S, and thus P′ has been assigned a magic solution over G. □

When two semiregular polygons have the same stream sets, we say they are *streamorphic,* (pronounced stree a morphic). Thus in Theorem 6.1 polygons P, P′ are streamorphic. (Also, any sequence of hexagon inversions as described in Chapter 1 will carry a semiregular polygon P to a *streamorphic image* P′.) The method of Theorem 6.1 was used to create the streamorphic pair in Figure 6.1.

In Section 5.5 the concept of a *slide transformation* was introduced and its use in restructuring semiregiular hexagons was left as a Conjecture, (Conjecture 5.10). These transformations are also streamorphisms whose importance comes again to light when considering the relationship between an integral octagon and its dual. In Figure 6.2(I) an integral O(2,3,2,3) octagon P is illustrated along with its dual P′, (O(2,3,3,2)) in Figure 6.2(II). P contains as kernel an H(1,2,3) hexagon bound in red. In P′ the same hexagon has been *exploded* into three parts also bound in red. The parts are formed by the stream intersections C∩(F,G,H}, C∩{I,J} and {F,G,H}∩{I,J}, and consequently can be reassembled to form a semiregular polygon, in this example, an integral H(1,2,3). We will argue that a series of slide transformations can be used to make this reassemblage.

Clearly permuting the order of parallel streams is a streamorphism. In our example, 6.2(III) is obtained from 6.2(II) by interchanging the parallel streams A,C.

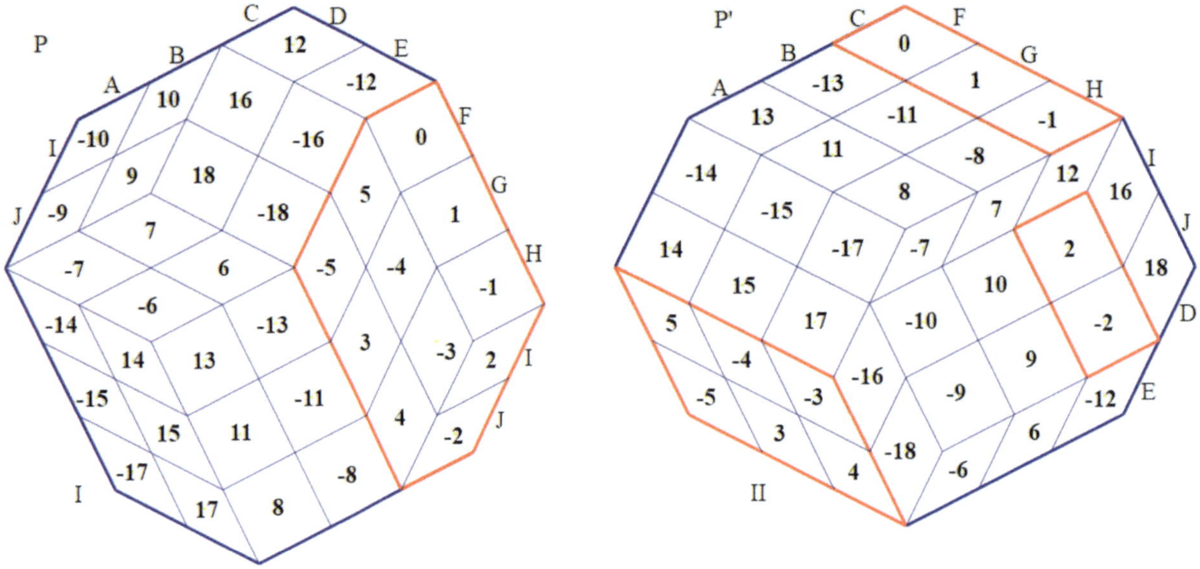

Figure 6.2

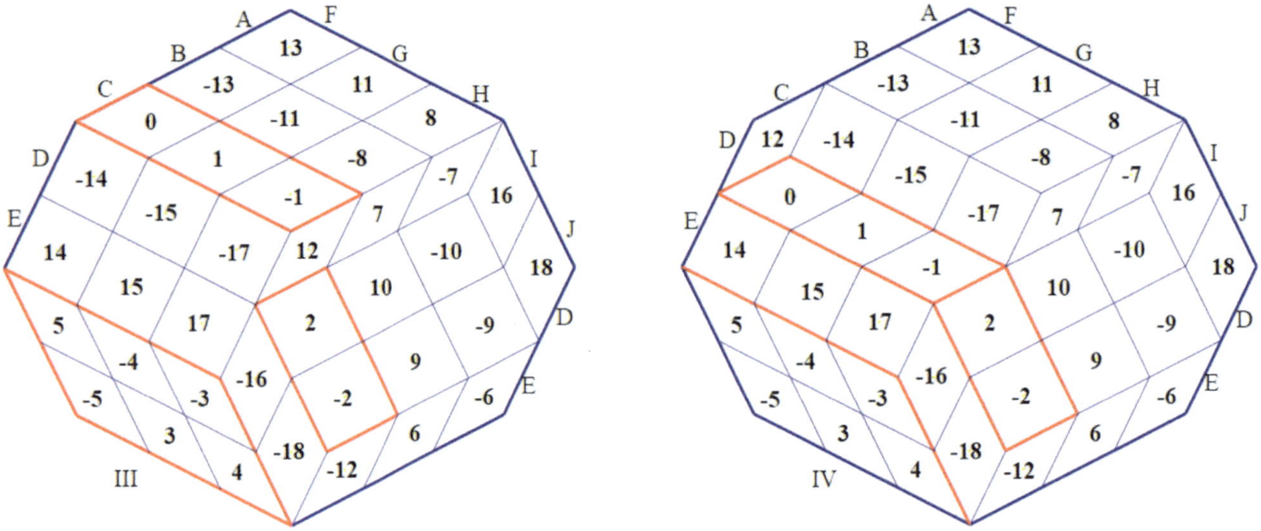

Figure 6.2

Moving from 6.2(III) to 6.2(IV) involves sliding tiles 0,1,-1 across the chasm consisting of tiles -14,-15,17,12. The emergent consists of these same tiles with the same shapes; consistent with the preservation of composition sequence stipulated hy Theorem 1.2. This particular slide transformation could also have been accomplished by three hexagon inversions; in fact, a hexagon inversion is itself a simple slide transformation. From 6.2(IV) to 6.2(V) involves sliding tiles 0,1,-1,2,-2 across the chasm consisting of tiles 14,15,17,-16,-18. This slide transformation could also be achieved by a series of five hexagon inversions.

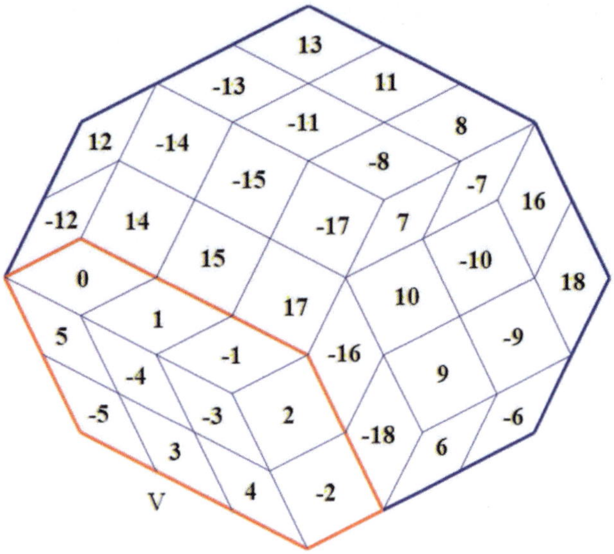

Figure 6.2

The end result is illustrated in 6.2(V) in which the H(1,2,3) inclusion is achieved. The transformation P→P′ from a semiregular polygon to its dual will be called a *peripheral streamorphism.* More generally, if P′ is any semiregular polygon with the same set of side vectors as P, then P→P′ is a *peripheral streamorphism.* Slide transformations will be referred to as *slide streamorphisms.* We have illustrated the following general Theorem whose proof we omit.

Theorem 6.2 *Let P be a semiregular polygon with a magic solution over an abelian group G and let P′ be obtained from P by a peripheral streamorphism. If P contains an inclusion over a subset H of G, then there exists a series of slide streamorhisms which transform P into a semiregular polygon P* which contains an inclusion over the same subset H.* □

6.2 Regular Octagons over Z(n) and Z*(n)

For a regular octagon we will use the notation O(e) = O(e,e,e,e) where e denotes the number of basic edges per side. Although basic solutions are possible, we will be concerned only with natural and integral solutions. In case e is odd neither of these types of solutions are possible.

Theorem 6.3 *If e is odd then O(e) has no natural solution over Z(n) and no integral solution over Z*(n).*

Proof. The order n of O(e) is given by $n = 6e^2$, and the number of streams in any parallelogram tiling is given by the semiperimeter s = 4e. Suppose O(e) has a natural solution over Z*(n) with stream sum S. Then $sS = 2\Sigma Z^*(n) = 2(n/2) = n$, and thus S = n/s must be an integer. But S = n/s = (3e)/2 is not an integer. On the other hand, if O(e) has a natural solution over Z(n) then $sS = 2\Sigma Z(n) = n(n-1) = 6e^2(6e^2-1)$, and thus $S = 6e^2(6e^2-1)/4e = 3e(6e^2-1)/2$ must be an integer. But this is not so since both 3e and $6e^2-1$ are odd. □

As a consequence of Theorem 6.3, it remains only to consider solutions for O(e) when e is even. An integral and a natural solution for O(2) are given in Figure 6.3.

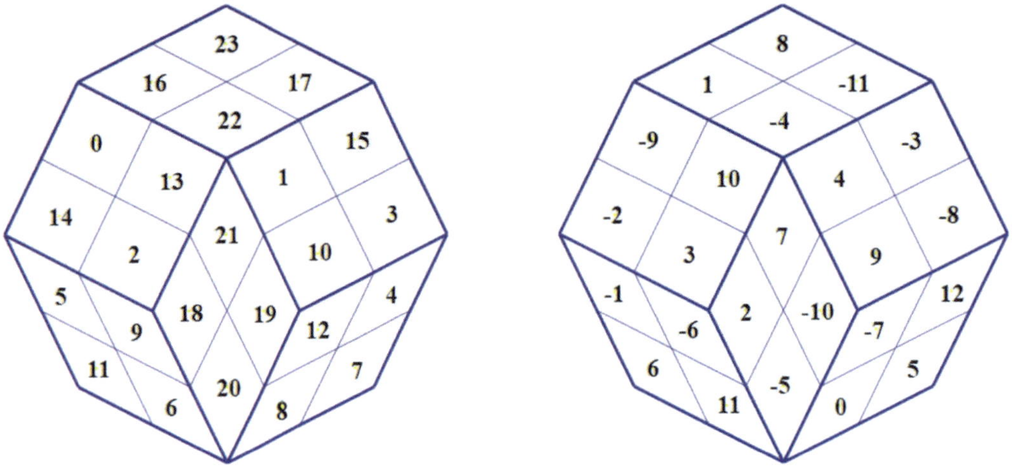

Figure 6.3

The stream sum for the natural solution in Figure 6.3 is 39 and the stream sum for the integral solution is 3. Note the various equivalences among sums in the indicated major parallelograms. In Figure 6.4 these sums are indicated by A,B,C,D,E,F. In case A=F then it is easily shown that C=E and B=D as in both examples in Figure 6.3. In fact these equivalences must always hold.

Theorem 6.4 *With major parallelogram sums as indicated in Figure 6.4, it must be that A=F, B=D and* C=E.

Proof. Let O(e) be a natural (or integral) octagon with stream sum S. Then

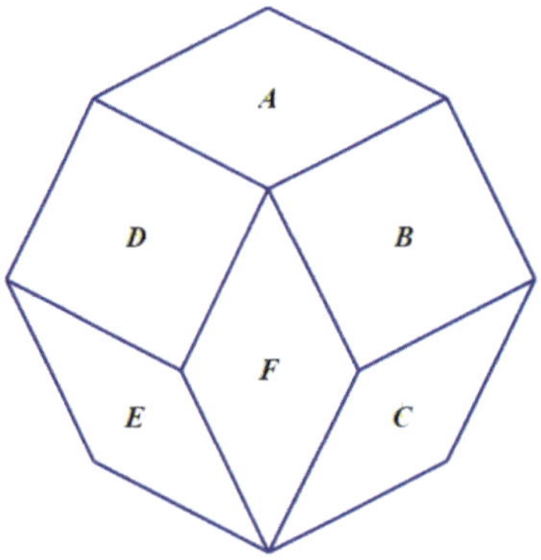

Figure 6.4

(1) A+B+C = eS

(2) A+D+E = eS

(3) B+F+E = eS

(4) D+F+C = eS.

From (2),(3) we obtain

(5) A+D = B+F,

and from (2),(4) we have

(6) A+E = F+C,

and from (1),(4) we obtain

(7) A+B = D+F,

and from (1),(3) we have

(8) A+C = F+E.

If A=F then (7) yields B=D, and (6) yields E=C. When equations (5),(6),(7),(8) are added we obtain

(9) 4A + (D+E+B+C) = 4F +(B+C+D+E,

Which yields A=F, and consequently B=D and E=C. □

It is also possible to find octagons all of whose major parallelogram sums are identical. In Figure 6.5 a natural and an integral O(2) are illustrated with common major parallelogram sums 46, 3 respectively.

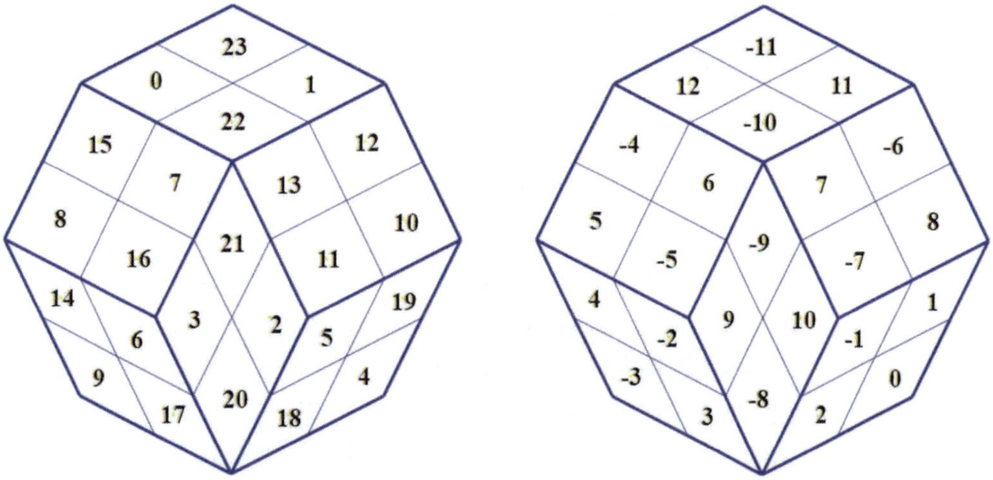

Figure 6.5

We will use the term *perfect* to describe any integral or natural magic polygon P all of whose major parallelograms have the same sum. The term only has significance in case P has 8 or more sides since, according to Theorem 5.1, all three major parallelograms of a magic hexagon *must* have the same sum. The triple of major parallelogram sums for a magic octagon P will be called the *signature of P.* The signature can be similarly defined in case P has more than 8 sides. If P is a semiregular octagon, then the *integral/natural signature set of P* is the multiset consisting of all possible signatures obtained by integral/natural solutions for P. The signature set provides a reasonable means for classifying solutions.

Suppose P is a magic polygon over the abelian group G and S_1, S_2 are two streams in P. Suppose further that $x_1, \ldots, x_m$ are elements in S_1 and $y_1, \ldots, y_m$ are elements in S_2 with $x_1 + \ldots + x_m = y_1 + \ldots + y_m$ and such that x_i, y_i lie in the same stream for each $i \in \{1, \ldots, m\}$. Then the interchanges $x_i \leftrightarrow y_i$, $i \in \{1, \ldots, m\}$, leave all stream sums unchanged thus preserving the magicality of P. This action we call a *simple interchange transformation,* and the sets $\{x_1, \ldots, x_m\}$, $\{y_1, \ldots, y_m\}$, *interchange sets.* An abundance of two element interchange sets can be found in the octagons in both Figures 6.3, 6.5; for example {6,17}, {21,2} in 6.5(I), and {6,-5}, {2,-1} in 6.5(II). All of these interchanges preserve the signature of the octagon.

From Figures 6.3, 6.5 we see that the signature set K of O(2) contains the triples {78,29,31}, {46,46,46}. By adding 4 to each element of 6.5(I) a natural solution with signature {14,62,62} is obtained, and by adding 4 to each cell of 6.3(I) a natural solution Ω with signature {46,45,47} is obtained. Finally, if 8 is added to each cell of Ω a natural solution with signature {14,61,63} results. These transformations will be called *natural algebraic equivalences.* The natural signature set for O(2) thus contains 5 triples. The full determination of the signature set for O(2) and other semiregular polygons is an interesting open question.

In Figure 6.6 an integral O(4) has been constructed as an extension of an integral O(2). Note the abundance of pairs of adjacent cells with sum 1. These *1-cells* are employed to insure that all streams which pass through the O(2) inclusion have the desired sum of 6. Various interchanges must then be made among the streams *not* passing through O(2) in order to insure that they also achieve a sum of 6. We were not successful in making the further extension O(4)→O(6) using this technique.

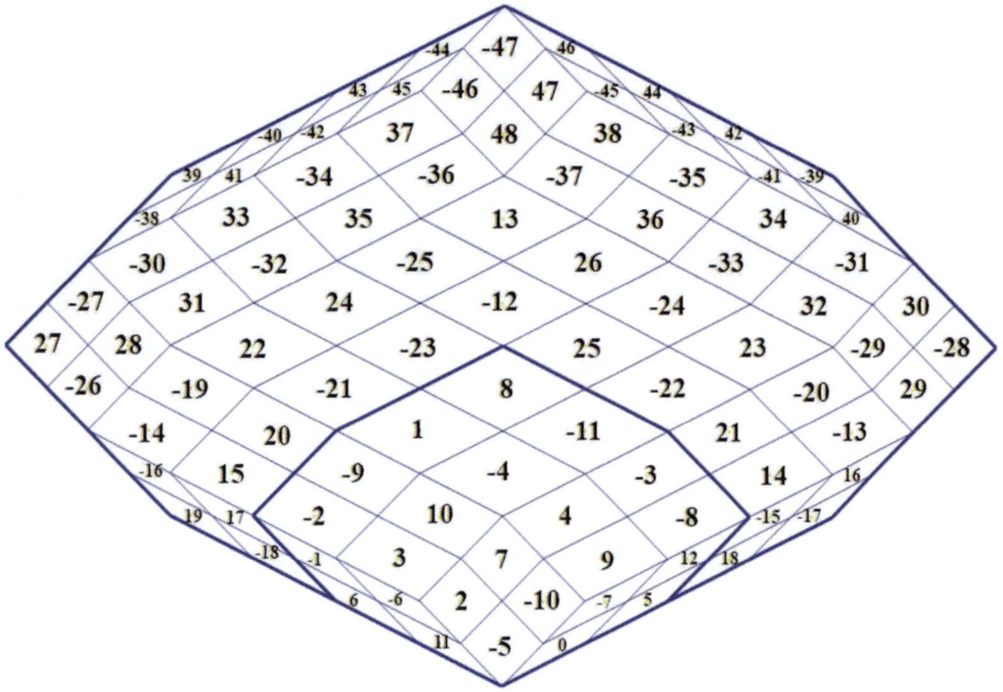

Figure 6.6

The following simple Theorem and Corollary make possible the construction of certain perfect integral polygons including integral regular octagons.

Theorem 6.5 *Let $n = 16t$, $t \in \{3,4,5, \ldots\}$ and let $K = n/2$. Then $Z^*(n)$ has a partition into t near magic squares of order 4, each with all rows and columns having sum 2.*

Proof. For A = 1, 5, 9, …, 4t-3 define the square M_A as in Figure 6.7. These squares are near magic with constant row and column sum 2. They also determine a partition of $Z^*(n)$ into t parts. □

A	*K*-*A*	1-*A*	*A*-*K*+1
K-*A*+1	*A*+1	*A*-*K*	-*A*
-*A*-1	*A*-*K*+3	*A*+2	*K*-*A*-2
A-*K*+2	-*A*-2	*K*-*A*-1	*A*+3

Figure 6.7

Corollary 6.6 *A perfect regular integral polygon P with e basic edges per side can be constructed whenever e is a multiple of 4.*

Proof. If e = 4m and P is a regular polygon with w sides, then the order n of P is given by $n = (4m)^2(1/2)q(q-1)$ where q = w/2. Thus n is a multiple of 16 and Theorem 6.5 provides $t = m^2(1/2)q(q-1)$ integral near magic squares of order 4 which determine a partition of Z*(n). These can be compressed into t major parallelograms and then used to tile P. □

In Figure 6.8 the construction indicated by Corollary 6.6 is used to create a perfect integral O(4). In fact, it is possible to construct perfect regular integral polygons whenever the number e of basic edges per side is even and larger than or equal to 4. The following construction requires the existence of even order squares with a *balanced integral solution.* An integral solution is *balanced* in case the number of positive values and the number of nonpositive values is the same in each row and column. The even order integral magic squares constructed in Theorem 3.21 all have balanced solutions.

Let M_n be a balanced integral square with n even, n≥4, and let K be a positive integer. Let $M_n \mp K$ denote the square obtained by (i) adding K to each positive value in M_n, and (ii) subtracting K from each nonpositive value in M_n . If M_n has constant row and column sum S, then $M_n \mp K$ will also have S as constant row and column sum. Furthermore the squares

$$\{M_n, M_n \mp n^2/2, M_n \mp n^2, M_n \mp 3n^2/2, M_n \mp 2n^2, M_n \mp 5n^2/2, \ldots, M_n \mp (t-1)n^2/2\}$$

Determine a partition of $Z^*(tn^2)$ into t squares of order n each with all rows and columns having the same sum S. These squares can be compressed into parallelograms and used to tile the major parallelograms which constitute a regular polygon. So we have the following result.

Theorem 6.7 *Let P be a regular polygon with an even number e of basic edges per side, (e≥4). Then P has a perfect integral solution which is composite in case e≥6.*

Proof. If P has w sides and q = w/2, then P has t = (1/2)q(q-1) major parallelograms each with order $n = e^2$. The balanced squares $\Pi = \{M_n, \ldots, M_n \mp (t-1)n^2/2\}$, listed above, can be compressed and used to tile the major parallelograms of P. The rows and columns of these squares all have the same sum and thus the streams constituting P will all have the same sum. Since all the rows and columns of every square in Π have the same sum, ΣA is also the same for ever square A in Π and thus P has a perfect integral solution.

Finally, for each n we use the balanced integral square M_n provided by the nesting proved in Theorem 3.21. This will ensure that for e≥6, our solution for P will be composite. □

In Figure 6.9 the construction described in Theorem 6.7 is used to create a perfect integral O(4). Note that it is composite with the top major parallelogram serving as kernel.

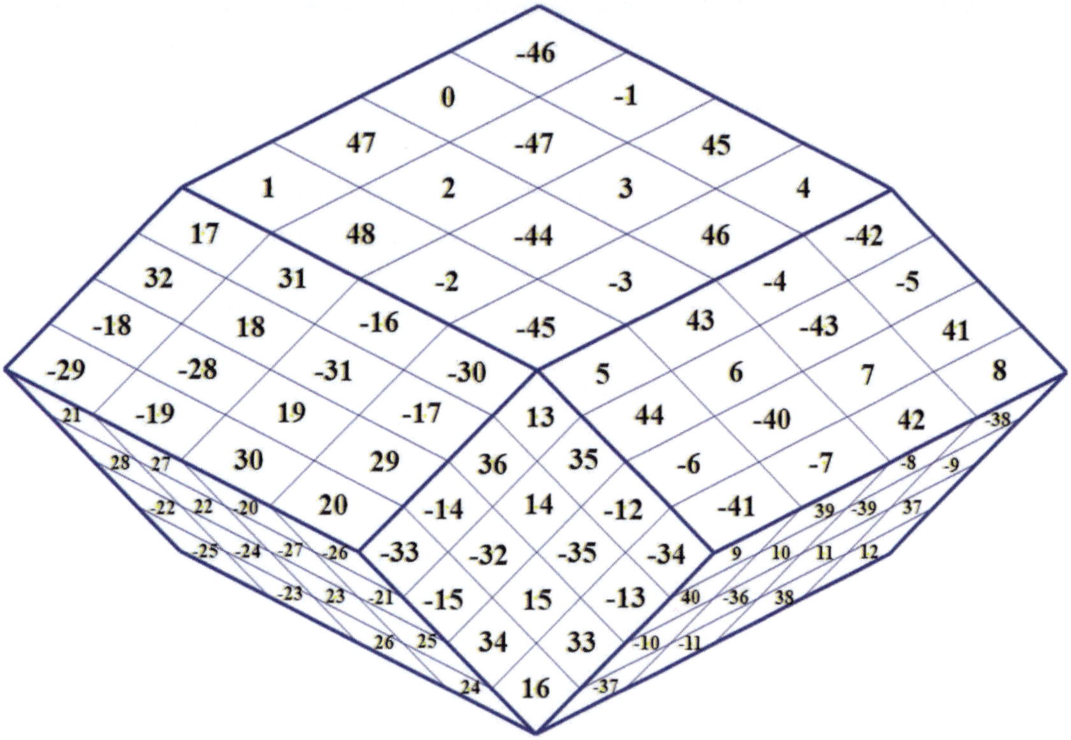

Figure 6.8

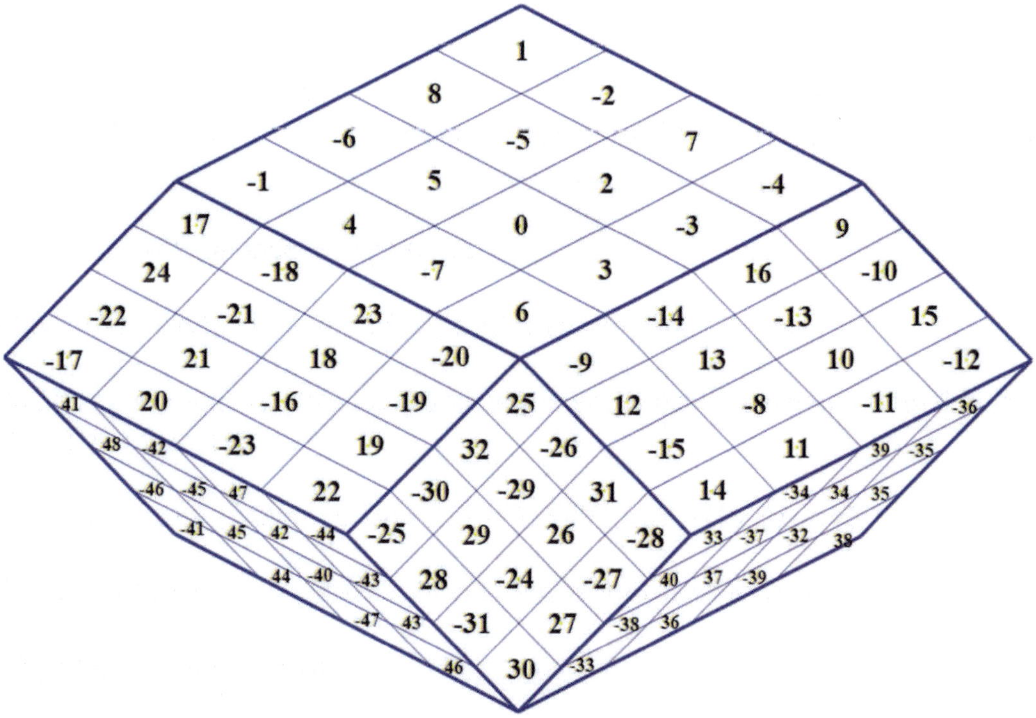

Figure 6.9

Theorem 6.8 *Let $n = 16t$, $t \in \{3,4,5, \ldots\}$ and let $K = n/2$. Then $Z(n)$ has a partition into t near magic squares of order 4, each with all rows and columns having sum $2n-2$.*

Proof. For A = 0,4,8, …, 4(t-1) define the square M_A as in Figure 6.10. These squares are near magic with constant row and column sum 2n-2. They also determine a partition of Z(n) into t parts. □

A	K-A-3	n-A-1	K+A+2
n-A-2	K+A+3	A+1	K-A-4
K-A-1	A+2	K+A	n-A-3
K+A+1	n-A-4	K-A-2	A+3

Figure 6.10

Corollary 6.9 *A perfect regular magic polygon P over Z(n) with e basic edges per side can be constructed whenever e is a multiple of 4.*

Proof. If e = 4m and P is a regular polygon with w sides, then the order n of P is given by n = $(4m)^2(1/2)q(q-1)$ where q = w/2. Thus n is a multiple of 16 and Theorem 6.7 provides t = $m^2(1/2)$ q(q-1) integral near magic squares of order 4 which determine a partition of Z(n). These can be compressed into t major parallelograms and then used to tile P. □

In Figure 6.10 the construction indicated by Corollary 6.9 is used to construct a perfect magic O(4) over Z(96).

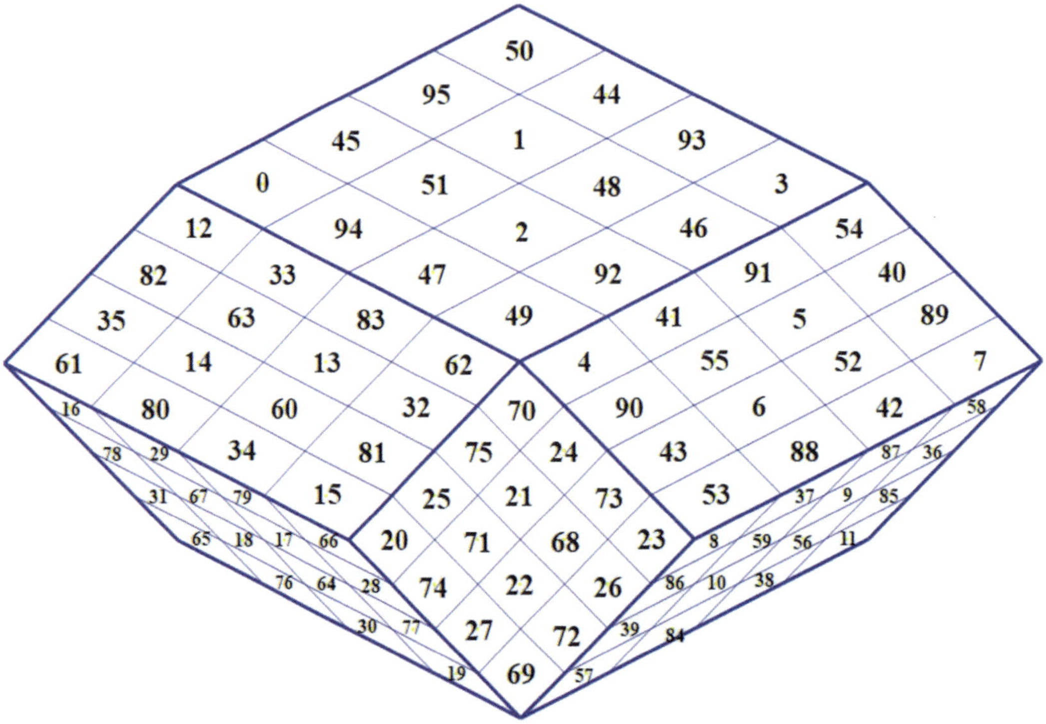

Figure 6.10

Yet another important method for constructing natural and integral solutions for a regular polygon involves the use of the two types of squares indicated in Figure 6.11.

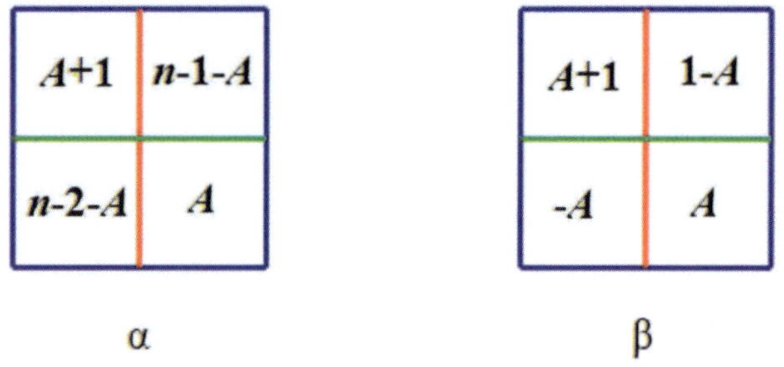

Figure 6.11

These 2 by 2 squares, as well as squares obtained by applying rotations or reflections to these, will be called *α-squares* and *β-squares.* If P is a regular polygon of order n with 2 basic edges per side, then for A = 0,2,4, …, (1/2)(n-4) the resulting α-squares determine a partition of Z(n). Similarly, for A = 1,3,5, …, (n/2)-1, the corresponding β-squares determine a partition of Z*(n). For an α-square the columns separated by a red line sum to n-1, and the rows separated by a green segment sum to n and n-2. For the β-squares the columns separated by red have sum 1, and the rows separated by green have sums 0 and 2.

In case P is a regular octagon, a solution over Z(n) or Z*(n) is indicated by a scheme for orienting 6 constituent α or β-squares, as illustrated in Figure 6.12(I). Corresponding actual solutions are given in 6,12(II,III). Many schemes may be possible; the only requirement is that each stream contains an even number of green segments which align with the direction of the stream. In this case the α or β-squares can be oriented so that all streams of P have the same sum.

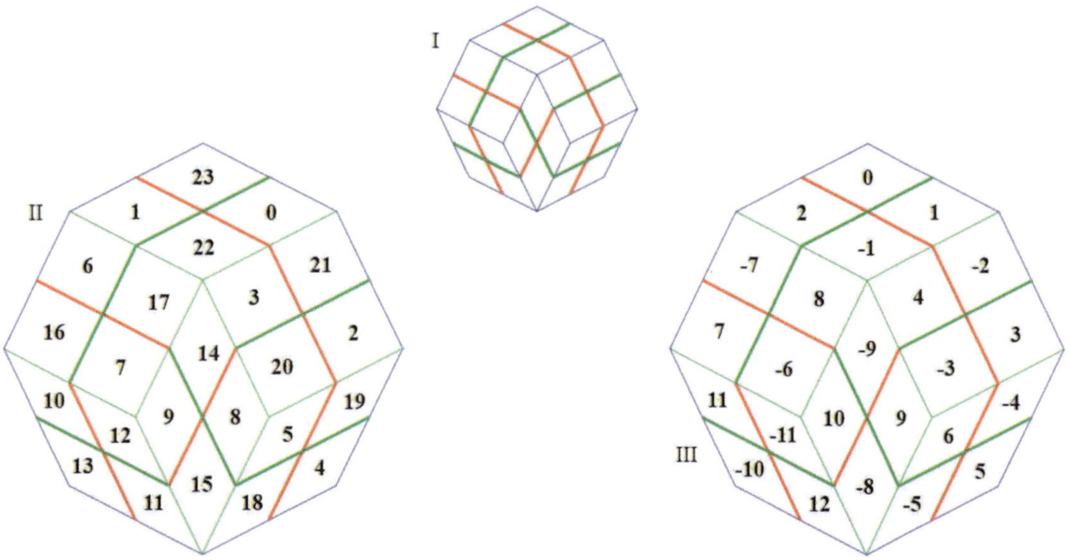

Figure 6.12

In case P is a regular octagon O(e) with e any even number, it is also possible to obtain a solution for P using α or β-squares. In Figure 6.13(I) a scheme is illustrated which can be used to create a 4 by 4 near magic square. If e is any multiple of 4, then 4 by 4 squares of this type can be combined to obtain either a natural or an integral solution for P. One such solution for O(4) is illustrated in Figure 6.13(II). The construction in this Figure illustrates the extension M(4)→H(4)→O(4) provided β-squares are used.

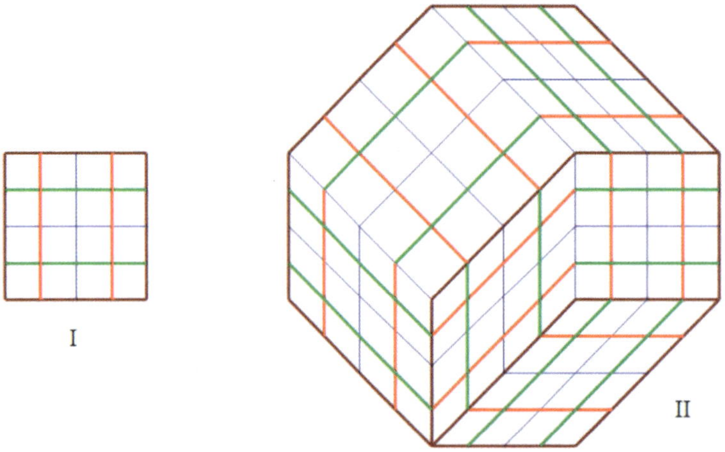

Figure 6.13

In Figure 6.14 a scheme for a solution, (integral or natural), for O(6) is illustrated. Note that it is unnecessary to draw the red bars; it is only needed that each stream contains an even number of green bars aligned with it.

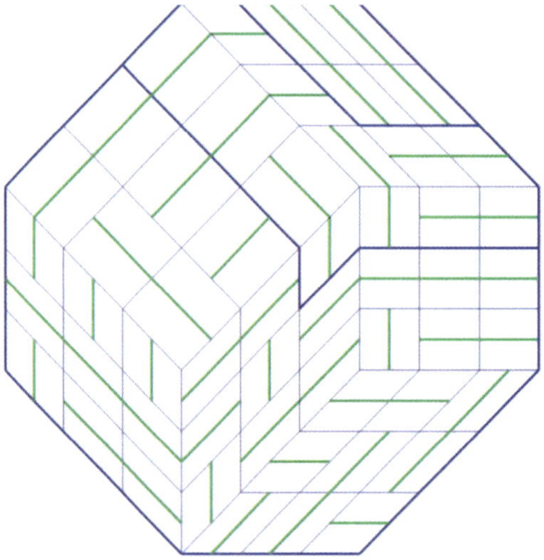

Figure 6.14

Also note that Figure 6.14 illustrates the extension M(4)→M(8)→O(6), provided β-squares are used. In Chapter 7 α and β-squares will be used to construct regular magic polygons with more than 8 sides.

6.3 Regular Octagons over Direct Product Bases

There can be quite a few different ways to represent a given abelian group using direct products. For example abelian groups with 24 elements have the following possible representations:

> Z(24), Z(2)×Z(12), Z(3)×Z(8), Z(2)×Z(2)×Z(6), Z(2)×Z(2)×Z(2)×Z(3)
> Z(2)×Z(3)×Z(4), Z(4)×Z(6).

Although all abelian groups of order 24 are isomorphic to one of the three groups Z(24), Z(2)×Z(12), Z(2)×Z(2)×Z(6), natural solutions for these do not guarantee that natural solutions exist for the remaining representations. In Chapter 5, for example, although natural solutions were found with base Z(12) and Z(2)×Z(6), no natural solution with base Z(3)×Z(4) exists.

To obtain a natural solution over a direct product representation it is necessary to obtain natural solutions for each of the component factors. In addition these natural solutions must be *orthogonal.* For example, to construct a solution over Z(2)×Z(12), a natural solution over $(Z(2))^{12}$ and an orthogonal natural solution over $(Z(12))^2$ must be found. These are provided in Figure 6.15.

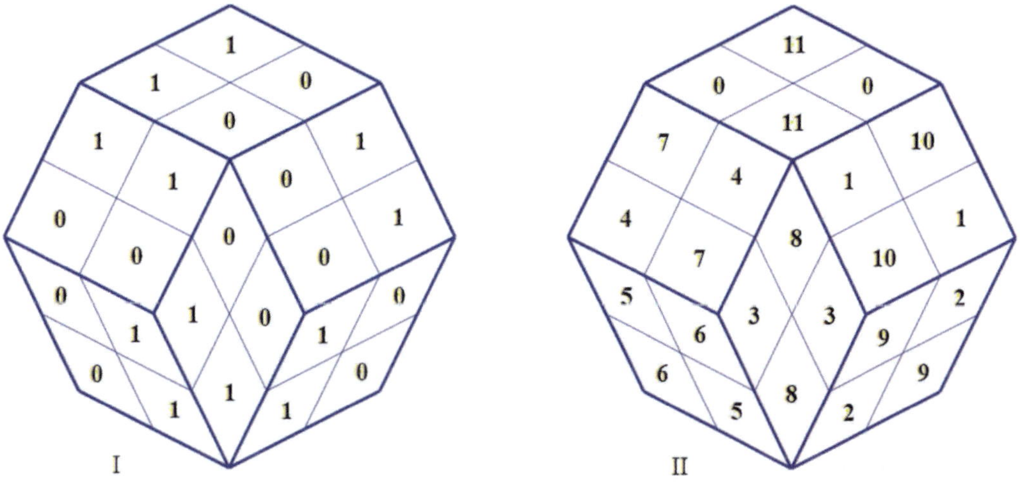

Figure 6.15

Note that the constructions in 6.15(I,II) are perfect and each employs 6 α-squares. For any regular octagon O(e) with e even, a solution over $Z(2)\times Z(3e^2)$ can be obtained similarly. In Figure 6.16 a magic O(2) over Z(3)×Z(8) is obtained via orthogonal $(Z(3))^8$ and $(Z(8))^3$ octagons, and in Figure 6.17 the same octagon is obtained but in this example both of the $(Z(3))^8$ and $(Z(8))^3$ octagons are perfect.

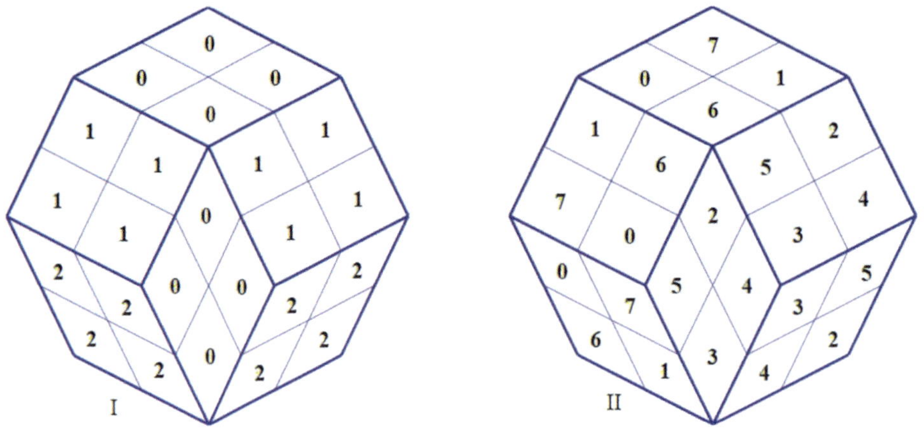

Figure 6.16

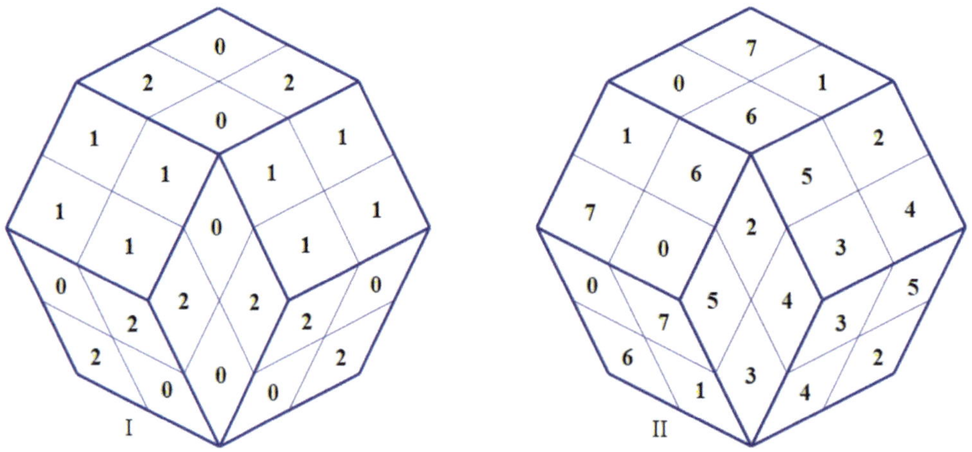

Figure 6.17

In Figure 6.18 a perfect O(2) with base Z(4)×Z(6) is obtained by the superimposition of the perfect $(Z(4))^6$ in 6.18(I) and the perfect $(Z(6))^4$ in 6.18(II).

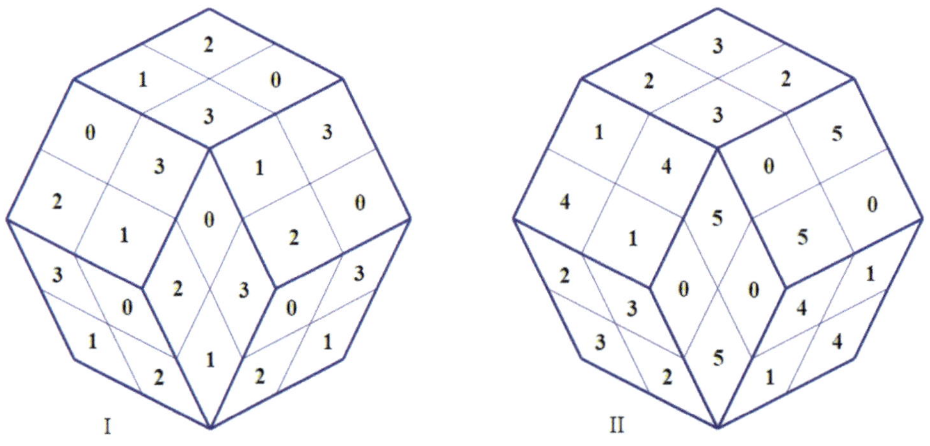

Figure 6.18

In 6.18(II), if the mapping ϕ:Z(6)→Z(2)×Z(3) defined by ϕ(x) = (x mod 2, x mod 3) is applied, a perfect O(2) with base ((Z(2)×Z(3))^4 is obtained. This octagon remains orthogonal to 6.18(I) and thus a perfect O(2) with base Z(2)×Z(3)×Z(4) is obtained.

Figure 6.19(I) is a perfect O(2) with base Z(8)×Z(3) obtained by superimposing 6.16(I,II). The first components of this octagon are then replaced by their base 2 equivalent to obtain the perfect O(2) with base Z(2)×Z(2)×Z(2)×Z(3) in Figure 6.19(II).

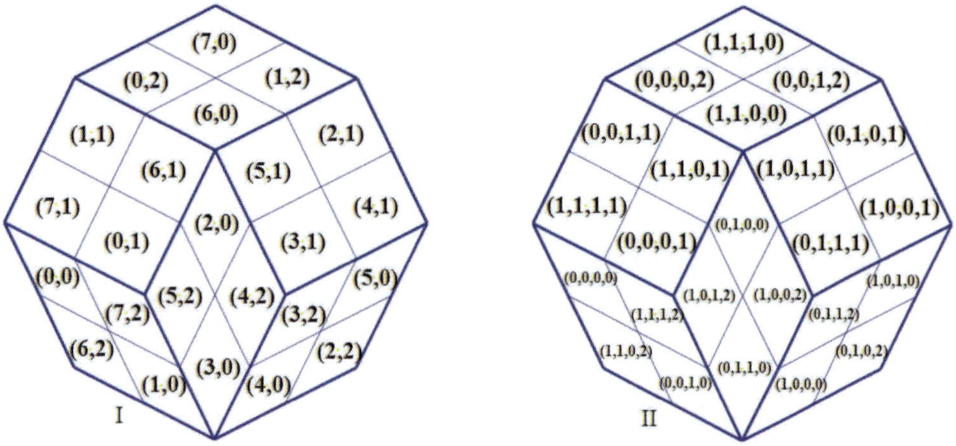

Figure 6.19

Finally, when the values in 6.18(I) are converted to base 2, a perfect O(2) with base (Z(2)×Z(2))^6 is created which remains orthogonal to 6.18(II). As a consequence, a perfect O(2) with base Z(2)×Z(2)×Z(6) is obtained. We have thus shown in this Section and in Figures 6.12(II,III), that O(2) is universally magic, universally perfect and integral.

Several of the O(2) constructions suggest procedures for constructing larger magic octagons. For example, the same pattern used to obtain the O(2) with base Z(3)×Z(8) in Figure 6.17(I,II) is used in Figures 6.20(I,II) to obtain a perfect O(4) over Z(3)×Z(32).

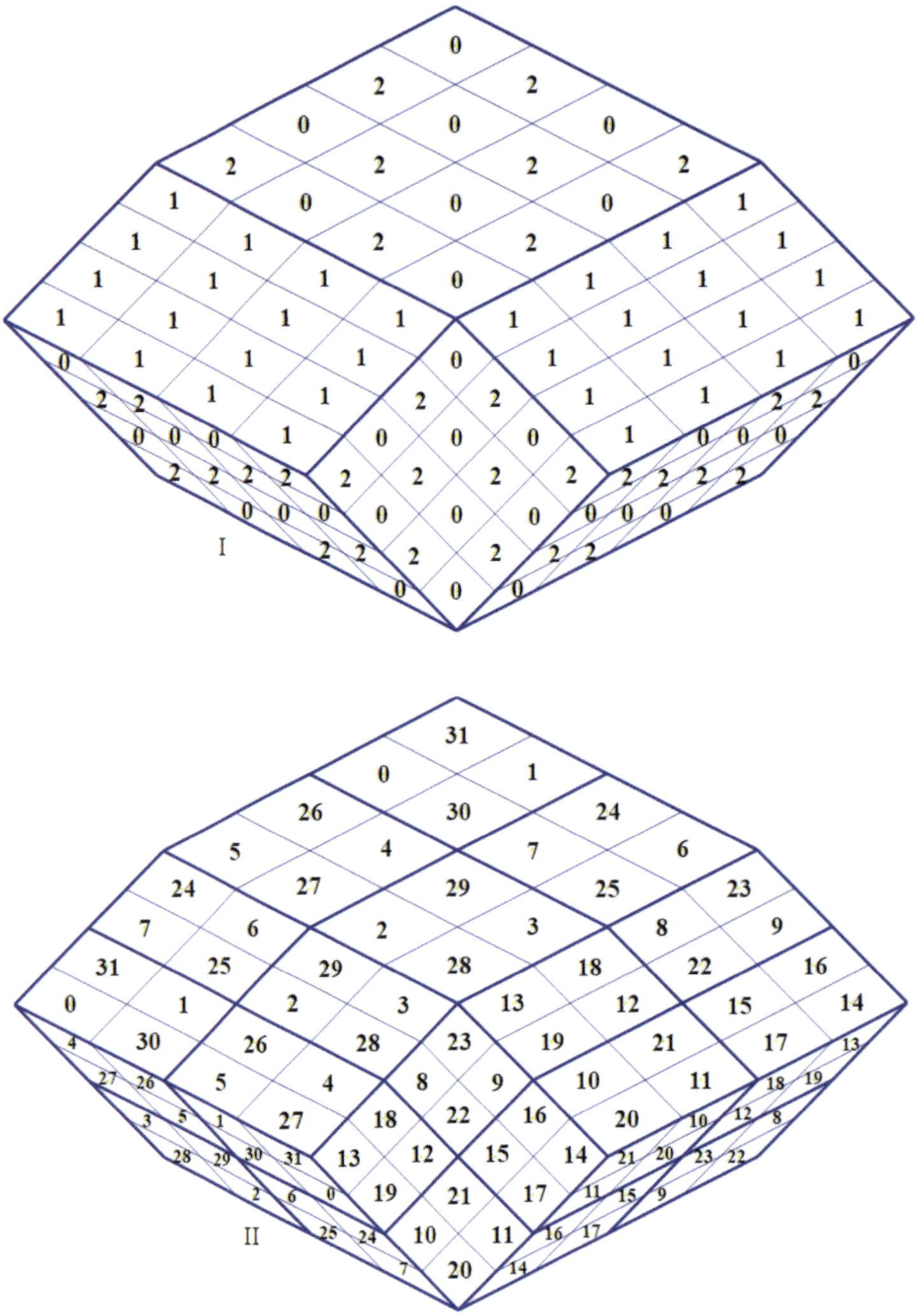

Figure 6.20

In Figures 6.21(I,II) the pattern established in Figures 6.18(I,II) is used to construct a perfect O(4) with base Z(4)×Z(24) via orthogonal octagons with bases $(Z(4))^{24}$ and $(Z(24))^{4}$.

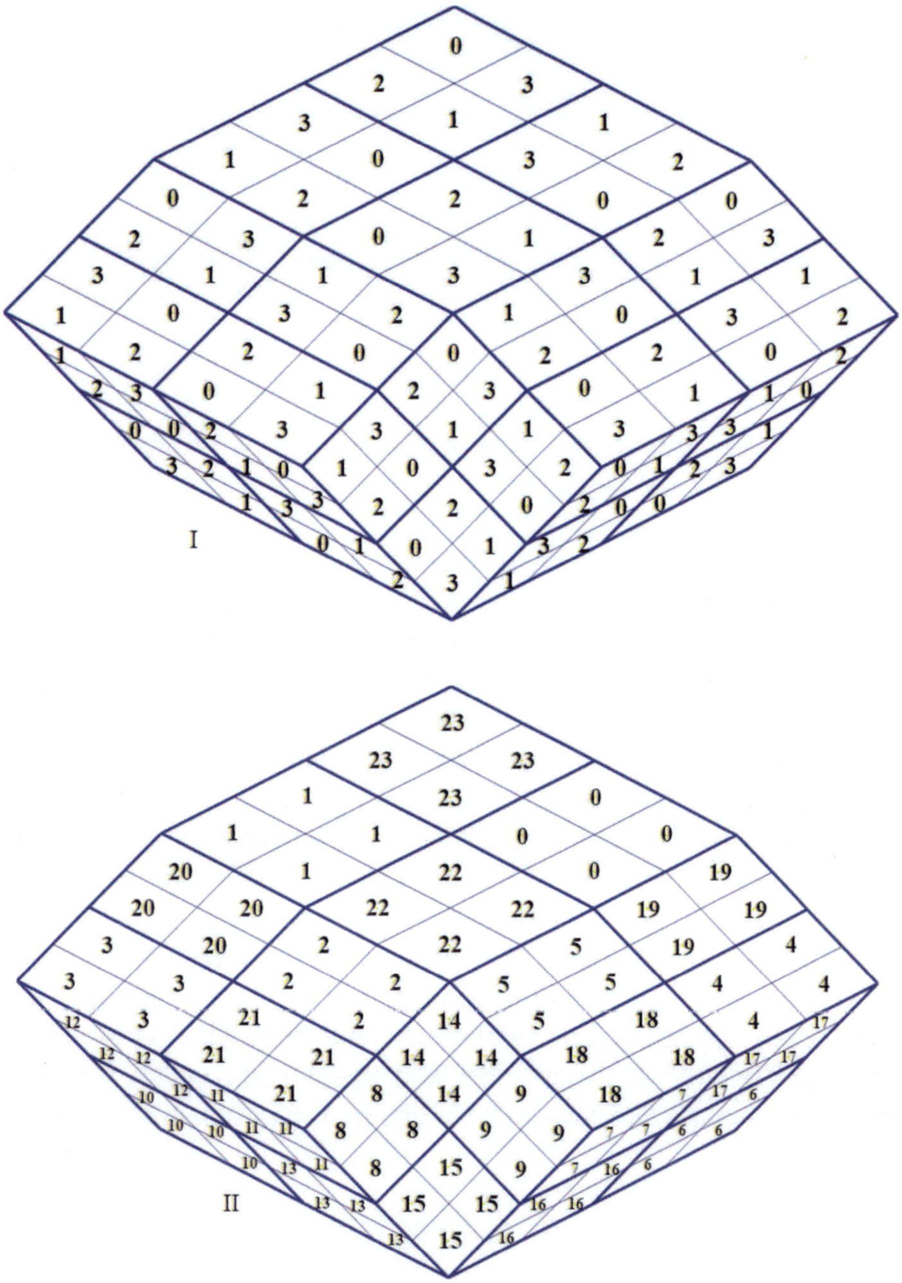

Figure 6.21

The order of O(e) is $6e^2$, and as e gets larger, (through even values), there will be numerous representations of $Z(6e^2)$ making it seemingly impossible to obtain solutions over all of them. We will then be satisfied with just one more construction in Figure 6.22 which also exhibits the extensive use of α-squares.

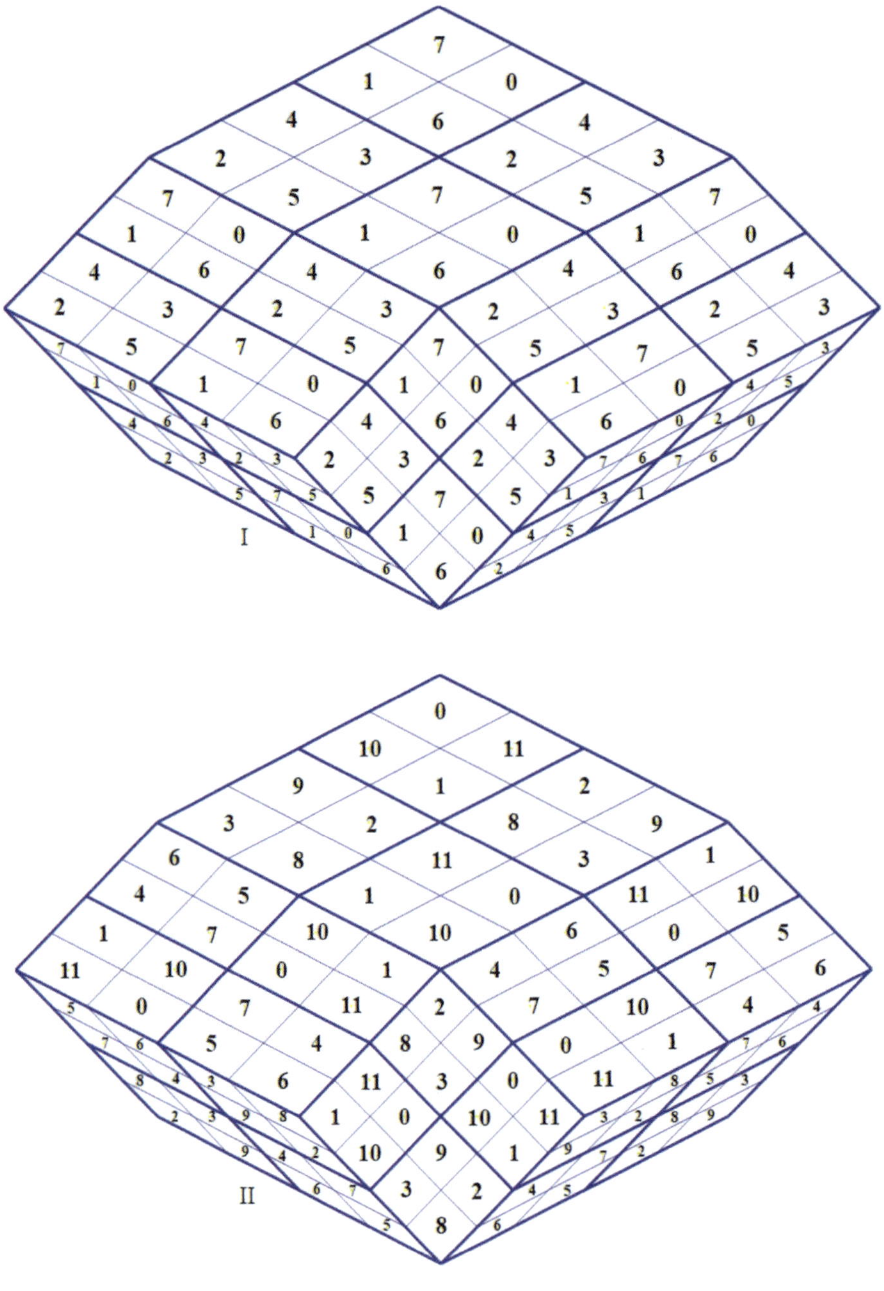

Figure 6.22

Figure 6.22(I) is a perfect O(4) over $(Z(8))^{12}$ and 6.22(II) is a perfect O(4) over $(Z(12))^{8}$. These are orthogonal, so their superimposition yields a perfect O(4) over Z(8)×Z(12). It is probably the case that O(4) is both integral and universally perfect, and even the following Conjecture may hold true.

Conjecture 6.10 *The regular octagon O(e) is both integral and universally perfect for every even side length e.*

6.4 Odd Order Integral Octagons

The order of the octagon O(p,q,r,s) is given by n = pq+pr+ps+qr+qs+rs which is odd in case either exactly one or exactly two of the parameters p,q,r,s are even. If exactly one parameter is even, we call O(p,q,r,s) *Class 1,* and if exactly two of the parameters are even, we say O(p.q,r,s) is *Class 2.* Our main objective will be to determine the primary integral octagons in either class, and then build larger integral octagons via extensions. We will start by examining Class 1 octagons in which p=q=r=1, and s is even. Examples with s = 2,4,6 are illustrated in Figure 6.23.

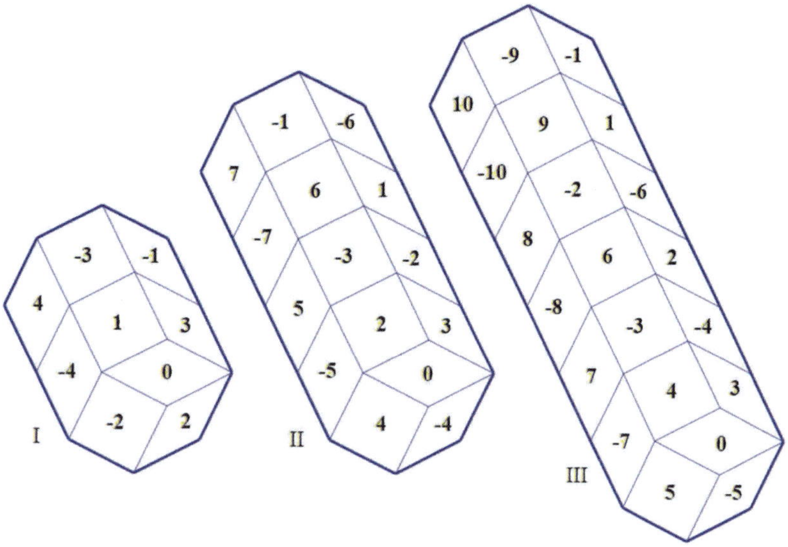

Figure 6.23

One larger example is provided in Figure 6.24. This is an O(1,1,1,14) with base Z*(45) which has been constructed according to the pattern given in Section 4.7 for constructing integral rectangles. Instead of a stream of length 3, the octagon has a simple hexagon at its base. To make the three long streams in the octagon have sum 0, the indicated interchanges must be made. For other large (even) values of s, octagons O(1,1,1,s) can be constructed similarly.

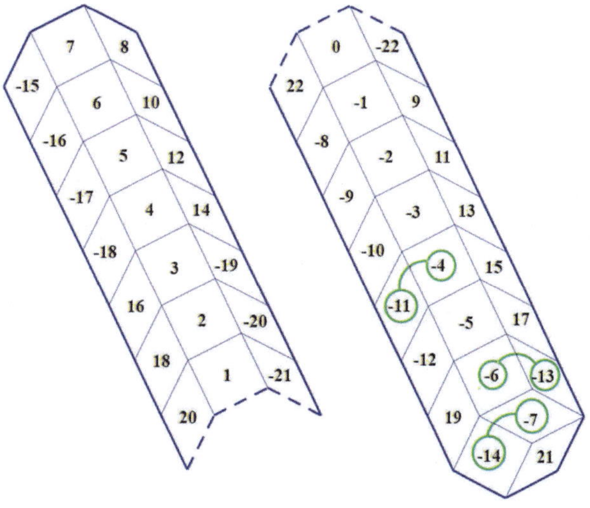

Figure 6.24

For small even s, octagons O(1,1,1,s) like the ones in Figure 6.23 are certainly primary, but for sufficiently large s composite ones can be constructed. The method described in Theorem 5.15 is used to create a 0-rectangle with three rows. This is

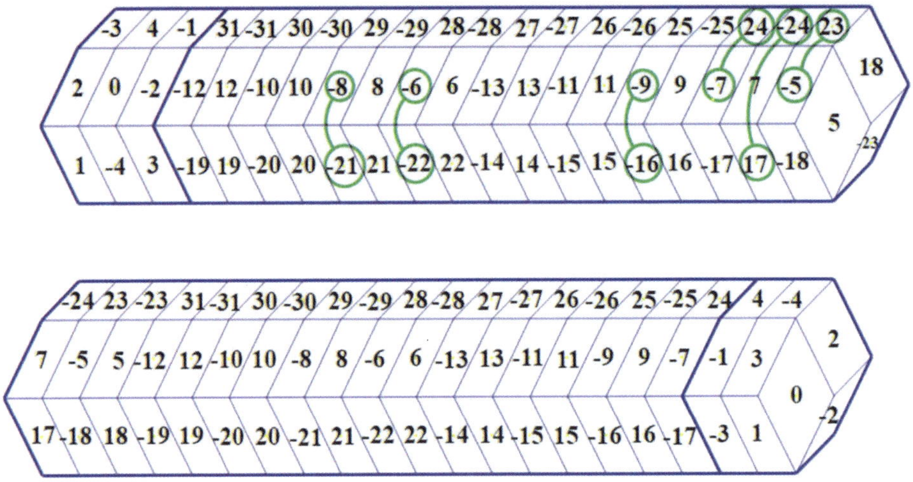

Figure 6.25

converted to an isorectangle and used to extend either a Lo Shu isosquare or a O(1,1,1,2) octagon as in Figure 6.25. The extension of the isosquare requires a number of interchanges as indicated. Note that according to Lemma 5.14, the construction of the isorectangle over a sequence of consecutive integers starting with a = 5 must end with the value b = 7a-4 = 31.

In Figures 6.26, 6.27 the only primary odd order integral octagons not in the set {O(1,1,1,x) : x is even} are illustrated.

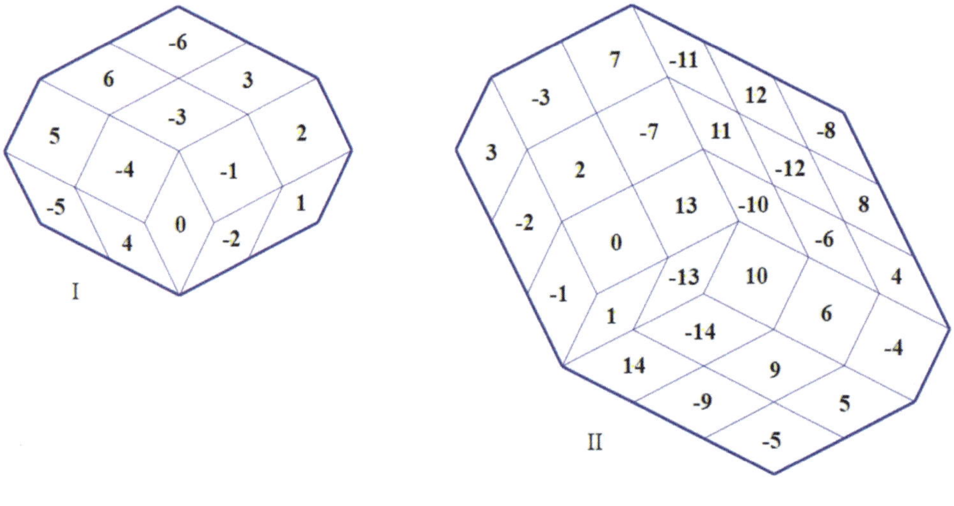

Figure 6.26

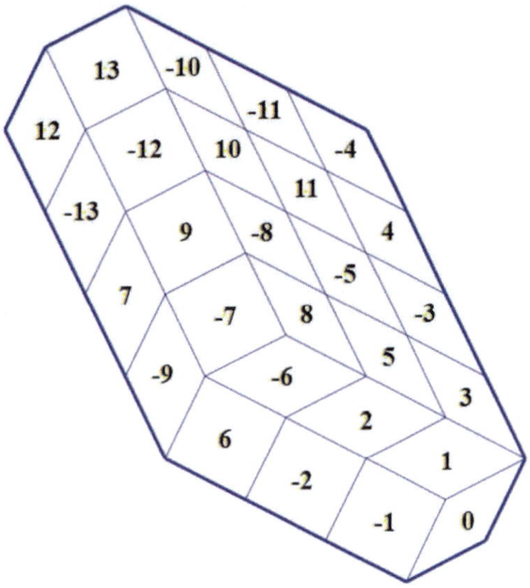

Figure 6.27

Integral octagons of the form O(1,1,3,x) with x even and x≥6 appear to be composite although the constructions are difficult. In Figure 6.28 (I,II) integral octagons O(1,1,3,8) and O(1,1,3,10) are represented each as extensions of O(1,1,2,3), and in 6.29 octagon O(1,1,3,12) is obtained as an extension of O(1,1,3,4). Note that O(1,1,3,6) and O(1,1,3,10) can be obtained via domino extensions O(1,1,1,6)→O(1,1,3,6) and O(1,1,1,10)→O(1,1,3,10).

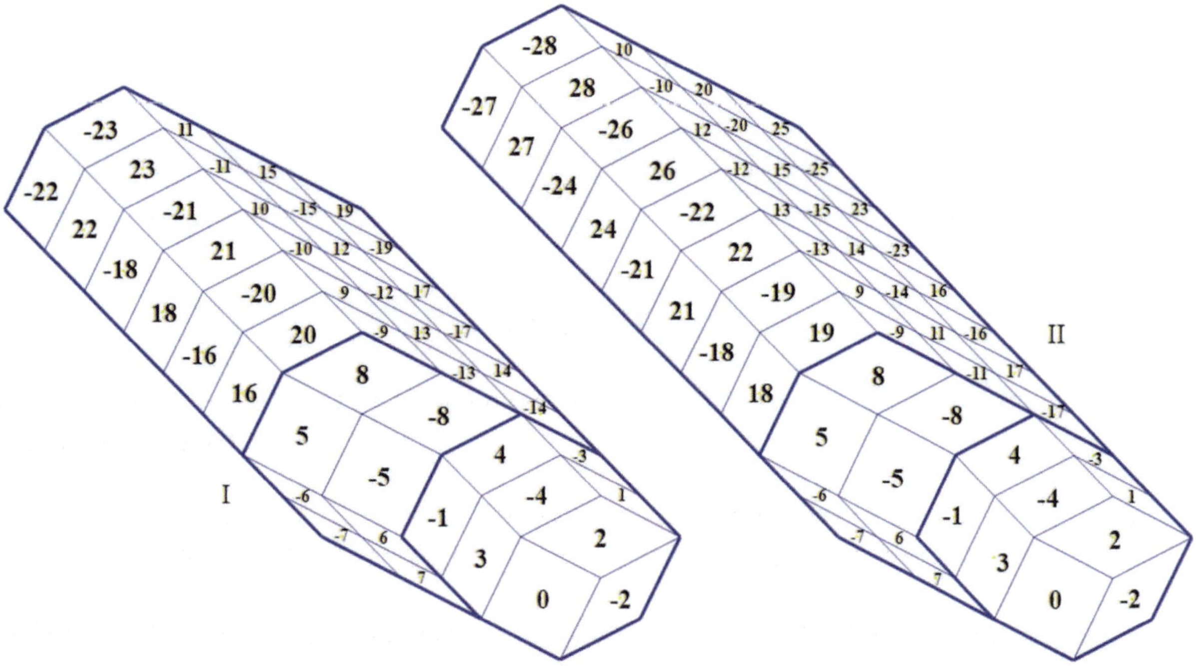

Figure 6.28

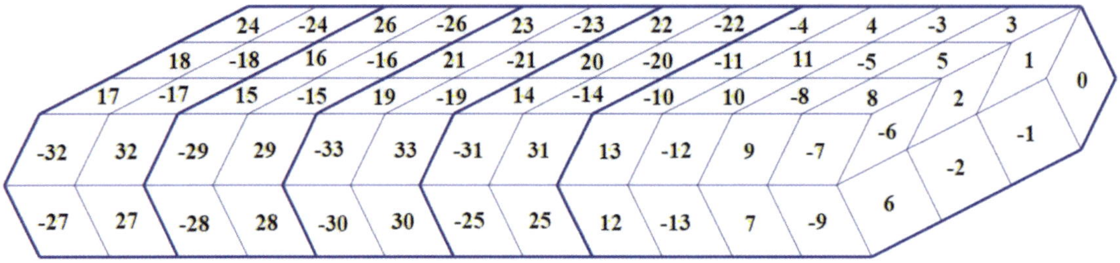

Figure 6.29

In constructing integral octagons via extensions, besides extending smaller octagons, it is also possible to achieve a construction by extending an integral isosquare, isorectangle, or isohexagon. In Figure 6.30(I) an O(1,1,2,5) is obtained as an extension of an H(1,1,3) isohexagon. An O(1,1,2,5) can also be obtained as an extension of an O(1,1,1,2) as in Figure 6.30(II).

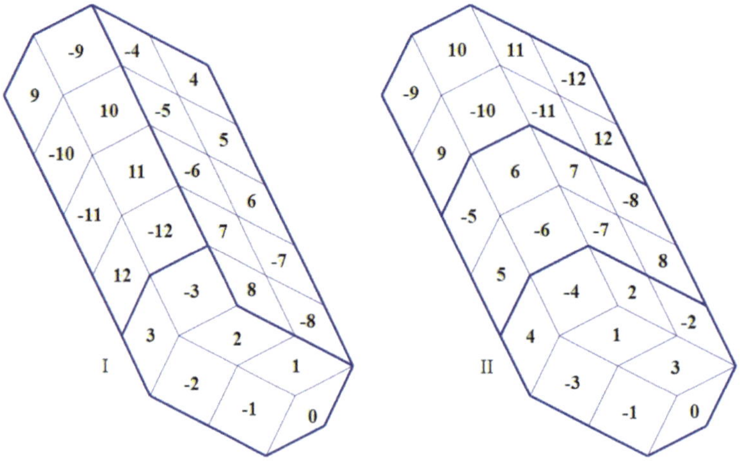

Figure 6.30

To show that O(1,2,2,3) is composite it is necessary to extend a Lo Shu isosquare as in Figure 6.31.

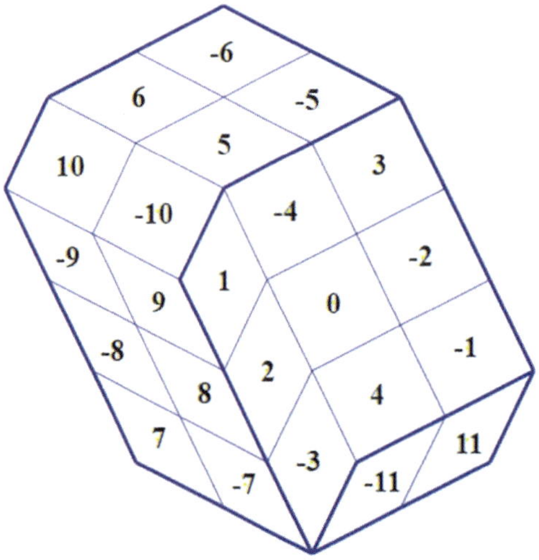

Figure 6.31

Suppose O(a,b,c,x) is a Class 1 integral octagon with a,b,c odd and x even. If 4|(a+b+x) then the extension O(a,b,c,x) → O(a,b,c+2,x) is accomplished by means of a 0-domino extension. If a+b+x has the form 4k+2 then the extension O(a,b,c,x) → O(a,b,c+4,x) can similarly be achieved. These observations yield the following two *reduction rules.*

(I) If c≥3 and 4|(a+b+x) then the extension O(a,b,c-2, x)→ O(a,b,c,x) can be made.

(II) If c≥5 and a+b+x has the form 4k+2, then the extension O(a,b,c-4, x) → O(a,b,c,x) can be made.

By means of these reduction rules and the observations regarding streamorphisms in Section 6.1, a Class 1 integral octagon can be reduced to one of the following four Types.

1. {O(1,1,1,x): x is even}
2. {O(1,1,3,x) : x is a multiple of 4}
3. {O(1,3,3,x) : x = 4k+2}
4. {O(3,3,3,x) : x is a multiple of 4}

Type (1) octagons were covered at the beginning of Section 6.3. There are many primary octagons of this Type, but it was shown that composite ones also exist. The primary Type (2) octagon O(1,1,3,4) is illustrated in Figure 6.27, and O(1,1,3,8) is illustrated in Figure 6.32 where it is shown to be an extension of O(1,1,2,3). We conjecture that the remaining Type (2) octagons are composite.

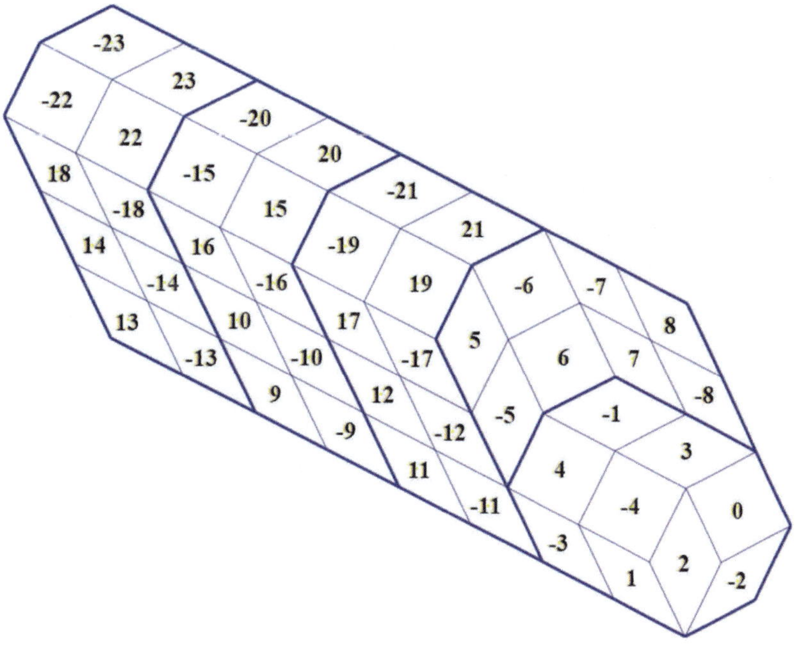

Figure 6.32

The Type (3) primary octagon O(1,3,3,3) is illustrated in Figure 6.26(II), and O(1,3,3,6) is shown to be an extension of hexagon H(1,3,3) in Figure 6.33.

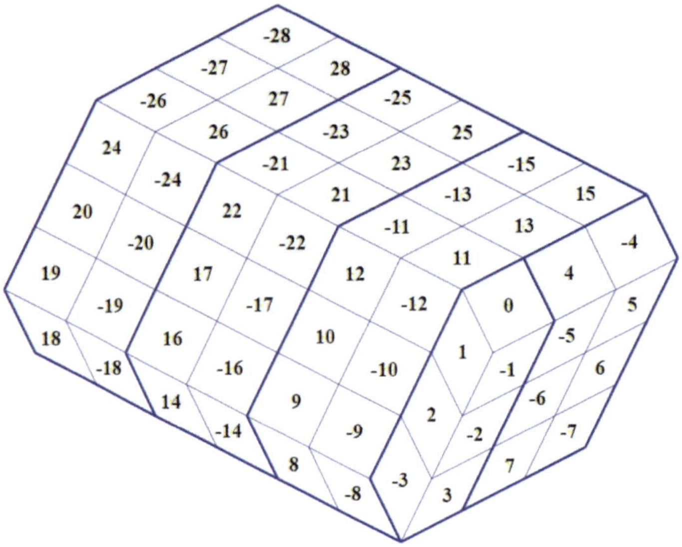

Figure 6.33

We conjecture that the remaining Type (3) octagons are composite. In Figure 6.34 the Type (4) octagon O(3,3,3,4) is shown to be an extension of O(1,1,3,2). We conjecture that the remaining Type(4) octagons are composite.

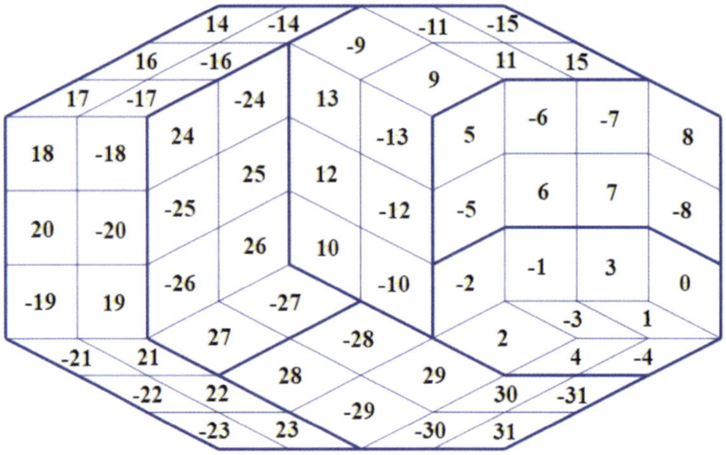

Figure 6.34

Now let O(a,b,x,y) represent a Class 2 integral octagon with a,b odd and x,y even. Suppose one of x, y is larger than or equal to 6. Say y≥6, then if 4|(a+b+x) the extension O(a,b, x, y-4)→O(a,b,x,y) can be accomplished using dominoes. On the other hand, if a+b+x = 4k+2, then the extension O(a,b, x, y-4)→O(a,b,x,y) can similarly be obtained.

Thus, in the hunt for primary integral octagons, we can suppose that x≤4 and y≤4. In case 4|(a+b+x) the extension H(a,b,x)→O(a,b,x,y) can be made using dominoes, and in case a+b+x = 4k+2 the extension H(a,b,x)→O(a,b,x,4) can be made similarly. From these observations it is seen that O(a,b,4,4) and O(a,b,2,4) are always composite. As a consequence only Class 2 octagons of the form O(a,b,2,2) can be primary. In Figure 6.35 the sequence of extensions O(1,1,2,2)→O(3,3,2,2)→ O(5,5,2,2)→O(7,7,2,2) is illustrated. The octagon O(1,1,2,2), illustrated in Figure 6.1 is primary, and the set of octagons {O(a,a,2,2) : a≥3 is odd} are likely all composite.

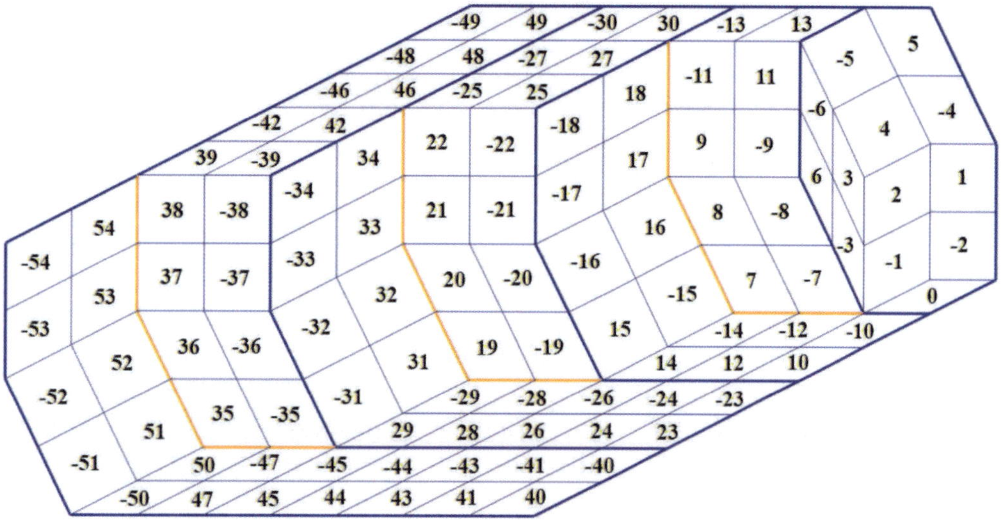

Figure 6.35

Besides the ones illustrated in Figure 6.35, there are of course many more octagons of the form O(a,b,2,2). In Figures 6.36, 6.37 we construct just two of these.

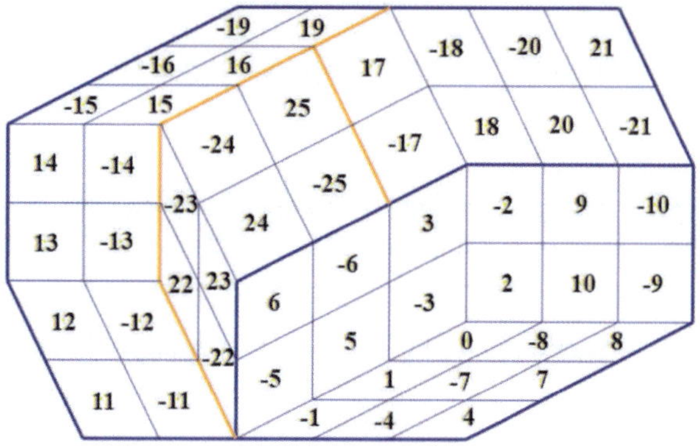

Figure 6.36

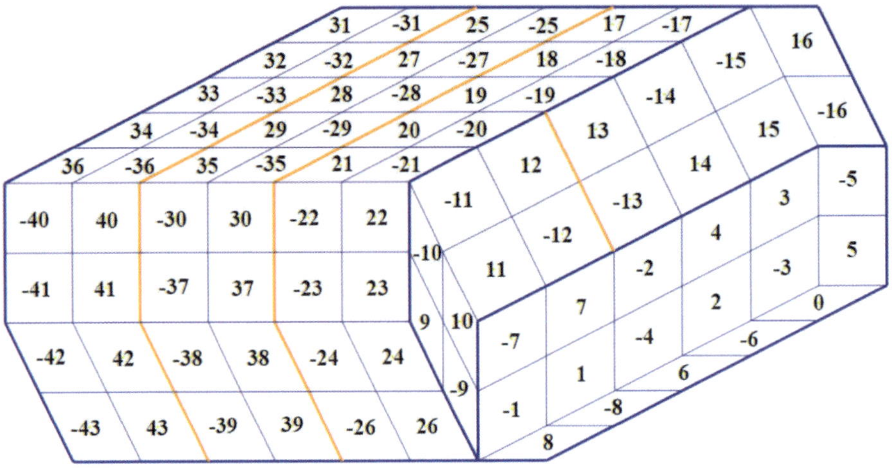

Figure 6.37

Figure 6.36 illustrates the extension H(2,3,3)→O(3,5,2,2), and Figure 6.37 illustrates the series of extensions H(1,2,5)→O(1,5,2,2)→O(5,7,2,2). The results of this Section strongly indicate the truth of the following Conjecture.

Conjecture 6.11 *Primary odd order integral octagons consist precisely of certain ones in the set {O(1,1,1,x) : x is even} as well as O(1,1,2,2), O(1,1,3,4) and O(1,2,3,3).*

6.5 Even order Integral Octagons

Let O = O(a,b,c,d) be an integral octagon with even order n, semiperimeter s = a+b+c+d and stream sum S. Then since s also represents the number of streams, we must have sS = 2ΣZ*(n) = 2(n/2) = n, and thus

S = n/s = (ab+ac+ad+bc+bd+cd)/(a+b+c+d) must be an integer.

Since n is even, the semiperimeter s must also be even, yielding two possibilities: a,b,c,d are all even, or a,b,c,d are all odd. The first possibility is satisfied by every regular octagon O(e) with e even. In Theorem 6.7 it was shown that O(e) is composite in case e≥6. If O is irregular, (but semiregular), then the conditions n/s an integer and s even, are satisfied by O(1,1,1,3), O(2,4,6,8) and O(3,3,3,9). These are the smallest such octagons and even for these, integral solutions have yet to be found.

CHAPTER 7

MAGIC POLYGONS WITH MORE THAN 8 SIDES

7.1 Constructions with α and β-squares

The 2 by 2 α and β-squares introduced in Section 6.2 are useful for constructing natural and integral regular magic polygons with more than 8 sides. Recall that for a natural polygon P with order n, the sum in the α-square in each column separated by a red bar is n-1, and the two sums separated by a green bar are n, n-2. For an integral polygon the sum in the β-square in each column separated by a red bar is 1, and the sums in the columns separated by a green bar are 0, 2. In order for all streams to have the same sum, the green bars must occur an even number of times in each stream. In Figures 7.1, 7.2 magic *regular* decagons with 2 basic edges per side are illustrated. (Note that we use the term *regular* in a nonstandard way to apply to any semiregular polygon with the same number of basic edges per side.)

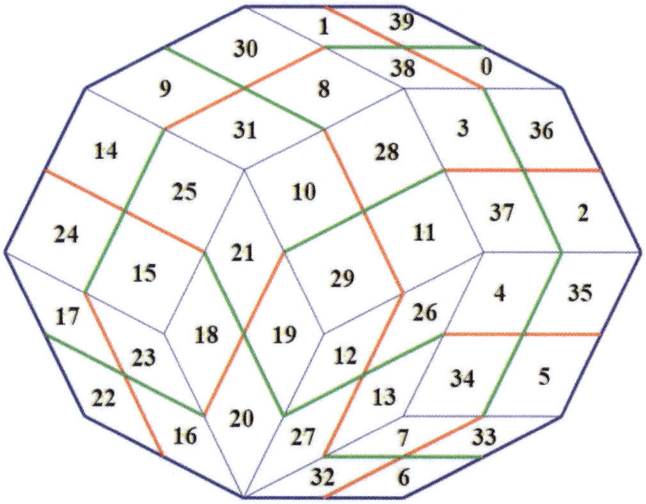

Figure 7.1

We will use the notation $P_w(e)$ to denote a regular polygon with w sides and e basic edges per side. The order n of $P_w(e)$ is given by $n = (1/8)e^2w(w\text{-}2)$. To start we will consider the case e = 2, then n = (1/2)w(w-2). As there must be an even number of α or β-squares, n must be a multiple of 8 and consequently w(w-2) must be a multiple of 16. Thus w ∈ {8, 10, 16, 18, 24, 26, 32, 34, …}. For all these choices for w, we will show that α and β-squares can be used to construct perfect magic $P_w(2)$ polygons.

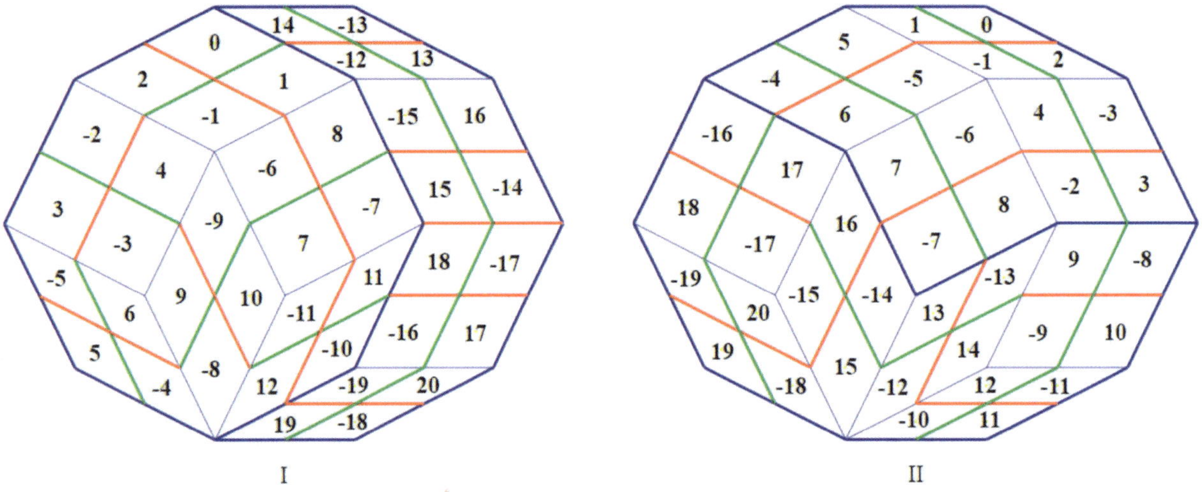

Figure 7.2

In Figure 7.2(I) an integral $P_{10}(2)$ is illustrated as an extension of the integral octagon $O(2) = P_8(2)$, and in 7.2(II), $P_{10}(2)$ is illustrated as an extension of the integral *isosquare* M*(4). (In general we will let M*(e) denote the isosquare which is isomorphic to a 0-sum square of order e.) Whenever e is a multiple of 4, it is possible to construct M*(e) using β-squares. In Figure 7.3, M*(8) is constructed as an extension of M*(4).

-11	11	10	-8	-3	3	2	0
12	-10	-9	9	4	-2	-1	1
-15	15	14	-12	-7	7	-4	6
16	-14	-13	13	8	-6	5	-5
-27	27	26	-24	-19	19	18	-16
28	-26	-25	25	20	-18	-17	17
-31	31	30	-28	-23	23	22	-20
32	-30	-29	29	24	-22	-21	21

Figure 7.3

A solution for $P_{16}(2)$ is illustrated in Figure 7.4. Each green banded block represents either an α-square or a β-square. If every block is an α-square then a natural solution is obtained, and if every block is a β-square then an integral solution is obtained. (Note that it is unnecessary to include the red bars.) The polar tiling in Figure 7.4 yields a partition into a 0-sum isosquare M*(8) and two $P_8(2)$ polygons each of which have the indicated β-square solutions. These regular polygons which contain the polar point p we call *polar polygons.*

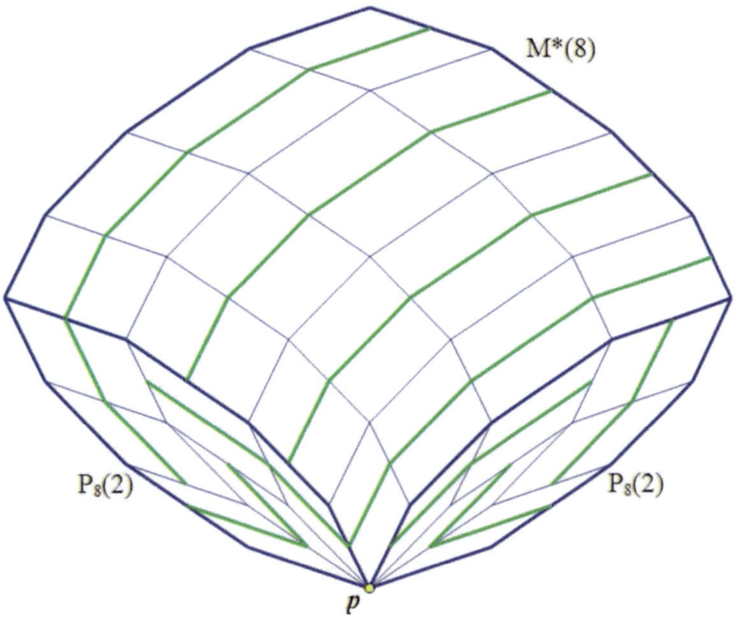

Figure 7.4

Note that either one of the polar polygons or the 0-sum isosquare can serve as an inclusion, and thus our construction of $P_{16}(2)$ is composite. In Figure 7.5 a solution for $P_{18}(2)$ is similarly obtained. In this case the dissection of the polar tiling of $P_{18}(2)$ consists of the isosquare M*(8) and two $P_{10}(2)$ polar polygons which share a single side. Two integral solutions for $P_{10}(2)$ are given in Figure 7.2. Again, either one of the polar polygons or the isosquare can be chosen as an inclusion.

To obtain solutions for every $P_w(2)$, (w(w-2) a multiple of 16), an inductive argument can be used. In Figure 7.6(I) a dissection of the polar tiling of $P_{32}(2)$ into 0-sum isosquares, (M*(8)), and smaller solvable polar polygons is illustrated. The collection of $P_w(2)$ polygons solvable using α or β-squares can be put into two *Categories*: (1) those in which w is a multiple of 8, and (2) those in which w = 8k+2. Figure 7.6(I) indicates the general procedure for solving Category (1) polygons.

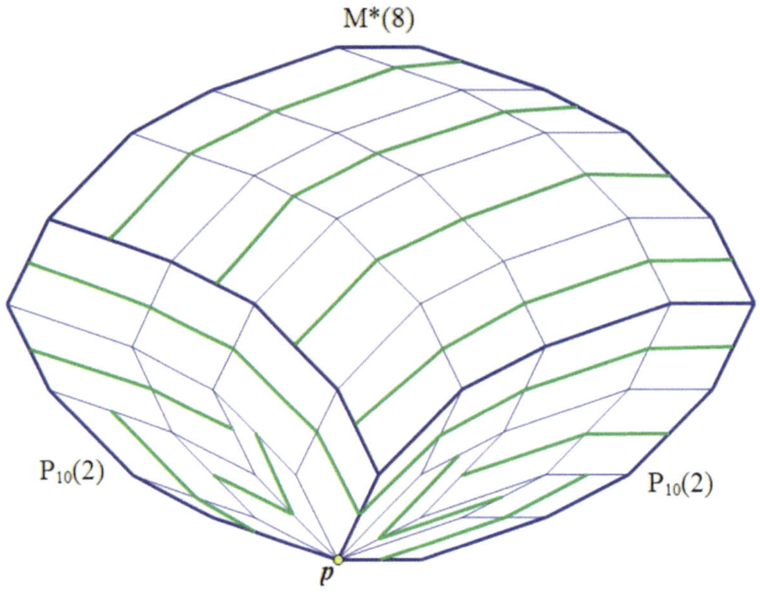

Figure 7.5

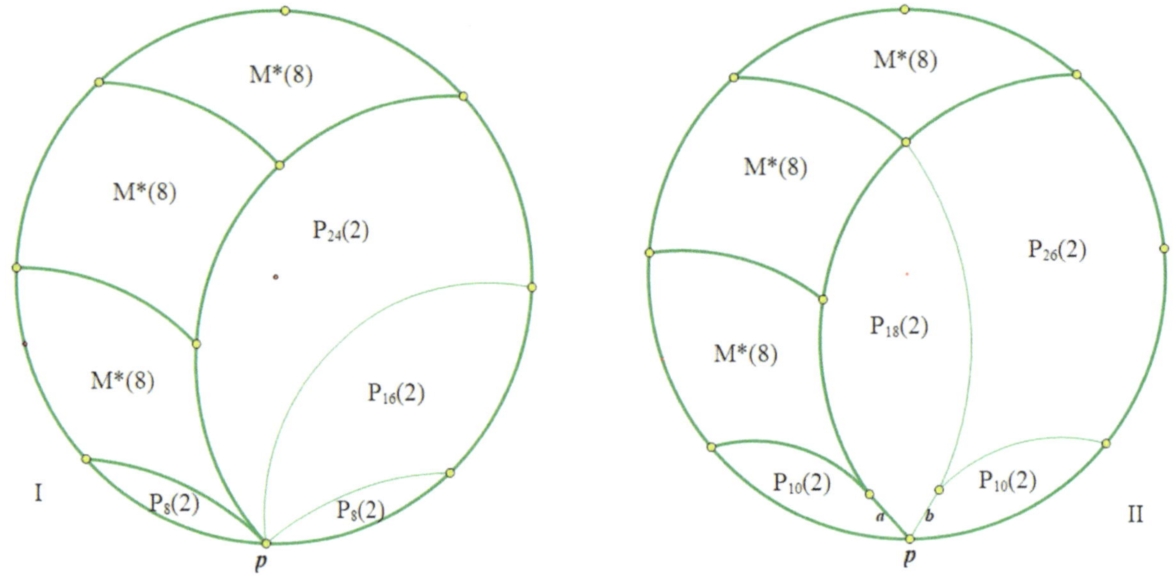

Figure 7.6

Figure 7.6(II) provides the scheme for solving $P_{34}(2)$. It is indicative of the general procedure which can be used to solve Category(2) polygons. Note that, unlike Category(1) polygons, the polar polygons for Category (2) have edges in common with their neighbors,(marked a, b in 7.6(II)). We have proved the following.

Theorem 7.1 *Every regular polygon $P_w(2)$ can be solved with α or β-squares provided w(w-2) is a multiple of 16. In case β-squares are used, then the integral $P_w(2)$ is composite whenever w≥10.* □

Theorem 7.1 should also be true for $P_w(e)$ whenever e is a multiple of 4 or if e is any even integer and the number of major parallelograms which constitute $P_w(e)$ is even.

7.2 Odd Order Integral Decagons

Odd order integral semiregular polygons with any (even) number of sides can be often obtained by extensions. For decagons we will just provide examples in Figures 7.7-7.11.

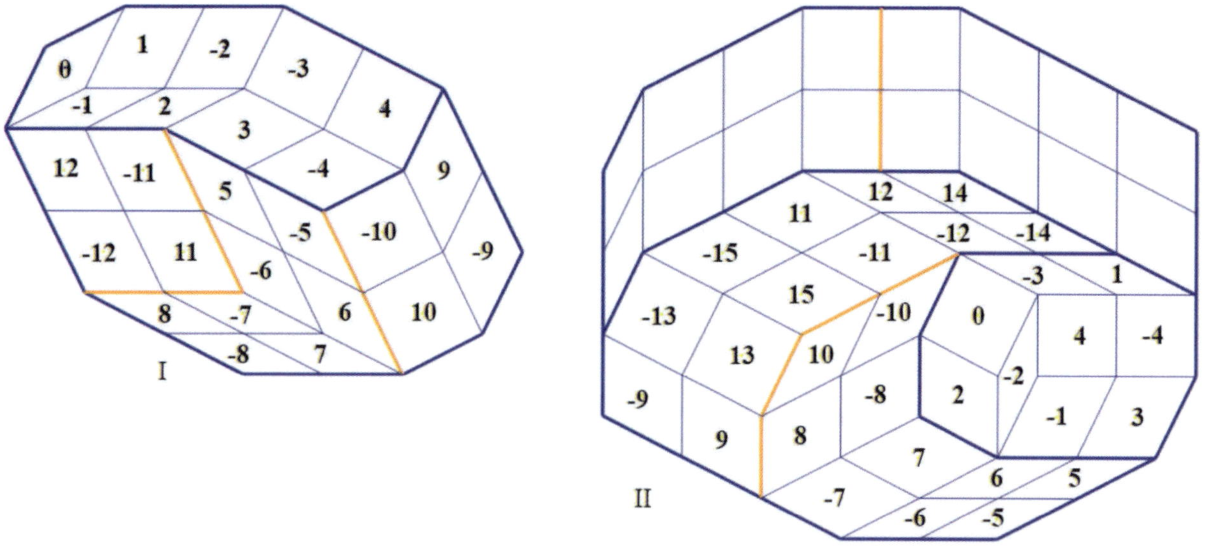

Figure 7.7

The notation *isoH(a,b,c)* will be used to denote an isohexagon which is isomorphic to a hexagon with semiperimeter consisting of three sides consisting of a, b and c basic edges. The notation D(a,b,c,d,e) will be used to similarly denote a semiregular decagon. In Figure 7.7(I) the extension: isoH(1,1,4)→D(1,1,2,2,2) is illustrated, and in Figure 7.7(II) the extensions: O(1,1,1,2)→D(1,1,2,2,3)→D(1,2,2,3,3) are illustrated. The numerical designations for the extending dominoes have been omitted since these can be easily supplied by the reader.

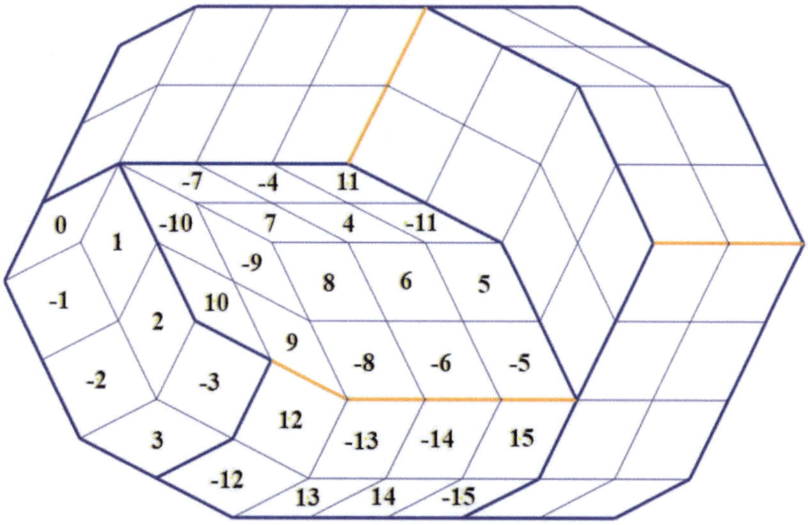

Figure 7.8

In Figure 7.8 the extensions: isoH(1,1,3)→D(1,1,3,2,2)→D(1,3,2,2,3)→D(1,5,2,2,3) are illustrated. In Figure 7.9 the extension: O(1,1,1,2)→D(1,1,2,1,6) is illustrated.

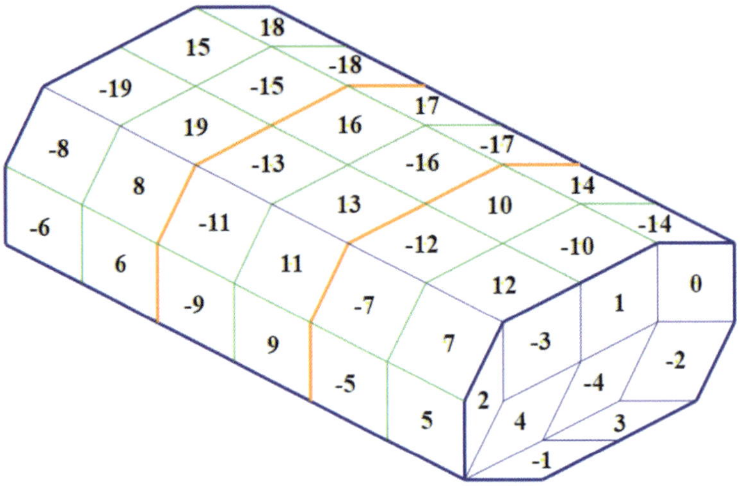

Figure 7.9

In Figure 7.10 the extension: O(1,1,2,2)→D(1,1,2,3,4) is illustrated. Note that the extension O(1,1,2,2)→D(1,1,2,2,4), (not illustrated), is easily obtained by a domino extension.

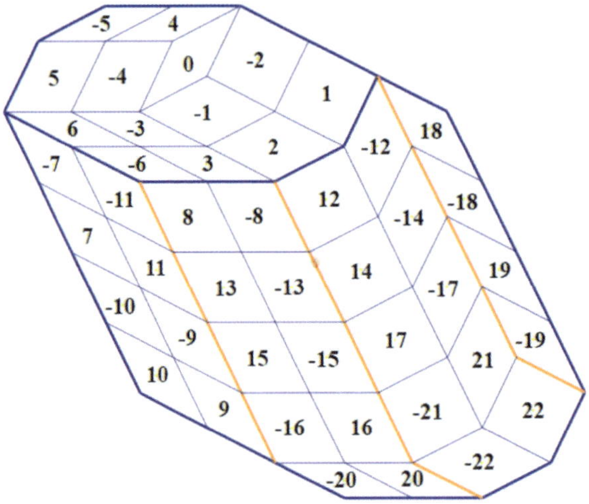

Figure 7.10

In Figure 7.11(III) a primary D(1,1,1,2,2) is constructed. In 7.11(II) a D(1,1,1,2,4) is illustrated which copies the method used in 7.11(III). In 7.11(I) the extension: isoO(1,1,1,6)→D(1,1,1,2,4) shows that D(1,1,1,2,4) is composite.

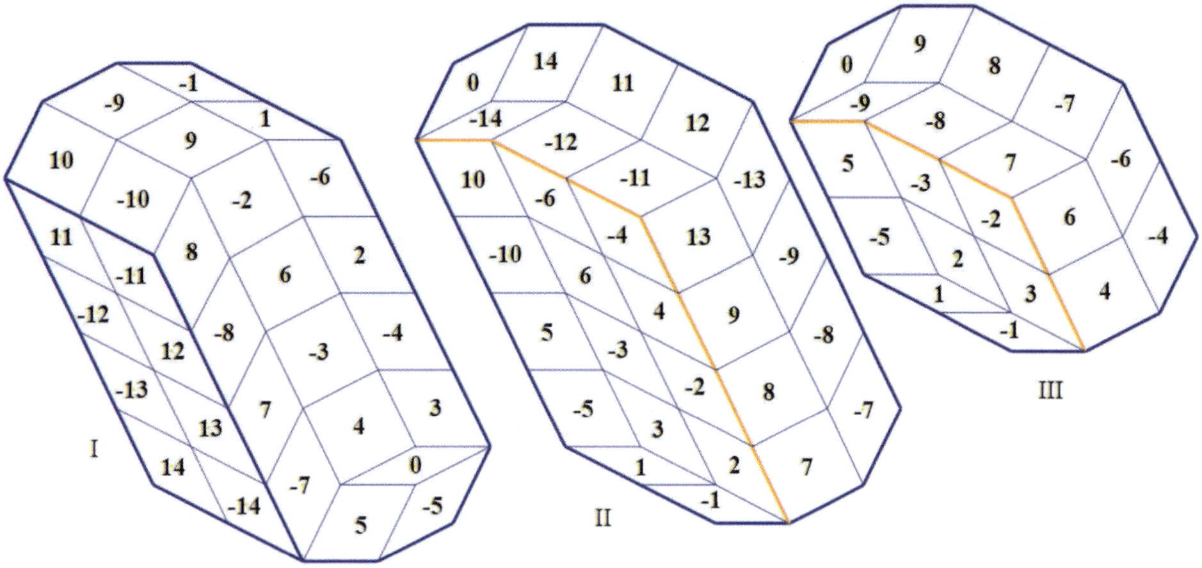

Figure 7.11

In Figure 7.12 we explore decagons of the form D(1,1,1,2,s), s even, a little further. Figure 7.12(I) illustrates the extension: O(1,1,1,4)→D(1,1,1,2,6), and in 7.12(II), the extension: isoO(1,1,1,10)→D(1,1,1,2,8) is shown. The octagons in these constructions are styled after the ones given in Section 6.3. It should be true that the D(1,1,1,2,s) octagons are all composite for s≥6, (s even). In fact, the following Conjecture may well hold true.

Conjecture 7.2 *Every odd order decagon has an integral solution; D(1,1,1,2,4) is primary and all other odd order integral decagons are composite.*

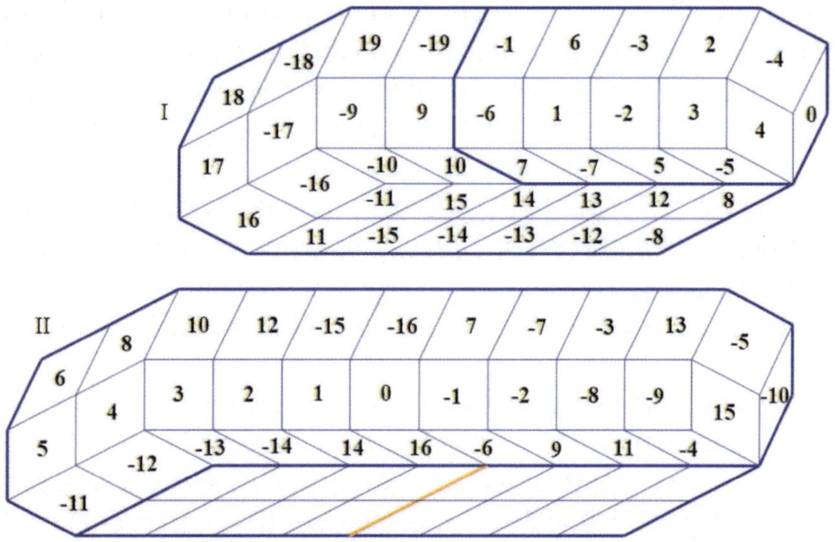

Figure 7.12

7.3 Dodecagons

In Chapter 8 a Latin dodecagon with an orthogonal mate will be used to construct a natural simple dodecagon with base Z(3)×Z(5) and a natural simple dodecagon with base Z(15). (See Figure 8.12). This is the smallest regular polygon with a natural solution. Beside the solution obtained in Figure 8.12(V), we have an additional solution in Figure 7.13 obtained by conversion of an integral dodecagon. Figure 7.13(II) represents a basic solution obtained by a direct conversion of 7.13(I) to base Z(15). The natural solution in 7.13(III) is obtained by adding 1 to each tile in 7.13(II). The indicated pair of polar hexagons can be shown by simple algebra to always have the same sum.

We will use the notation DD(a,b,c,d,e,f) to denote a dodecagon with parameters a,b,c,d,e,f. In Figure 7.14(I) the extension: O(1,1,1,2)→DD(1,1,1,2,2,2) is illustrated, and in Figure 7.14(II) the extension D(1,1,2,2,2)→DD(1,1,2,2,2,2) is achieved using dominoes. Thus even the smallest irregular integral dodecagons are composite.

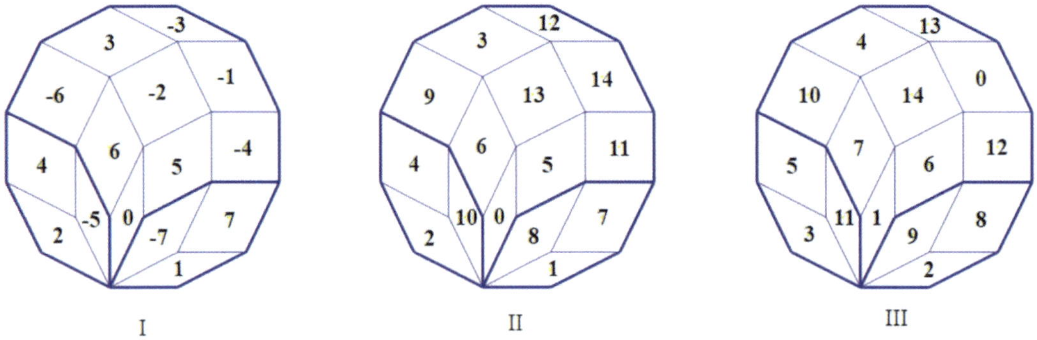

Figure 7.13

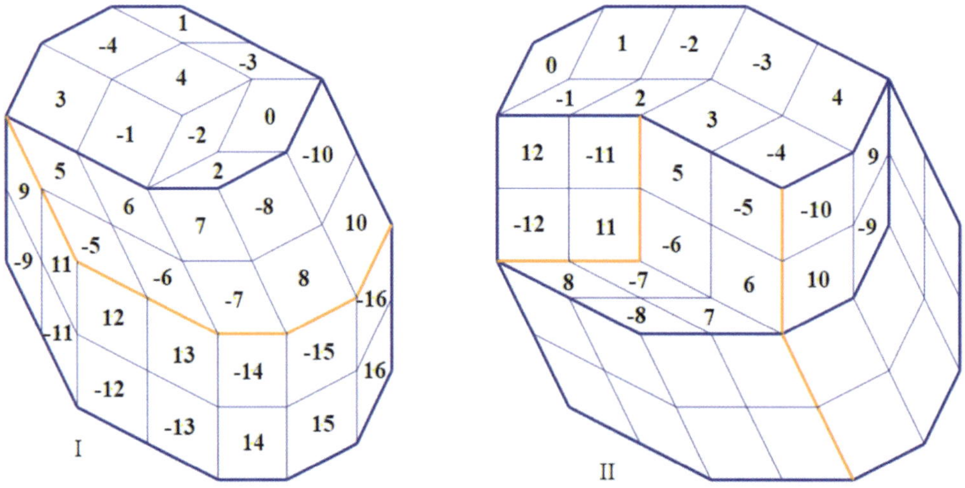

Figure 7.14

In Figure 7.15 the natural and integral dodecagons of Figure 7.13(I. III) have been retiled. These retilings yield some interesting observations regarding sums of subpolygons. Regarding the labeling of tiles in Figure 7.16 the following 6 equations are obtained.

1. L+G+D+F+N = 0
2. L+I+A+C+P = 0
3. J+A+B+M+N = 0
4. J+I+K+E+F = 0
5. C+B+K+G+H = 0
6. P+M+E+D+H = 0.

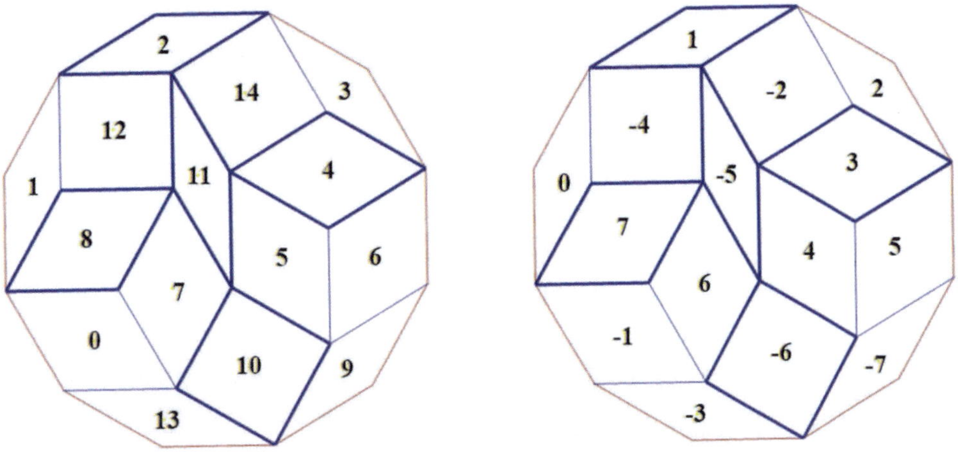

Figure 7.15

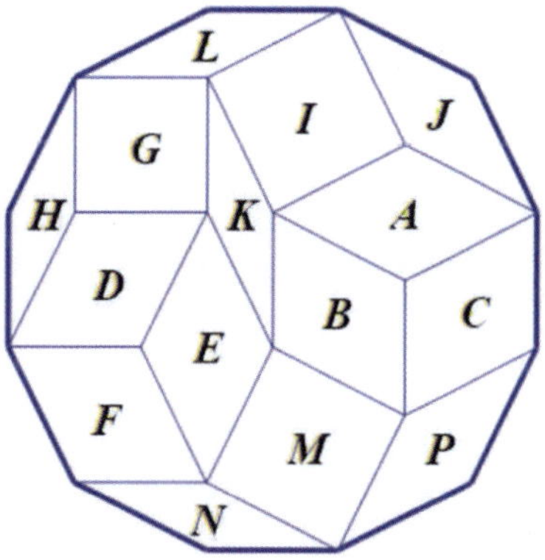

Figure 7.16

The letters in Figure 7.16 represent values from the base group Z*(15) and the equations represent stream sums, (all 0), for the integral version. Let Δ = (L+G+N) + (J+I+K) + (P+M+H). Then summing equations 1, 4, 6 we obtain

7. 2(E+D+F) + Δ = 0,

and summing equations 2, 3, 5 yields

8. 2(A+B+C) + Δ = 0.

From 7, 8 we obtain

9. A+B+C = E+D+F.

A similar analysis yields

10. H+G+D = J+I+A.

For the natural version change 0 to 5 in equations 1, ..., 6. The analysis proceeds similarly and equations 9, 10 are again obtained. More equivalences can be found in the examples in Figure 7.15, such as: L+K+E+B+M = 0 and L+K+A+D+M = 0 but these do not hold in general.

7.4 Basis

The polar tiling is useful for constructions and even for determining some sum equivalences between major parallelograms. But it is interesting that alternative tilings can reveal otherwise hidden properties in any magic assignment. The dodecagon above is one example; the H(2) flower in Figure 5.1(III) is another. In the H(2) flower three opposite pairs of hexagons can be found and it can be shown using stream equations that each member of a pair must have the same sum. A more impressive example occurs in considering the *O(2) flower* illustrated in Figure 7.17. The construction was obtained from the integral O(2) in Figure 6.3.

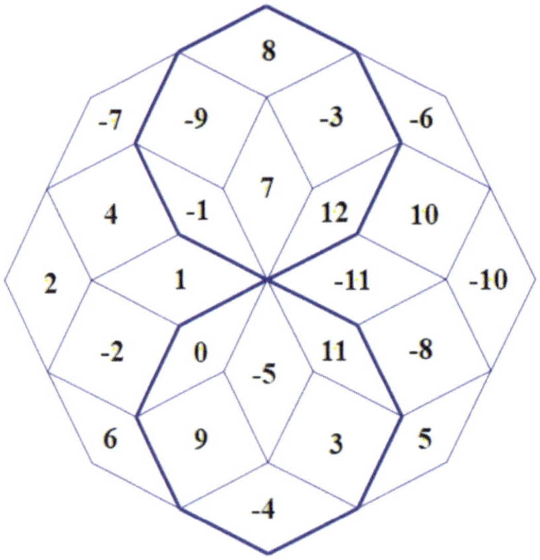

Figure 7.17

To show that the pair of octagons outlined in Figure 7.17 have the same sum it is not necessary to resort to a list of stream equations due to the following simple observation. Suppose P is a semiregular polygon with a parallelogram tiling and a magic solution over some abelian group G, and suppose P contains two tile disjoint polygons P_1, P_2. Require: (1) no stream in P passes through both P_1 and P_2, and (2) If T is any tile not in $P_1 \cup P_2$ then one of the streams through T passes through P_1 and the other passes through P_2. (3) the number of streams through P_1 is equal to the number of streams through P_2. **Note:** in case P has a 0-sum integral solution then (3) is **not** a requirement. If (1) and (2) are satisfied and P has 0 as common stream sum then $\{P_1, P_2)$ is called a *basis* for P, or if (1),(2) and (3) are satisfied then $\{P_1, P_2)$ is a basis for P.

Theorem 7.3 *Let P be a semiregular polygon with a parallelogram tiling and an integral or natural magic assignment over some abelian group G, and suppose* $\{P_1, P_2\}$ *is a basis for P. Then* $\Sigma P_1 = \Sigma P_2$.

Proof. Let Π denote the collection of tiles of P not in $P_1 \cup P_2$. Let S_1, S_2 denote respectively the sets of streams which pass through P_1, P_2. Then $\Sigma S_1 = \Sigma\Pi + 2\Sigma P_1$ and $\Sigma S_2 = \Sigma\Pi + 2\Sigma P_2$. Suppose P does not have a 0-sum integral solution. By requirement (3), S_1, S_2 contain the same number, say m, of streams. If s is the common stream sum of P then $\Sigma S_1 = ms = \Sigma S_2$. We then obtain $2\Sigma P_1 = 2\Sigma P_2$ and this implies $\Sigma P_1 = \Sigma P_2$.

In case P has a 0-sum integral solution, then $\Sigma S_1 = 0 = \Sigma S_2$ regardless of how many streams pass through P_1 or P_2, and again $\Sigma P_1 = \Sigma P_2$ is obtained. □

The outlined pair of octagons in Figure 7.17 form a basis for O(2), and so by Theorem 7.3, they must have the same sum. Two more such pairs can be found in the Figure. In Figure 7.16 the hexagons {(A,B,C), (D,E,F)} form a basis, and in the integral version in Figure 7.15, the polygon consisting of the single tile {A}, and the octagon {D,E,F,G,H,K} form a basis.

Magic squares require no retiling to find a basis. In Figure 7.18 the order 5 magic square of Figure 3.23 is illustrated along with several pairs of rectangles each of which form a basis.

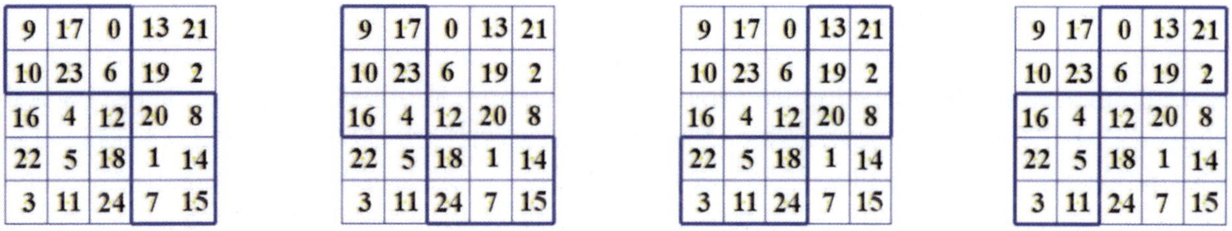

Figure 7.18

Hexagons on the other hand, do require special tilings in order to obtain a basis. The smallest Class 2 hexagon is H(1,1,2). Although this hexagon has no integral or natural solution, it does have a basis and it serves as a generator for tilings of Class 2 hexagons which admit a basis.

Theorem 7.4 *Every Class 2 hexagon has a basis.*

Proof. The basis for H(1,1,2) consists of just two parallelograms as indicated at the top of Figure 7.19 . Every Class 2 hexagon is obtained by starting with H(1,1,2) and progressively appending two basic edges either horizontally or at 60° or 120° degree angles. Whichever direction is chosen, *one* basic edge is added in that direction to the basis polygons. The result is a basis for the enlarged polygon. □

In Figure 7.19 the construction indicated in Theorem 7.4 is illustrated.

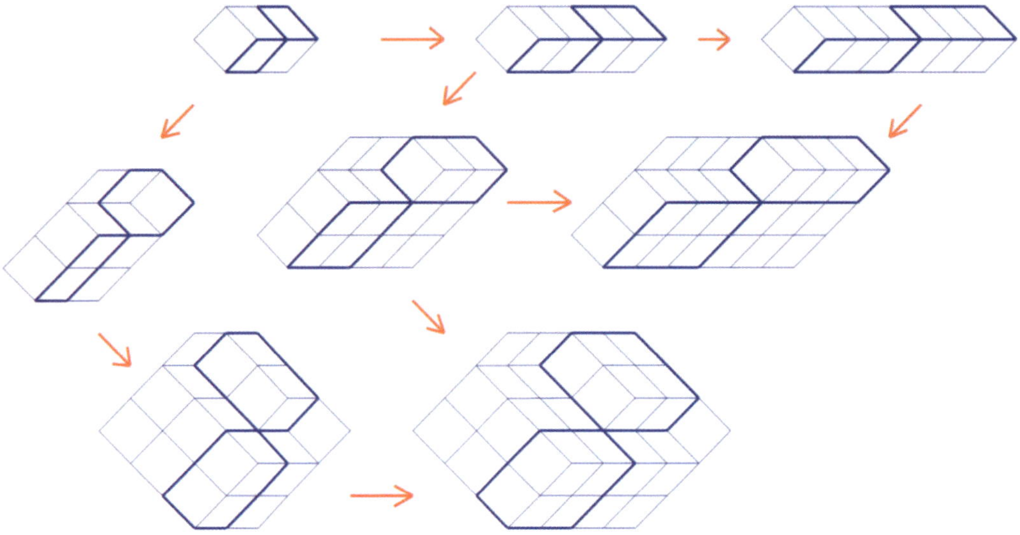

Figure 7.19

Owing to the following simple result, no Class 1 hexagon with a natural solution can have a basis.

Theorem 7.5 *If a semiregular polygon with a natural solution has an odd number of basic edges in its semiperimeter, then it can have no basis.*

Proof. Let P be a semiregular polygon with a basis $\{P_1,P_2\}$. Then the collection S of streams in P has a partition into two sets S_1, S_2, where S_1 denotes the collection of streams through P_1, and S_2 denotes the collection of streams through P_2. Since P_1,P_2 have the same semiperimeter, $|S_1| = |S_2|$. Then $|S| = |S_1| + |S_2| = 2|S_1|$ is even and thus the semiperimeter of P must contain an even number of edges. □

The remaining class of hexagons to consider are those of the form H(p,q,r) in which each of p,q,r is even. Many of these hexagons have no integral solution as can be determined by Theorem 5.23. The procedure for determining a basis is the same as for Class 2 hexagons except that the regular hexagon H(2) is taken as generator. Figure 7.20 illustrates the genesis of a basis for H(4) and H(2,2,8). The intermediate hexagons in 7.20(II, III, IV) have no solution by Theorem 5.23.

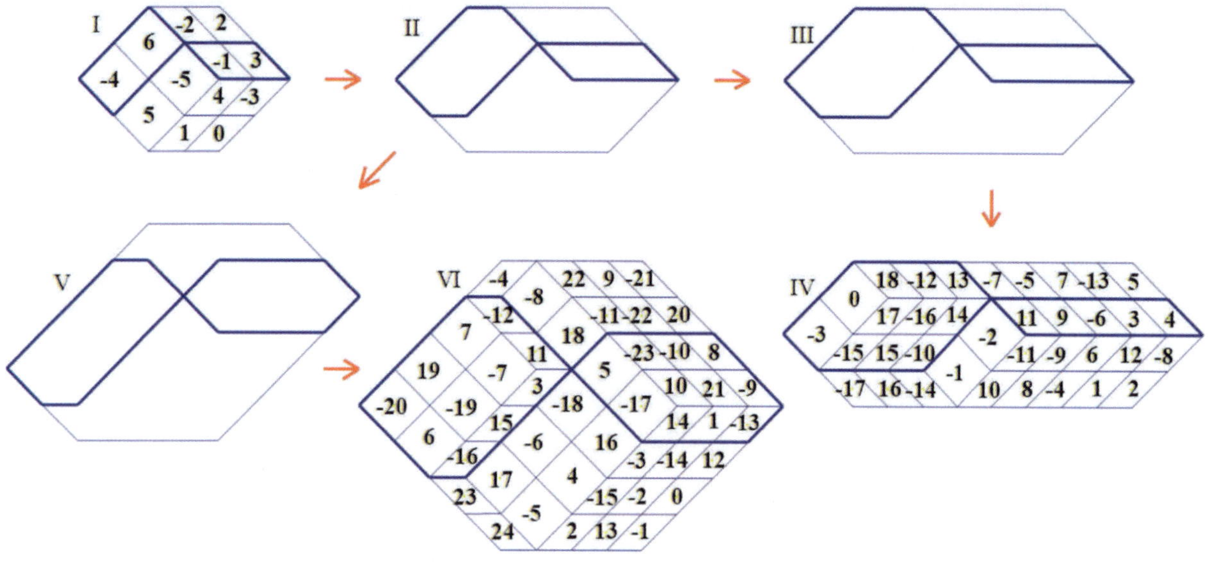

Figure 7.20

It is clear that the method illustrated in Figures 7.19 and 7.20 can be generalized to construct a basis for any semiregular polygon with semiperimeter of even length. It is only necessary to find a basis for the generating polygons, (H(1,1,2) and H(2) for hexagons). For octagons there are three generators: O(1,1,1,1), O(1,1,2,2) and O(2). These are illustrated, each with a basis, in Figure 7.21.

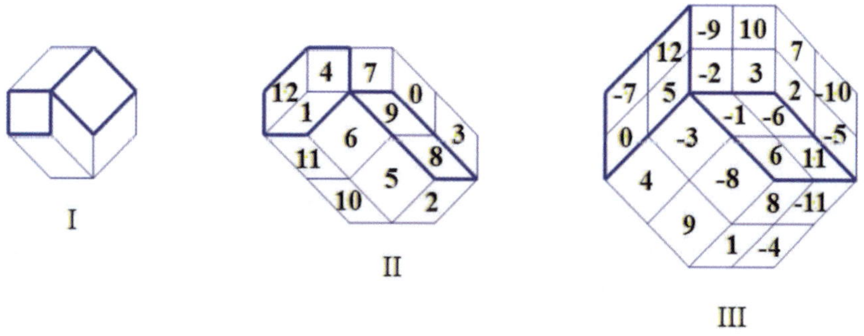

Figure 7.21

Notice that in 7.21(I) and 7.21(III) there are three pairs which constitute a basis. Compare 7.21(III) with Figure 7.17 in which three different basis pairs can also be found. Note also that the result of Theorem 6.4 which was proved using stream equations, follows immediately from diagram 7.21(III) and Theorem 7.3.

For decagons the generators are: D(1,1,2,2,2), D(1,1,1,1,2) and D(2) = D(2,2,2,2,2). Each one of these with a corresponding basis is illustrated in Figure 7.22.

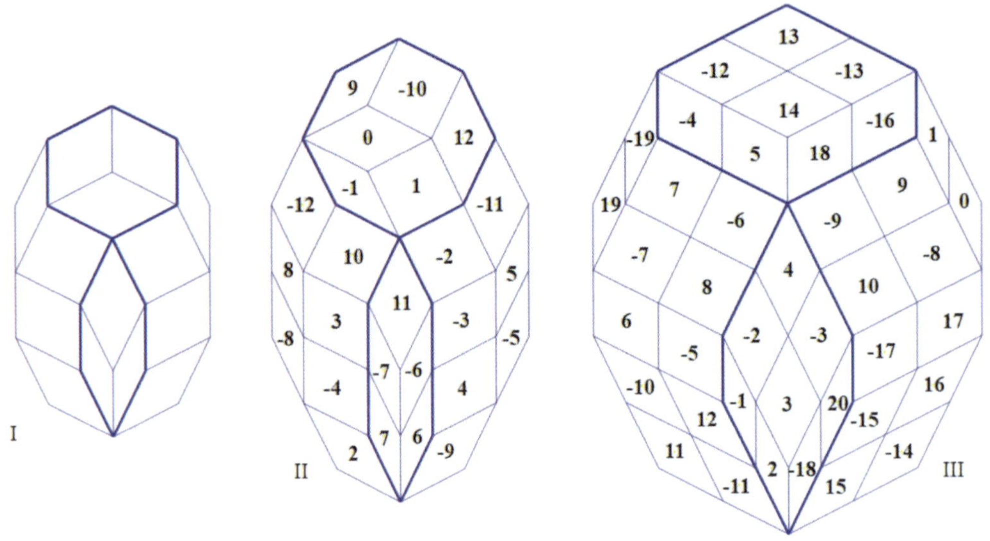

Figure 7.22

For 7.22(I) there is no magic solution since n/s = 14/6 is not an integer. (Here n is the order and s the semiperimeter.) The construction in 7.22(III) has the special property that the pair of basis polygons as well as the pair of complimentary polygons all have sum 5. For any semiregular polygon with an even length semiperimeter there are in fact many ways to construct a basis.

Theorem 7.6 *Let P be a semiregular polygon with a semiperimeter of length s = 2m. Then a basis for P can be constructed consisting of two semiregular polygons which share one common point and which have semiperimeter of length m.*

Proof. Choose any m consecutive basic edges $\{\mathbf{v}_1, \ldots, \mathbf{v}_m\}$ on the boundary of P and construct a semiregular polygon P_1 in the interior of P using the chosen edges as semiperimeter as in Figure 7.23. Order the m edges clockwise and let $\mathbf{v}_0$ denote the boundary edge of P which precedes $\mathbf{v}_1$ and let $\mathbf{v}_{m+1}$ denote the boundary edge of P which directly follows $\mathbf{v}_m$. Let q be a point on the boundary of P_1 which lies in the interior of P and construct a second interior semiregular polygon

$P_2 = (q, \mathbf{v}_{m+1,\ldots,} \mathbf{v}_{2m}, -\mathbf{v}_{m+1,\ldots,} -\mathbf{v}_{2m,} q)$.

The polygons P_1, P_2 are semiregular, each with semiperimeter consisting of m basic edges. Let q′ denote the point on the boundary of P_1 opposite q, and let q″ denote the point on the boundary of P_2 opposite q. Then q′, q″ are opposite points on the boundary of P, and as a consequence, every stream in P must pass through either P_1 or P_2. The parallelogram tiling of P is uniquely determined by the positioning of P_1, P_2, since every edge in the tiling must be parallel either to an edge of P_1 or an edge of P_2. Also P has 2m streams, m of which pass through P_1 and m of which pass through P_2. No stream therefore can pass through both P_1 and P_2, and we can conclude that $\{P_1, P_2\}$ is a basis for P. □

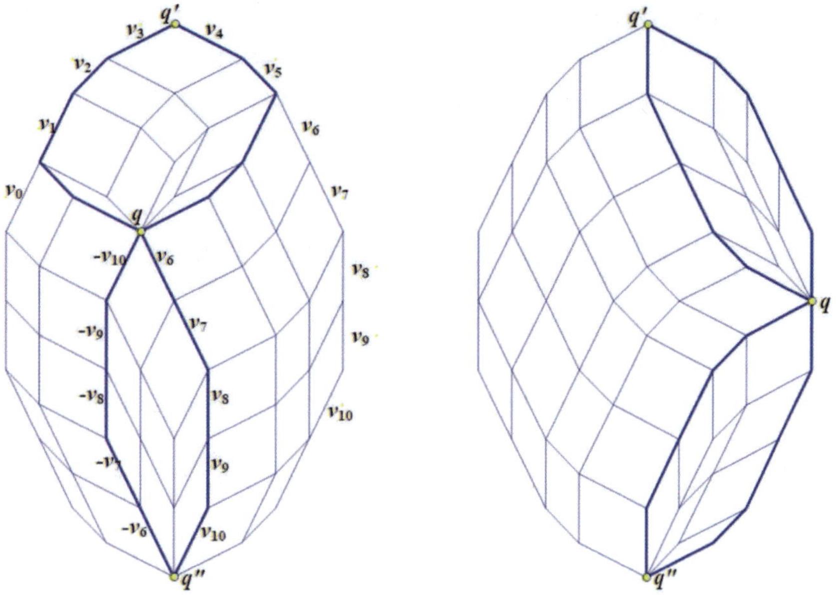

Figure 7.23

In the Theorem the contact point q is chosen interior to P. We call q the *eye* of the basis. This need not always be the case as illustrated at right in Figure 7.23. Neither is it essential that P_1 have m edges on the boundary of P as illustrated in the two examples in Figure 7.24.

If our attention is confined to 0-sum integral polygons. Requirement (3) is not needed, so basis pairs can be constructed with different semiperimeters. To construct basis pairs in semiregular polygons it is only necessary to determine a basis for the simple polygon in the class. The same kind of extensions used to analyze hexagons can be used for semiregular polygons with more sides, but it is only necessary to extend by one unit. As an example, consider the genesis of basis pairs for octagons and decagons illustrated in Figure 7.24. We have left unlabeled those polygons with no solutions. Note that, in the case of octagons, the simple generating octagon has 3 basis pairs which form a dissection. This property is preserved under any unit extension, thus every semiregular octagon has a tiling with such a dissection.

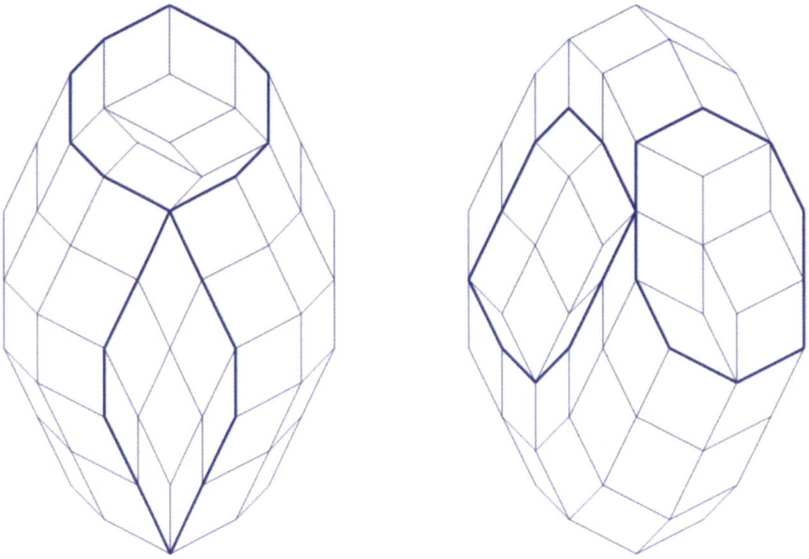

Figure 7.24

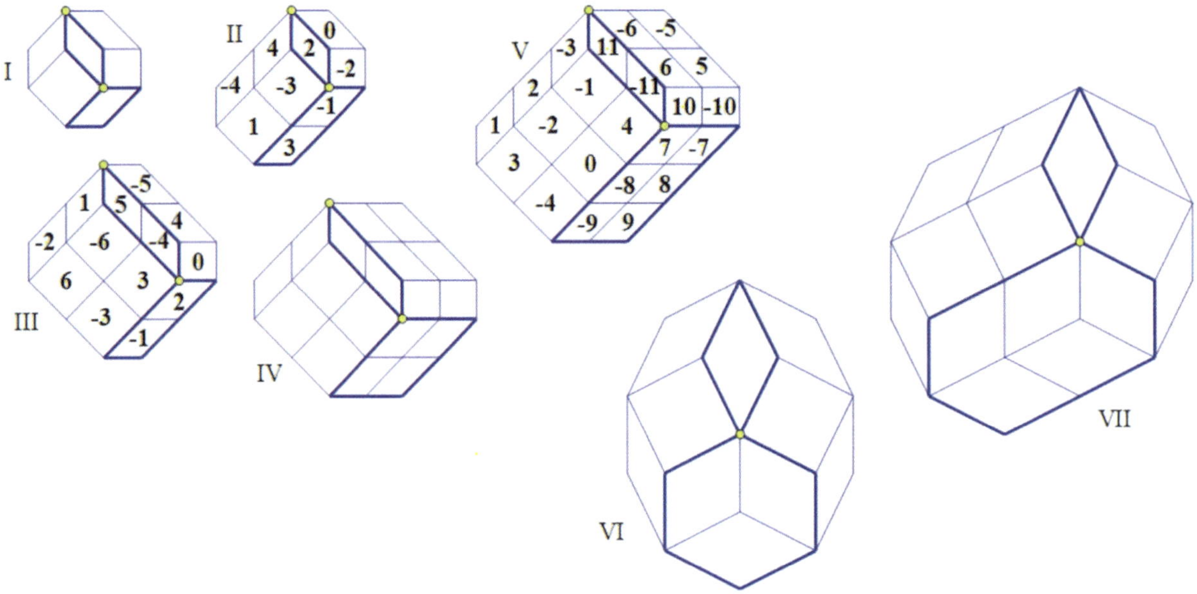

Figure 7.25

Note also that the decagon in 7.25(VI) has three overlapping basis pairs all with the same eye. In 7.25(VII) it can be seen that the unit extension also contains three basis pairs with a common eye.

Using the unit extension idea, the following Theorem has an easy proof.

Theorem 7.7 *Every semiregular polygon except the simple hexagon has a basis.*

Proof. The simple hexagon has no basis, but all subsequent semiregular hexagons can be obtained from H(1,1,2) by unit extensions. Since H(1,1,2) has a basis, (see Figure 7.19), it follows that all semiregular hexagons have a basis. For semiregular polygons with more than 6 sides, it suffices to exhibit a basis in the simple member of the class. For the simple octagon and simple decagon basis pairs are illustrated in Figure 7.25. In Figure 7.26(I) a general procedure for constructing a basis in simple polygons is indicated. In fact, the usual polar tiling suffices: any boundary point is selected and then covered with a single tile. This tile and the region lying between the two boundary crescents form a basis. One or more basis pairs are also formed by polar polygons with the pole as common eye. The *capacity* of a tiling is the number of basis pairs that it contains. Thus every polar tiling, with the exception of the simple hexagon, has two eyes and capacity at least 2.

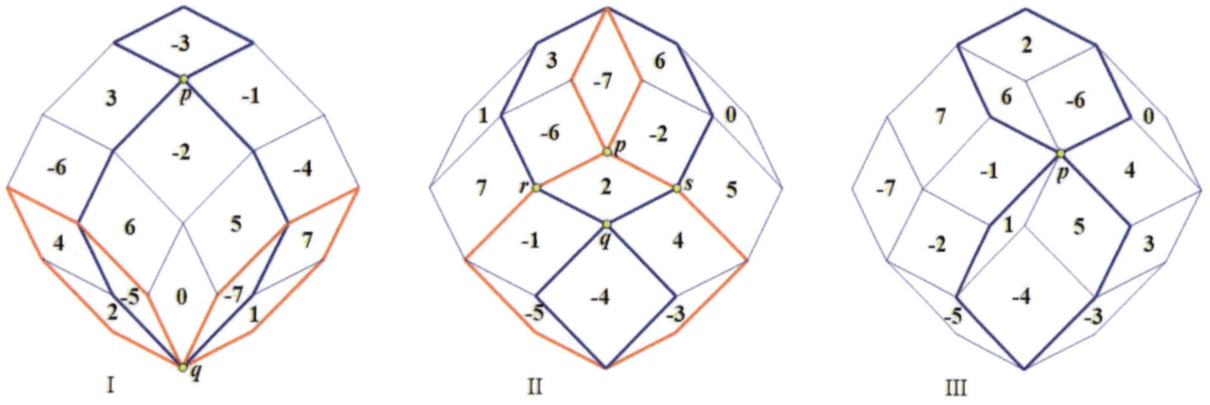

Figure 7.26

Note that besides the two basis pairs indicated in 7.26(I), there are two more:

{(0,5,-4,7,-7,1), (2)}   {(-6,6,0,4,-5,2), (1)}.

In the dodecagon in 7.26(II) there are 4 eyes labeled p,q,r,s, and besides the two basis pairs indicated, there are:

{(-7,-2,6), (4,-4,-3)}   {(3,-7,-6), (-1,-4,-5)}.

The dodecagon in 7.26(III) has one eye and a capacity of 6. The 6 basis pairs are:

{(2,6,-6), (1,5,-4)}   {(7,6,-1), (4,5,3)}

{(6,7,-7,1,-1,-2), (4)}   {(1,3,4,5,-3,-4), (6)}

{(0,4,-6), (1,-1,-2)}   {(1,-1,-2,5,-5,-4), (-6)}.

We define the *capacity of a semiregular polygon P* to be the maximum capacity among all parallelogram tilings of P. Thus the capacity of the simple dodecagon is larger than or equal to 6, and is most likely 6. Let P be a semiregular polygon. Some interesting pure tiling questions are (1) What is the capacity of P ? (2) Does a maximum capacity tiling of P always have one eye ? (3) How many eyes can a tiling of P have ?

7.5 Tetradecagons

For a tetradecagon we will use the notation F(a,b,c,d,e,f,g). In Figure 7.27 a simple integral tetradecagon is illustrated at left, and the extension D(1,1,1,2,2)→F(1,1,1,2,2,2,2) is illustrated at right. In Figure 7.28 the extension: O(1,1,2,2)→F(1,1,2,2,2,2,2) is illustrated. Finally, in Figure 7.29 a solution is given for the possibly primary tetradecagon F(1,1,1,1,1,1,2).

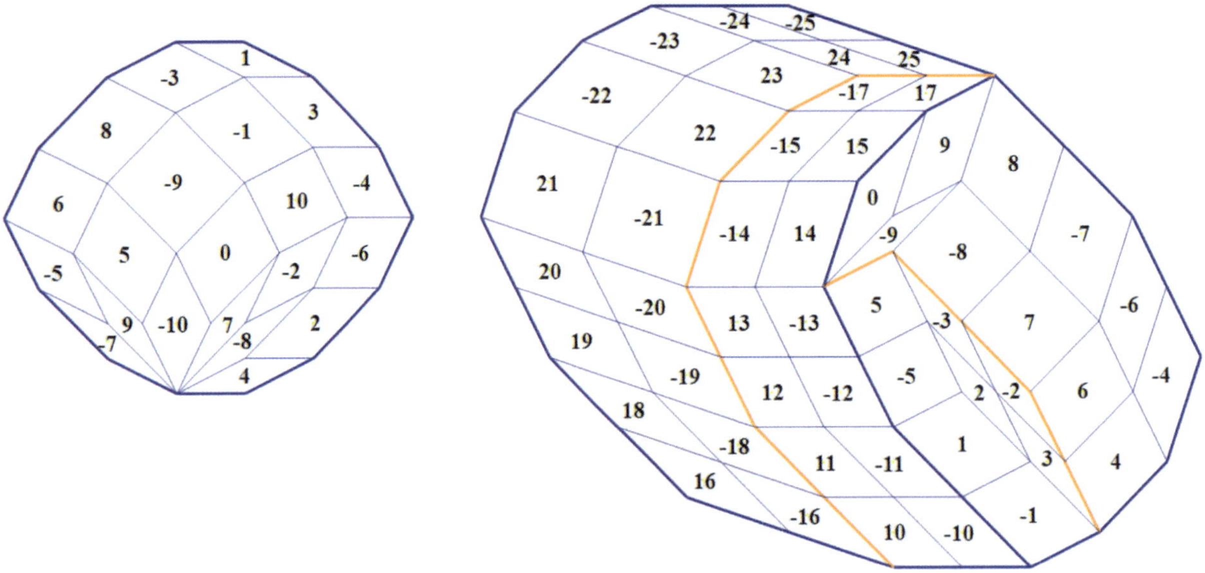

Figure 7.27

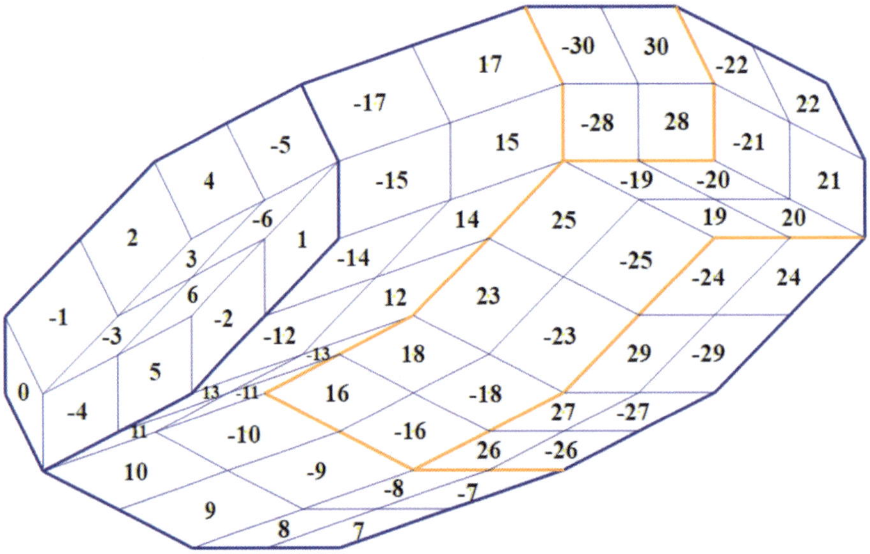

Figure 7.28

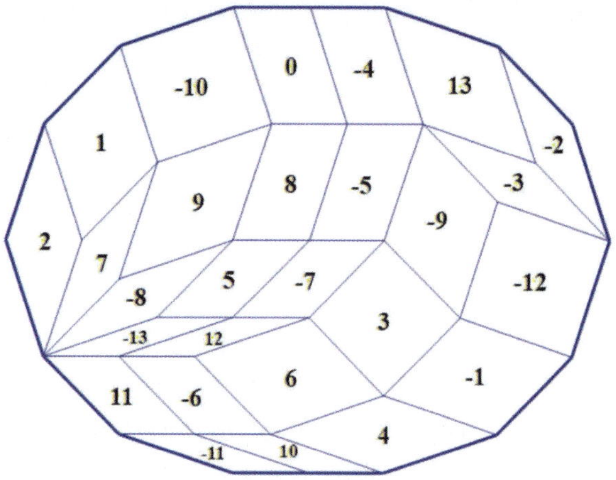

Figure 7.29

In Figure 7.30 a one-eyed tetradecagon derived from the simple integral tetradecagon in Figure 7.27 is illustrated. T has 7 basis pairs:

{(8,-5,0), (3,4,5,-1,-2,-6)} {(0,2,-4), (-1,-2,-3,5,6,-7)}

{(2,3,-9), (-2,-3,-5,6,10,-10)} {(3,5,-6), (0,8,-3,-5,10,-8)}

{(5,-1,-2), (0,1,2,-5,-4,8)} {(6,-2,-3), (0,2,3,9,-9,-4)}

{(10,-3,-5), (2,3,5,-6,-9,7)}

This construction indicates strongly that the capacity of the simple tetradecagon is 7.

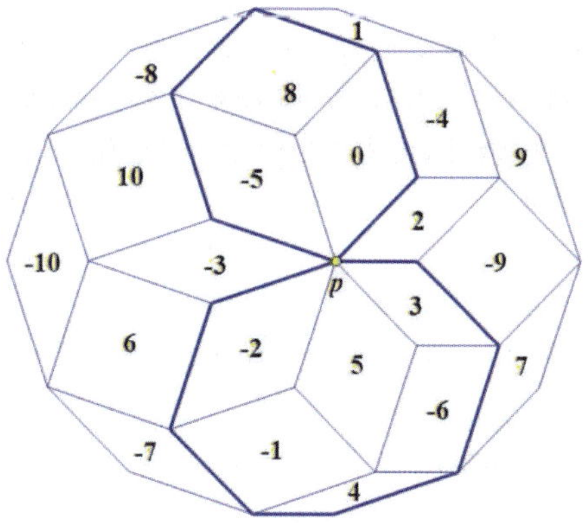

Figure 7.30

In Figure 7.31 the 14-gon in 7.30 is redesigned in order to easily see the symmetries. Also included is an 18-gon featuring the same symmetric construction.

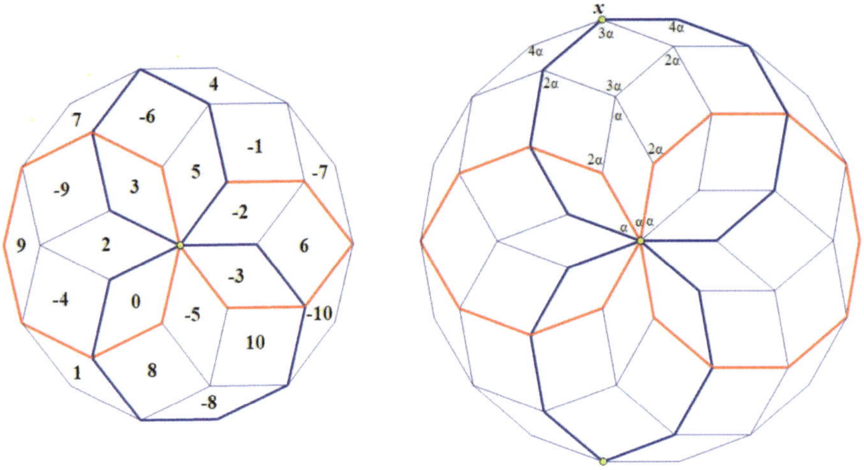

Figure 7.31

Let α be an angle such that mα = 360∘ for an integer m≥5, and let S(α) denote the star polygon formed by m α-rhombs surrounding a single central point p. Let F(α) denote the smallest regular polygon which contains S(α) at its center and is tiled entirely with rhombs employing only one edge length. We call F(α) the *flower generated by α.* In Figure 7.31 the flowers F(360∘/7), F(40∘) are illustrated. From the rotational symmetry, it can be easily seen that the capacity of F(α) is equal to m, *provided m is odd.*

In case m is even a different formula for the capacity is obtained. Consider the examples in Figure 7.32.

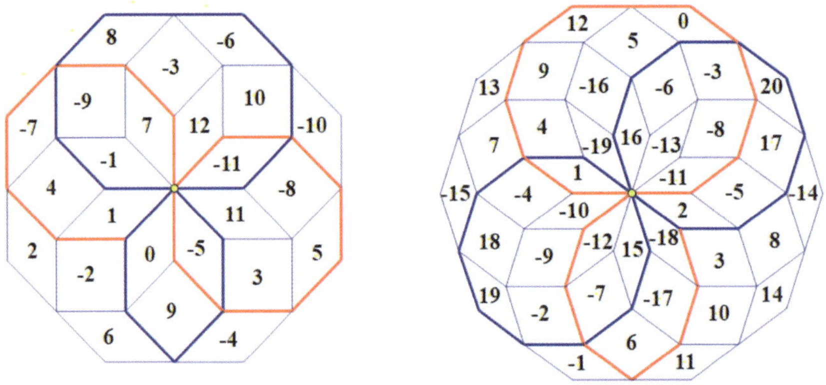

Figure 7.32

In this case, if S(α) consists of m α- rhombs, then F(α) is an m-gon with 2 basic edges per side. Two types of basis pairs emerge, one consisting of two simple semiregular m-gons and the other consisting of an (m-2)-gon and an (m+2)-gon. There are m/2 of the first type and m of the second type yielding a capacity of 3m/2. These formulas for capacity likely give the capacity of the corresponding regular polygon.

7.6 Flower Polygons

The following more general definition of a *flower polygon* is in order. Let V be a set of basic vectors positioned so that all tail points coincide at a *center point* x. Thus the vectors in V form a star. Require that the angles made at x between consecutive vectors are all less than 180°. Complete this configuration to a star polygon S consisting of rhombs with common point x. Extend this configuration by completing to a rhomb each pair of adjacent boundary edges of S. Repeat this process until a convex polygon Γ(S) is achieved. We call Γ(S) the *flower polygon generated by S.* In Figure 7.33 two examples of flower polygons are illustrated.

The basic properties of flower polygons are determined by the set V of generating basic vectors: (1) If |V| = m, then Γ(S) has the m basic edges determined by V in its boundary; (2) Γ(S) is semiregular; (3) Let k denote the number of pairs of the form {**v**,-**v**} in V. Then the number of sides of Γ(S) is given by the formula 2(m-k); (4) If V has no pairs of the form {**v**,-**v**} and x is not a boundary point or a point adjacent to a boundary point, then the capacity of Γ(S) is m; (5) If V contains k pairs of the form {**v**,-**v**}, then the capacity of Γ(S) is m+k, provided again that x is not a boundary point or a point adjacent to a boundary point.

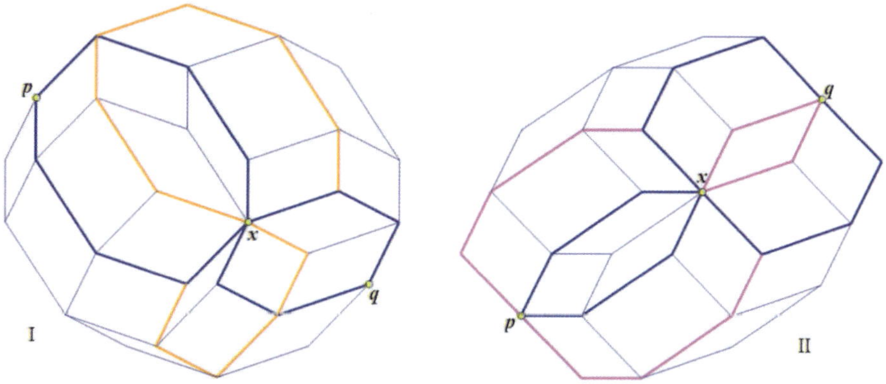

Figure 7.33

In 7.33(I) V consists of 7 vectors no two of which have opposite directions and Γ(S) is a semiregular 14-gon. Each basis pair consists of two semiregular polygons which touch at x. The points p, q which lie in these polygons and are opposite to x, are opposite vertices of Γ(S) . To each pair of opposite points of Γ(S) there corresponds a basis pair all with x as eye. The capacity of Γ(S) is thus m = 7.

The generating vector set V in 7.33(II) contains k = 2 pairs of the form {**v**, -**v**}. These pairs correspond to sides of Γ(S) which have two basic edges. Thus every such pair reduces the number of sides by 1 yielding a semiregular polygon with 2(m-k) = 10 sides. Every pair of opposite corner vertices of Γ(S) correspond to a basis pair with x as eye; this accounts for m-k distinct basis pairs. In addition, there are two basis pairs which correspond to opposite middle vertices of the sides with length 2. One such couple of basis pairs is illustrated in 7.33(II). Thus we obtain a total of 2k additional basis pairs, yielding a grand total of m+k distinct basis pairs in Γ(S).

The following Conjecture is very likely true.

Conjecture 7.8 *Let V be a set of m vectors which contains no pairs of the form {v,-v}. Then among all simple semiregular 2m-gons, the flower polygon generated by V has the greatest capacity. If P is a semiregular polygon with k sides of length 2 and m-k sides consisting of a single basic edge. Then the tiling of P which achieves the maximum capacity is obtained by the flower generated by a set of m vectors containing k pairs of the form {v, -v}.*

Suppose P is a simple regular polygon with semiperimeter m. Since every basis pair joins two opposite points on the boundary of P and the flower tiling yields capacity m, in order to exceed this capacity, it must be the case that some pair of opposite points of P are joined by more than one basis pair. This can be accomplished as in the example in Figure 7.34. In this 18-gon the opposite points p, q correspond to no less than 4 basis pairs only two of which are indicated. The points x,y,z,w are eyes. These are the only basis pairs so we fall far short of the 9 basis pairs which exist in a flower tiling.

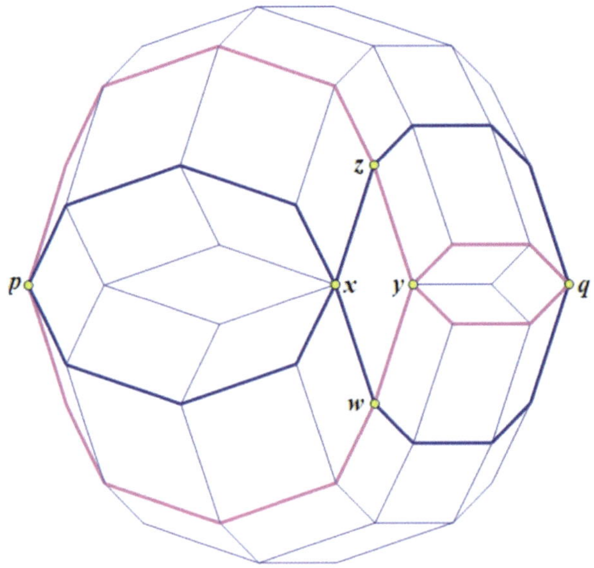

Figure 7.34

Another tiling of interest is the *bipolar tiling.* This tiling consists of a set of four crescents: two outer ones which envelop two inner ones. Let Π be an integral regular polygon with semiperimeter s and boundary point x. The outer crescents run along the boundary of Π extending from x in opposite directions as in Figure 7.35. These crescents enclose a semiregular polygon Δ. Let y denote the point of Δ opposite x and tile the boundary of Δ with two inner crescents extending from y in opposite directions. The inner crescents enclose, in turn, a regular polygon W as indicated in 7.35(I).

Let A, B denote the set of tiles which constitute the outer crescents and let

$$(A_1, \ldots, A_{s-2}, Q), (B_1, \ldots, B_{s-2}, Q)$$

denote respectively the integer values assigned to the sequence of tiles belonging to A, B. The value Q is assigned to the common endtile of A, B. Let C, D denote the set of tiles belonging to the inner crescents and let

$$(C_1, \ldots, C_{s-4}, R), (D_1, \ldots, D_{s-4,} R)$$

denote respectively the sequence of integer values assigned to the tiles belonging to C, D. The common endtile has value R and contains x as a vertex. Let Γ_C denote the regular polygon formed by the union of W and the tiles $\{C_1, \ldots, C_{s-4}\}$ and let Γ_D denote the regular polygon formed by the union of W and $\{D_1, \ldots, D_{s-4}\}$. Then we have the following result.

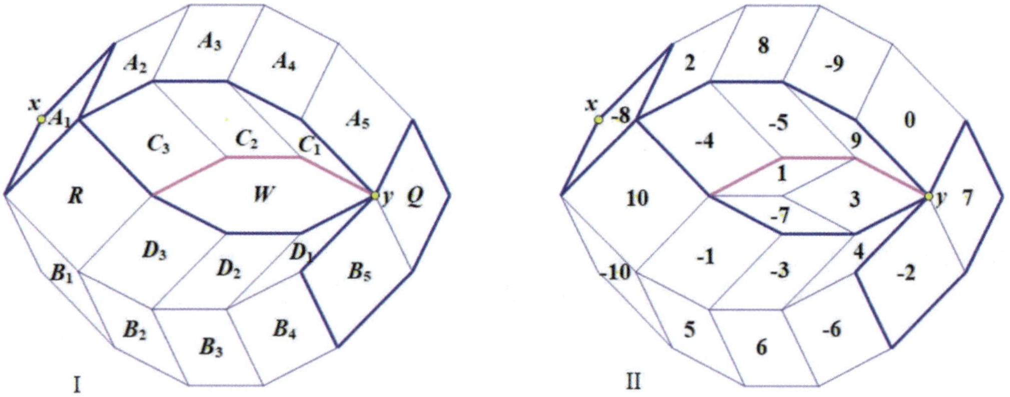

Figure 7.35

Theorem 7.9 *(i)* $A_1 + Q + B_{s-2} = \Sigma\Gamma_C$ *and (ii)* $B_1 + Q + A_{s-2} = \Sigma\Gamma_D$ *.*

Proof. From one of the streams through A_1 we have

$$(1)\ A_1 + R + (D_1 + \ldots + D_{s-4}) + B_{s-2} = 0,$$

and thus

$$(2)\ A_1 + Q + B_{s-2} = Q - R - (D_1 + \ldots + D_{s-4})\ .$$

Since $\{\Delta, Q\}$ form a basis for Π, we have $\Sigma\Delta = Q$ by Theorem 7.3. Thus

$$(3)\ R + \Sigma W + (C_1 + \ldots + C_{s-4}\} + (D_1 + \ldots + D_{s-4}) = Q.$$

Using (3) to substitute for Q in the right hand side of (2) yields

$$(4)\ A_1 + Q + B_{s-2} = \Sigma W + (C_1 + \ldots + C_{s-4}\} = \Sigma\Gamma_C\ .$$

From symmetry we also obtain $B_1 + Q + A_{s-2} = \Sigma\Gamma_D$. □

CHAPTER 8

LATIN POLYGONS

8.1 Definitions and Examples

Let P be a semiregular polygon tiled with parallelograms and let S be a set with r elements. P is *Latin over S* in case the tiles of P are labeled with the elements from S so that no two elements of S lie in the same tile stream. Thus each stream of P contains the full set of elements from S and consequently every stream of P must have the same length. The insertion of an element x from S in a tile T renders both streams intersecting at T unavailable for further insertions of x. As a consequence P must have an even number s of tile streams and the element x must occur in exactly s/2 positions. We call s/2 the *replication number for P.* If the elements of S constitute an abelian group G or a subset of G then all tile streams will have the same sum, and thus P is magic over $S^{(s/2)}$.

If the boundary of P is a regular polygon then we call P a *Latin polygon.* The construction of Latin polygons is the main concern of this Chapter. Latin squares of any order n are easily constructible; just take the Cayley Table for an abelian group of order n. But the construction of Latin hexagons, Latin octagons, Latin decagons, etc, can be quite challenging. In Figure 8.1(I) a Latin hexagon is illustrated, and in Figures 8.1(II,III) a simple Latin octagon and Latin dodecagon are given.

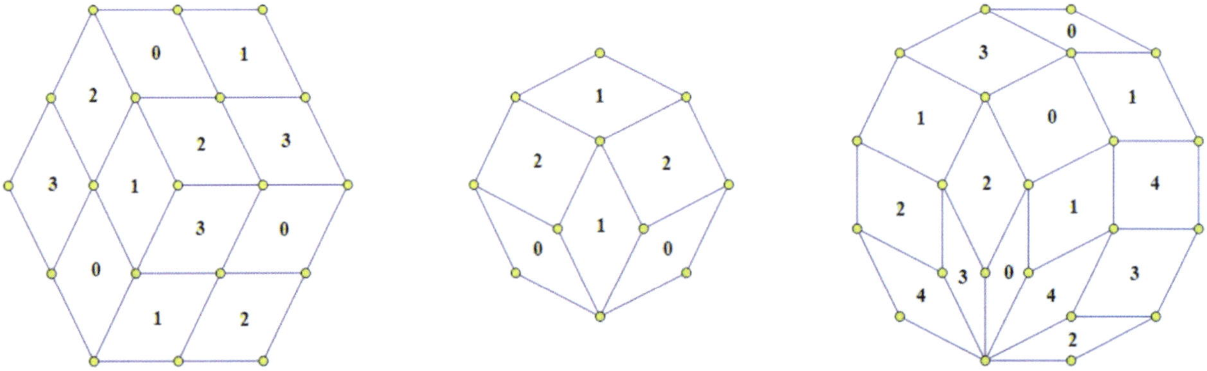

I II III

Figure 8.1

8.2 Latin Substitution

The Latin polygons in Figure 8.1 may be regarded as *primary* since no smaller Latin polygon in the same class exists. A Latin hexagon cannot have an odd number of streams and thus two edges per side is the minimum. There is an easy way to create larger Latin polygons in a given class provided a primary member in the class has been constructed. The method involves substituting a Latin square for each tile in a Latin polygon.

Theorem 8.1: *Let P be a Latin polygon over Z(n), and let k≥2 be an integer. Then a Latin polygon P' over Z(kn) can be constructed.*

Proof. Let Π= {S_0, S_1, …, S_{n-1}} be a partition of Z(kn) into n sets each with k elements, and for each i∈Z(n) let L(S_i) denote a latin square over S_i. Replace each tile in P with label i by L(S_i). The result will be a Latin polygon *P'* over Z(kn) with the same number of sides as P, but with the edgecount per side multiplied by k. □

The Latin octagon in Figure 8.2 was obtained from the simple Latin octagon in Figure 8.1(II) by the method just described.

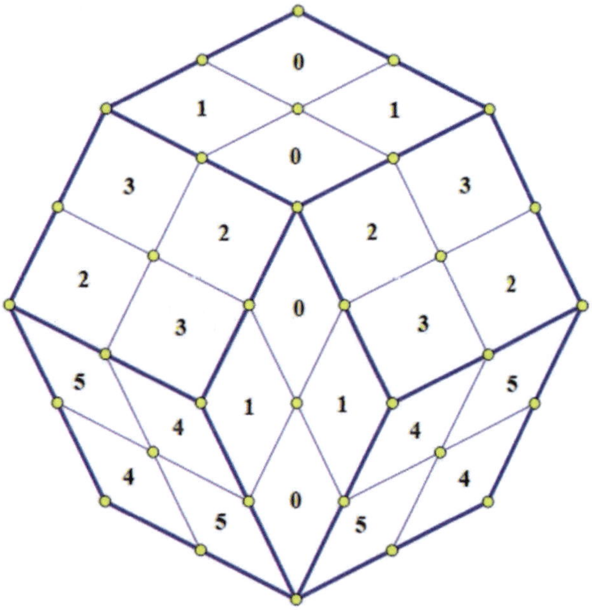

Figure 8.2

8.3 More Primary Latin Polygons

In Figures 8.3, 8.4 and 8.5 three more primary Latin polygons are illustrated.

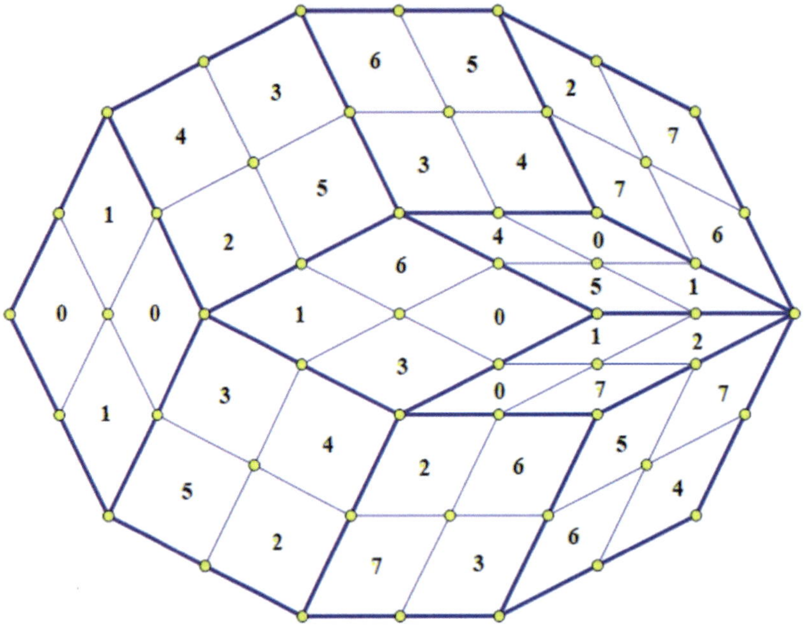

Figure 8.3

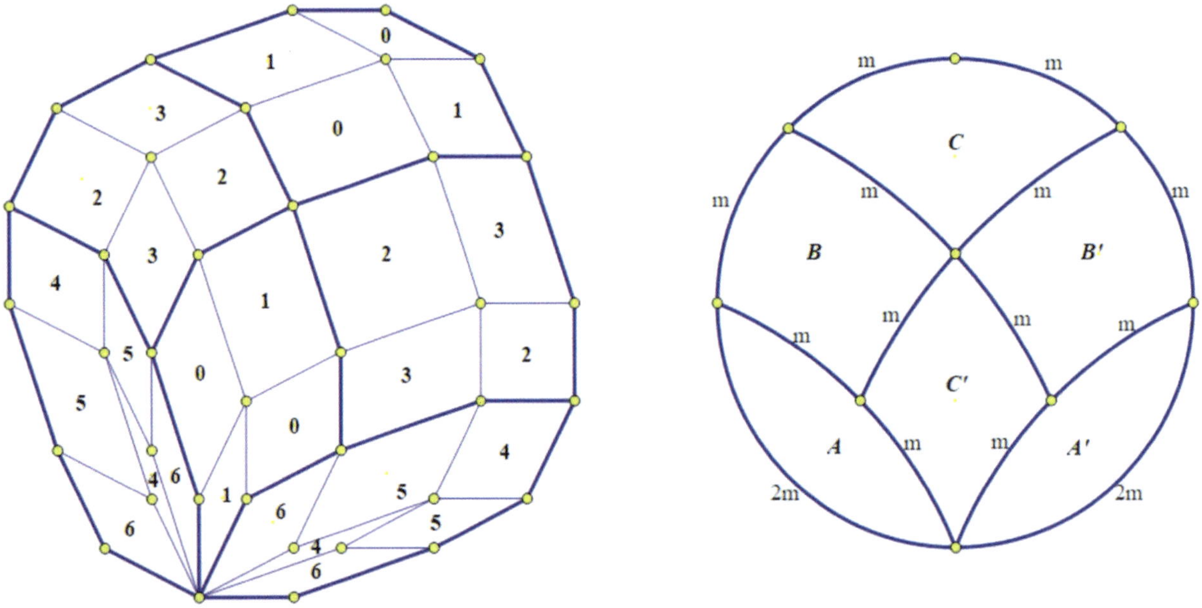

Figure 8.4

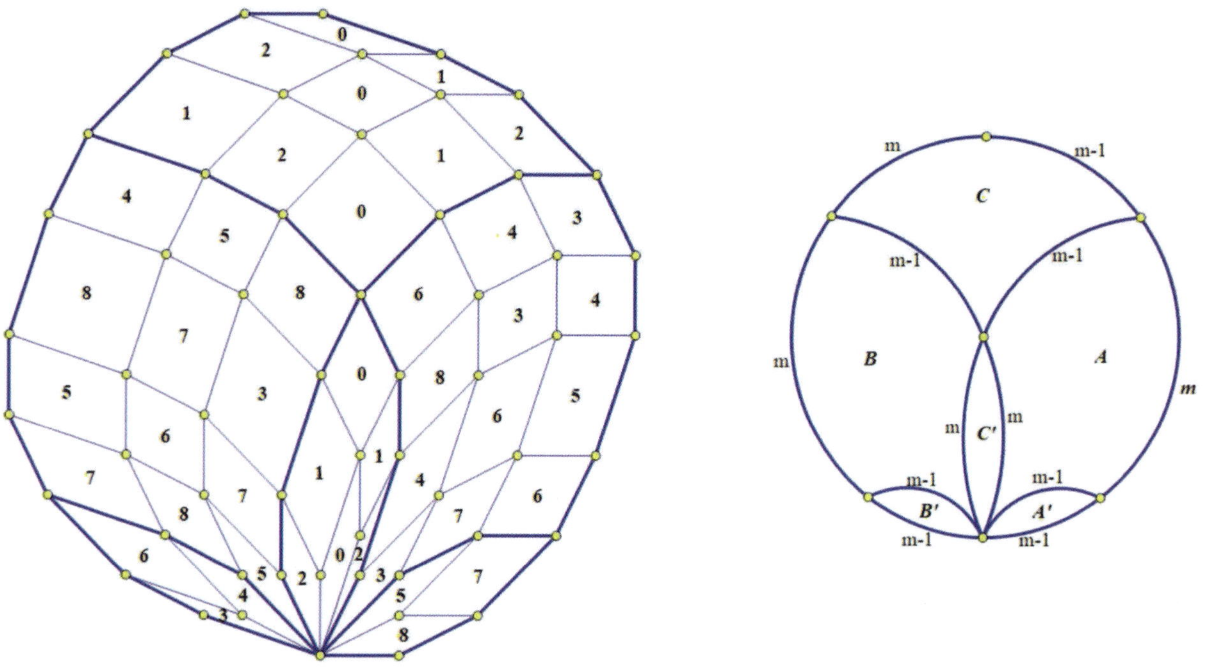

Figure 8.5

Figure 8.4 features a primary (simple) Latin 16-gon which indicates a general procedure for obtaining simple Latin polygons whose boundary contains 8m edges for some integer m. The general structure of such a Latin polygon is illustrated in the schematic diagram at right in Figure 8.4. The label on each arc gives the number of edges which constitute the arc. The labels C, C',B, B' each represent regions which are graphically equivalent to Latin squares, and A, A' represent simple Latin polygons with 4m boundary edges. All streams have length n = 4m-1. The set Z(n) is partitioned into three subsets {S, T, R} with |S|=|T|= m, and |R|=2m-1. Regions C, C' are Latin squares over S; regions B,B' are Latin squares over T, and A,A' are Latin 4m-gons over R. The result is a Latin 8m-gon over Z(n). The construction depends on the ability to obtain a Latin 4m-gon, but we can say the following.

Theorem 8.2: *A simple Latin 2^t - gon can be constructed for every integer t≥3.*

Proof. By induction on t: The construction for t = 3 is given in Figure 8.1(II), and for t = 4, the construction is given in Figure 8.4. Let P be a 2^k-gon for some integer k>3 and assume inductively that a Latin 2^n-gon can be constructed for every integer n with 3≤n<k. Let t = k-2 and let m =2^{t-1} in Figure 8.4. In this diagram sections A, A' will be Latin 4m-gons, i.e., Latin 2^{k-1}-gons. By the inductive hypothesis, a Latin 2^{k-1}-gon can be constructed. Then the construction given in Figure 8.4 produces a Latin assignment for P. □

In our constructions we have preferred a polar tiling. It is important to observe that hexagon inversions preserve Latinicity. It then follows from Theorem 1.5 that once a particular tiling of a polygon P yields a Latin assignment, then every possible parallelogram tiling of P must also have a Latin assignment. The polar tiling is especially useful due to possibilities for dissection. Latin assignments which fit the pattern given in Figure 8.4 will be called *Type I.* For Type I polygons the polar tiling yields a dissection into four Latin squares and two Latin polygons.

Figure 8.5 illustrates a typical *Type II* Latin polygon. The polar tiling yields a dissection consisting of a Latin square C, a Latin polygon C', and several other parts of a more complex nature. The semiperimeter s = 3m-2 gives the number of streams which must be even; thus m must be even. The tiles in sections C, C' are assigned values from the same set of m-1 elements from the base group Z(s-1). Each value in the base group is replicated r = s/2 times. The sections A,A',B,B' can be extracted from Figure 8.5 to form the Latin polyparallelogram illustrated in Figure 8.6 which we shall call a *double octagon.*

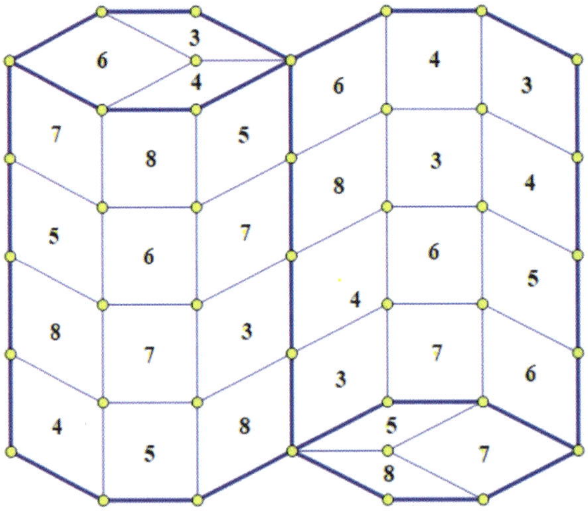

Figure 8.6

The construction of *double polygons* is the key to Latin assignments for Type II polygons. A *double polygon* consists of two semiregular polygons which share a common side, and for which the streams of a parallelogram tiling all have the same length. In Figure 8.7 a double dodecagon with a Latin assignment is illustrated, and in Figures 8.8(I,II) a double 16-gon with a Latin assignment is given. (stitch 8.8(I) and 8.8(II) together at the indicated points P,Q.) These double polygons can be used to construct corresponding Type II Latin polygons by following the pattern given in Figure 8.5. We have not been able to find a general method for obtaining Latin assignments for double polygons, and thus there are many Type II Latin polygons yet to be discovered.

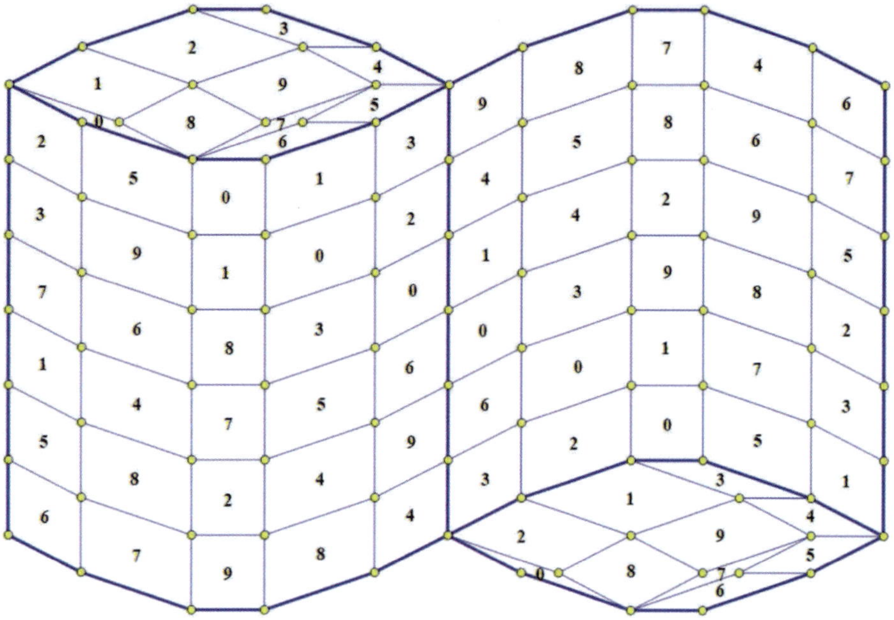

Figure 8.7

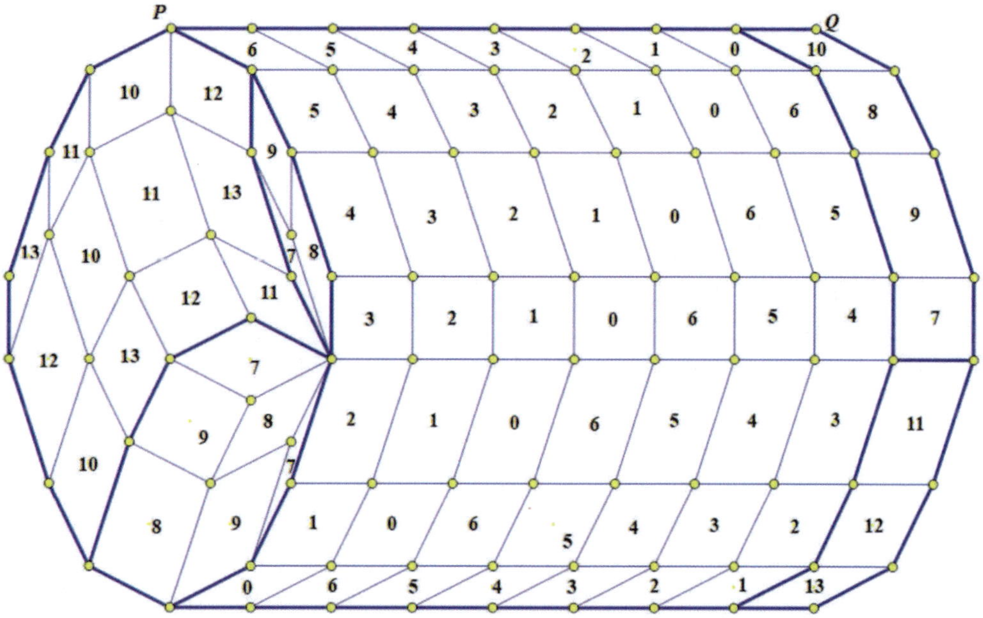

Figure 8.8(I)

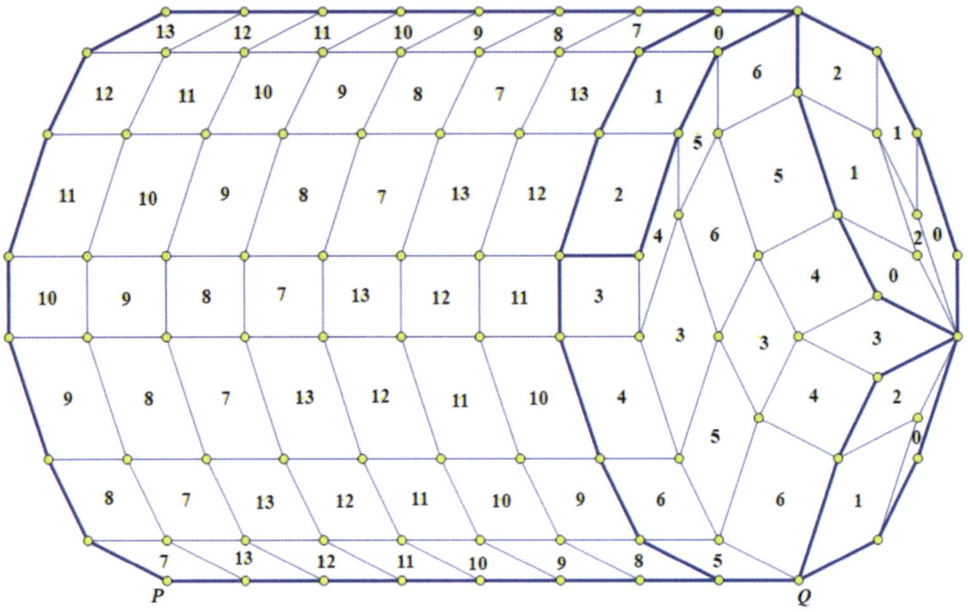

Figure 8.8(II)

In Figure 8.9 a typical Type III Latin polygon is illustrated. It is a primary 14-gon which must necessarily have two of edges per side to insure that there are an even number of streams. The base is Z(12) and the elements of Z(12) are partitioned into the three sets S_1 = {0,1,2,3}, S_2 = {4,5,6,7} and S_3 = {8,9,10,11}. Region C is a Latin square over S_1 ; region C' is a Latin hexagon over S_1 ; region B is a Latin square over S_2, B'∪B" is a *broken* Latin hexagon over S_2 ; region A is a Latin square over S_3, and region A' is a Latin hexagon over S_3.

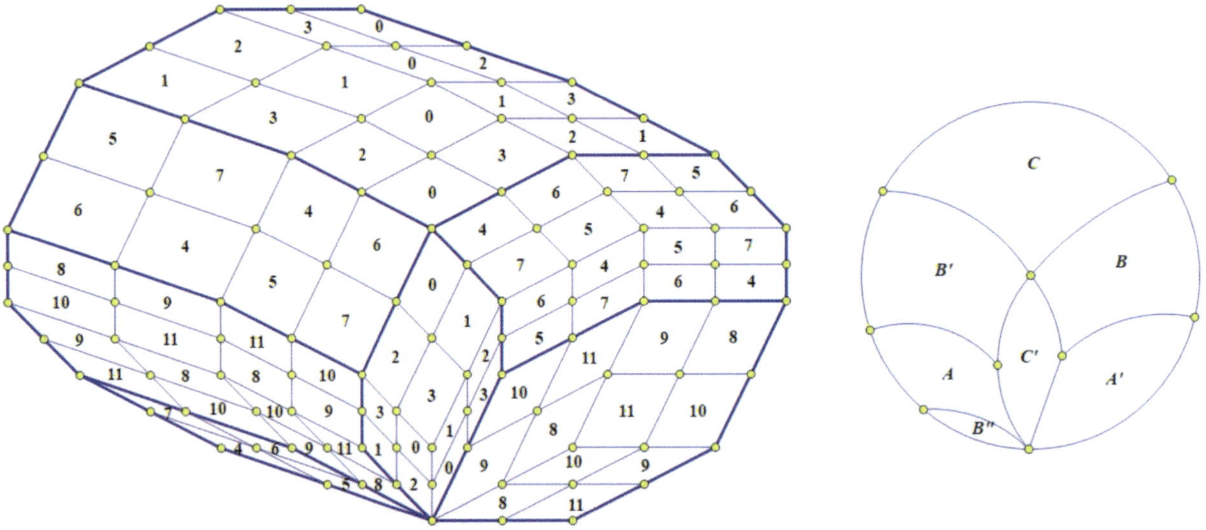

Figure 8.9

There are many more Latin polygons awaiting construction; the last one we will construct is the simple 28-gon in Figure 8.10(I, II) below. It is depicted in three sections with labeled vertices showing the boundaries of significant subtilings and indicating how the sections should be joined. Section AFEB is a Latin square over {0,1,2,3}; section ABFG is a Latin square over {5,6,7,8}; section

CDQP is a Latin square over {9,10,11,12}; section BCPH is a Latin dodecagon over {0,1,2,3,4}; sections BEDC and tiles 5,6,7,8 in octagon PQ when combined form the equivalent of a Latin hexagon over {5,6,7,8}; sections GHPR and tiles 9,10,11,12 in octagon RP combine to form the equivalent of a Latin hexagon over {9,10,11,12}. Note the four occurences of element 4 in octagons PQ and PR. A fairly complicated construction ! We suspect that similar complex constructions exist for even larger Latin polygons, so we make the following Conjecture.

Conjecture 8.3: *Every regular polygon with a parallelogram tiling and an even number of streams has a Latin assignment.*

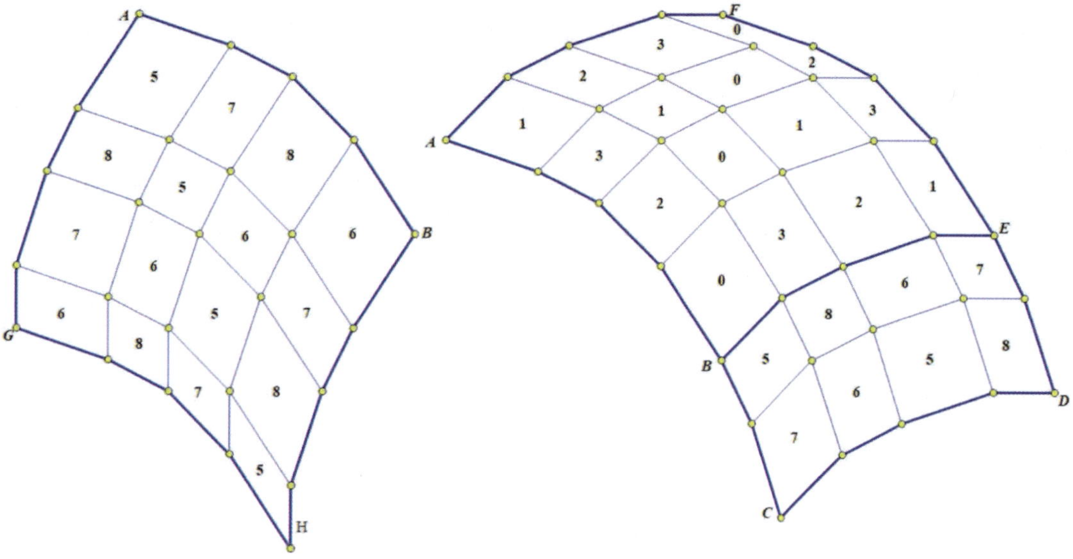

Figure 8.10(I)

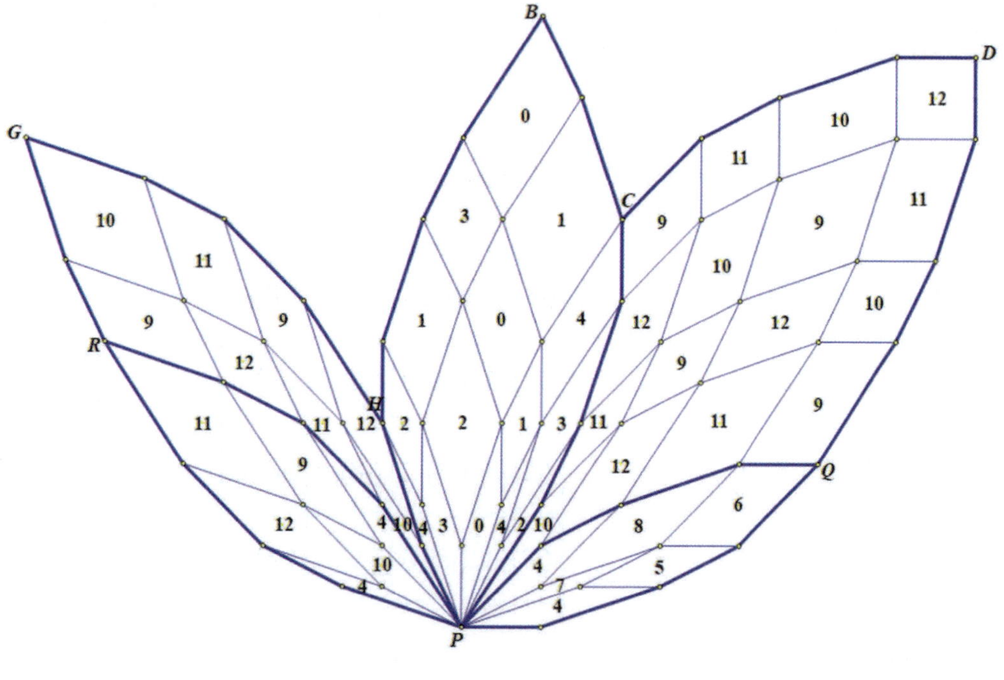

Figure 8.10(II)

8.4 Orthogonal Mates

The smallest Latin polygon with an orthogonal mate is the simple octagon illustrated in Figure 8.1(II) and repeated below in Figure 8.11(I) along with an orthogonal mate in Figure 8.11(II). Figure 8.11(I) is a $(Z(3))^2$ magic octagon, and the orthogonal mate in 8.11(II) is a $(Z(2))^3$ magic octagon. Superimposition of these two octagons yields the magic Z(2)×Z(3) octagon in Figure 8.11(III). Since 2, 3 are relatively prime, the mapping ϕ:Z(6)⟶Z(2)×Z(3) defined by ϕ(x) = (x mod 2, x mod 3) is an isomorphism which can be retracted to obtain the basic magic Z(6) octagon in Figure 8.11(IV).

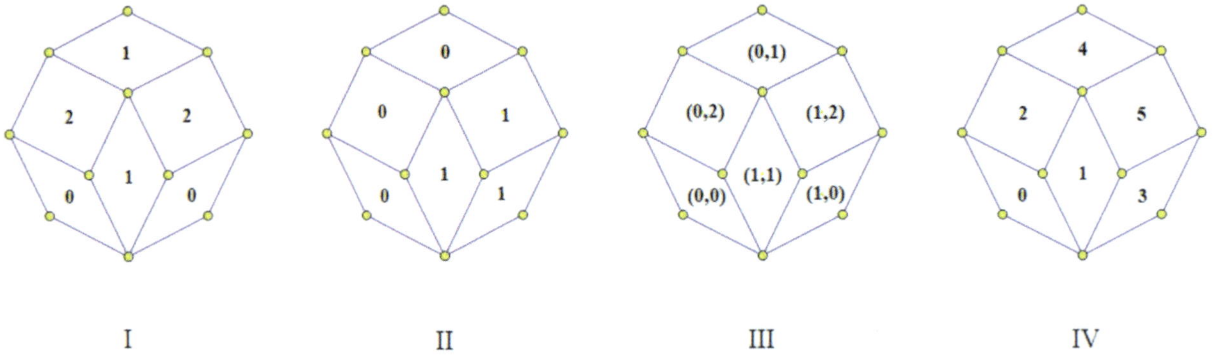

Figure 8.11

Two things are important to note regarding Latin polygons in general and orthogonal mates. These are summarized in the following Theorem.

Theorem 8.4 *(i) hexagon inversion preserves Latinicity as well as magicality; (ii) If P, Q are orthogonal magic polygons with the same parallelogram tiling T, then orthogonality is preserved by any hexagon inversion.* □

Much work has been done, e.g., in [5] regarding orthogonal Latin squares and systems of mutually orthogonal Latin squares. It is known [5] that every Latin square except those of orders 2, 6 has an orthogonal mate, (which is also a Latin square). For a nonsquare Latin Polygon there cannot exist an orthogonal mate which is also Latin. This is due to the fact that the replication number is not equivalent to the stream length. Orthogonal mates can be found however which are magic but not Latin as in Figure 8.11(I, II) and in Figure 8.12(I, II) below. In Figure 8.12(III) Figures I and II have been superimposed to create a magic dodecagon over Z(3)×Z(5). Since 3, 5 are relatively prime the mapping ϕ: Z(15)→Z(3)×Z(5) is an isomorphism which is retracted to obtain the basic magic dodecagon over Z(15) in Figure 8.12(IV). Finally by adding 9 to each tile in 8.12(IV) the natural magic dodecagon of Figure 8.12(V) is obtained.

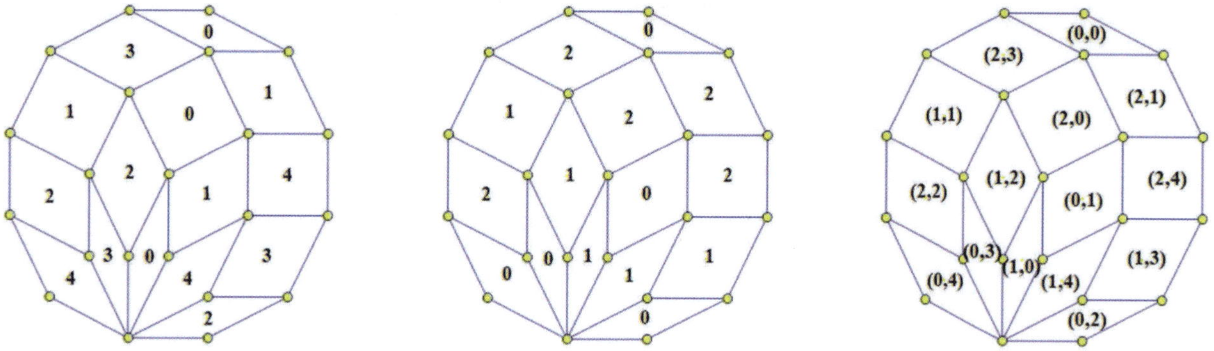

I II III

Figure 8.12

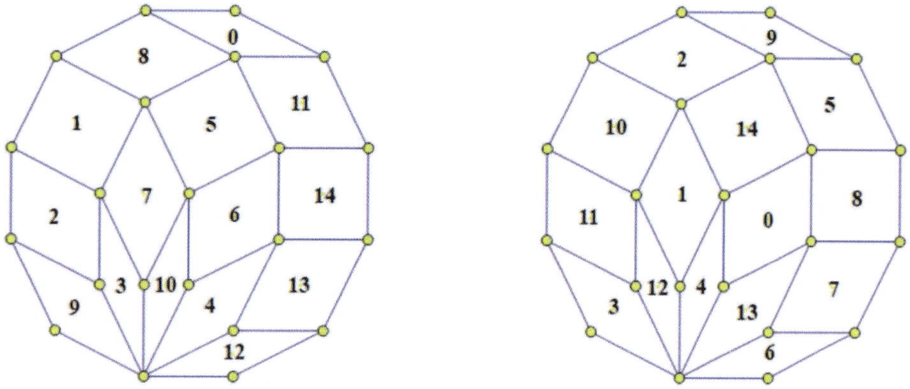

IV V

Figure 8.12

Not all Latin squares have orthogonal mates. In 1779 Euler [5] showed that no Latin square which derives from the Cayley table of an even order cyclic group can have an orthogonal mate. Later it was shown, (see [5], [9]) that for all n ≠1, 3 Latin squares of order n exist which have no orthogonal mate. Such squares have been called *bachelor squares* by Rees [8]. It can be shown that the Latin hexagon in Figure 8.1(I) has no orthogonal mate, hence it can also be termed a *bachelor.* It can also be shown that every Latin assignment for this hexagon is isotopic to the one given in Figure 8.1(I), and it follows that no Latin assignment for this hexagon has an orthogonal mate. The Latin octagon in Figure 8.2 does have an orthogonal mate which is illustrated in Figure 8.13(I). When superimposed with Figure 8.2, the Z(4)×Z(6) magic octagon in Figure 8.13(II) is obtained.

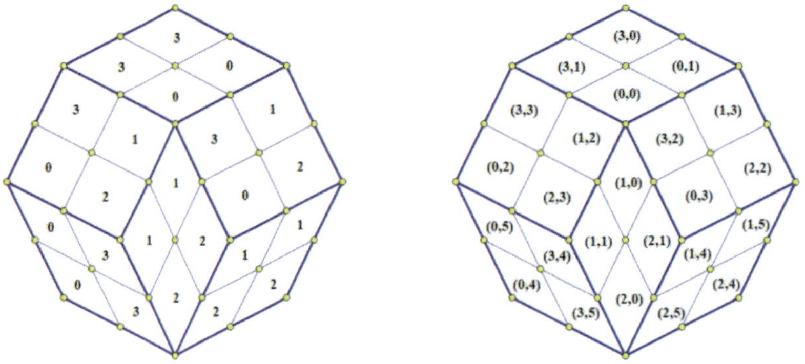

I

II

Figure 8.13

Finding orthogonal mates for Latin polygons seems to be quite difficult in general. We provide just one more example in Figure 8.14. Figure 8.14(I) is a Latin, (or $(Z(8))^6$ magic), hexagon, and in Figure 8.14(II) an orthogonal mate over $(Z(6))^8$ is illustrated. The superimposition of I and II yields a natural magic hexagon, (not illustrated), with base $Z(6)\times Z(8)$.

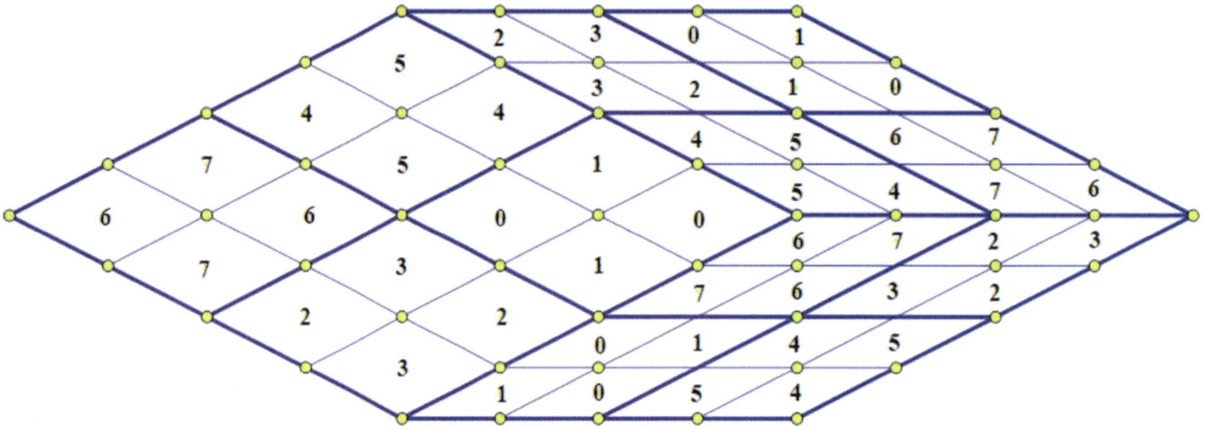

Figure 8.14(I)

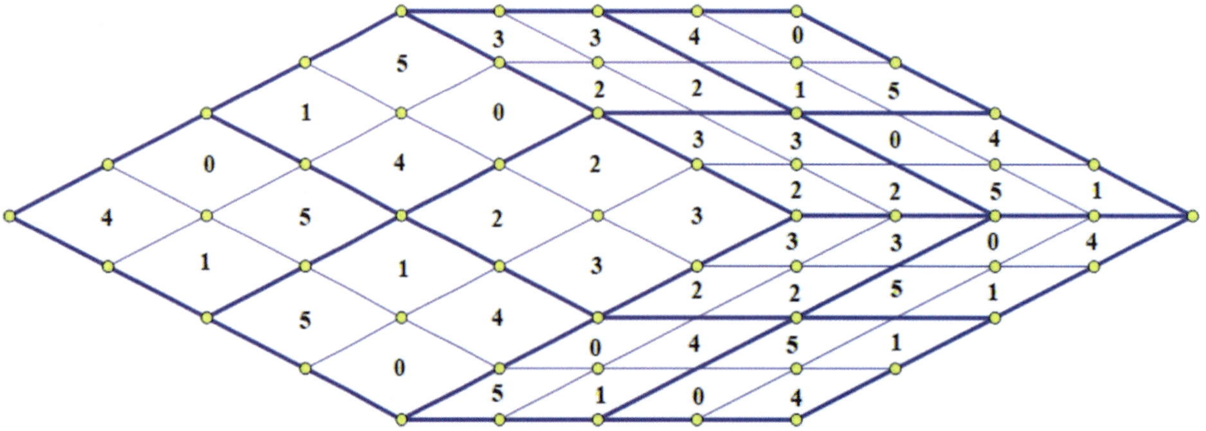

Figure 8.14(II)

CHAPTER 9

MAGIC CYLINDERS AND CIRCLE SYSTEMS

9.1 Definitions and examples

A *cylinder* P is a three dimensional polyparallelogram with top and bottom consisting of two congruent regular polygons P_1, P_2 and with lateral surface consisting of a spooled rectangular polysquare R. Let s denote the semiperimeter of P_1, P_2 and let e denote the edgecount, i.e., the number of basic edges per side for P_1, P_2 . Then the dimensions of R are 2s and e. The tile stream(s) which lie entirely in R are called *lateral* and have length 2s. The remaining tile streams pass through both P_1 and P_2 and occupy 2e tiles belonging to R. These we call *transversal streams.* Since the portion of a transversal stream which lies in P_1 or P_2 contains s-e tiles, the total length of such a stream is given by 2(s-e) + 2e = 2s. Thus all streams belonging to P have the same length 2s. The number of streams is given by m = s+e and the total number of tiles in P is given by m(m-1).

If n = m(m-1) and G is an abelian group with n elements, then P is *magic over G* in case the tiles of P are labeled with the full set of elements from G in such a way that all tile streams have the same sum in G. Our main concern will be the case in which G = Z(n) and P has a natural magic assignment over Z(n). In this case the common sum S of all tile streams must be s(n-1). There are several ways to represent P: in Figure 9.1(I) the cylinder with all its labeled tiles is illustrated; In Figure 9.1(II) a graph G(P) is given whose points represent tiles and whose edges are determined by tiles which share a common side. Since each tile of P shares a side with exactly four neighbors, G(P) is regular of degree 4. The tile streams of P correspond to cycles in G(P) and the m streams belonging to P produce a factorization of G(P) into m cycles of length 2s. In Figure 9.1(III) a circular representation of P is provided: each cycle in G(P) is represented by a circle in which the order of elements in any given cycle of G(P) is preserved.

In Figure 9.2(I) a second hexagonal magic cylinder P is depicted along with its corresponding graph G(P) in Figure 9.2(II). This cylinder differs from the first only in that the lower hexagon is inverted. As in the theory for *plane* magic polygons, such inversions preserve stream sets and thus preserve magicality. The order of values is altered as can be seen in the circular representation given in Figure 9.2(III).

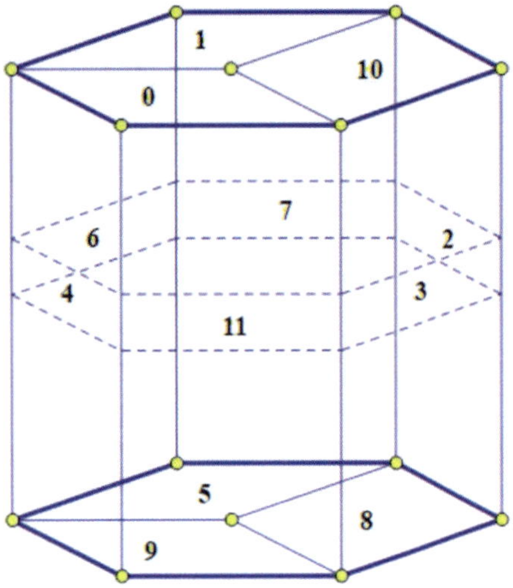

I

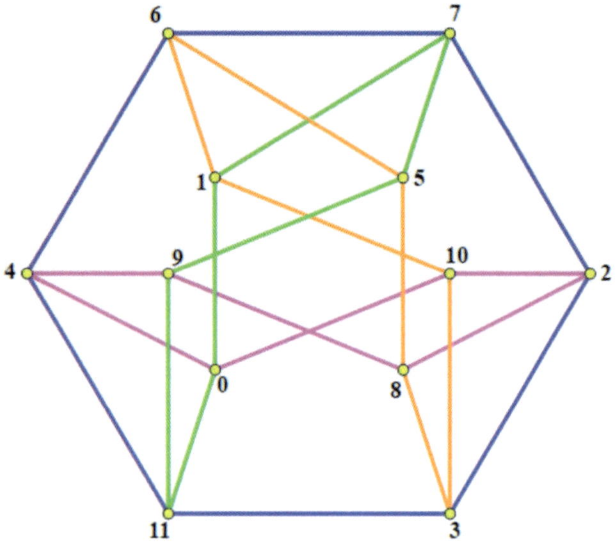

II

Figure 9.1

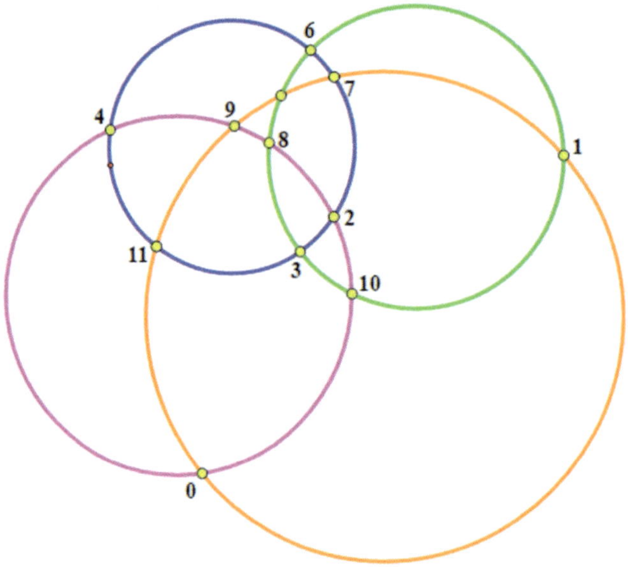

III

Figure 9.1

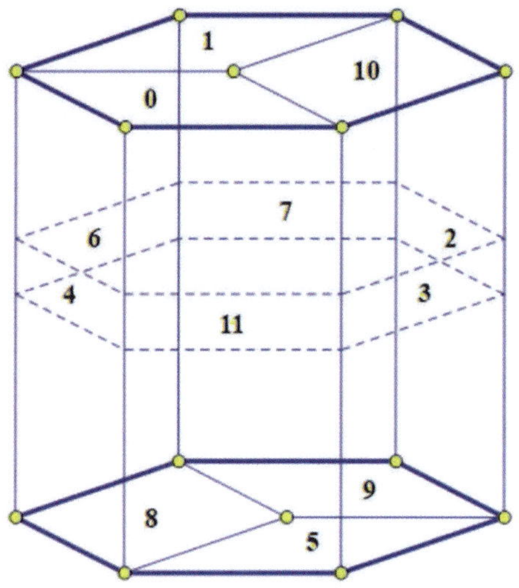

I

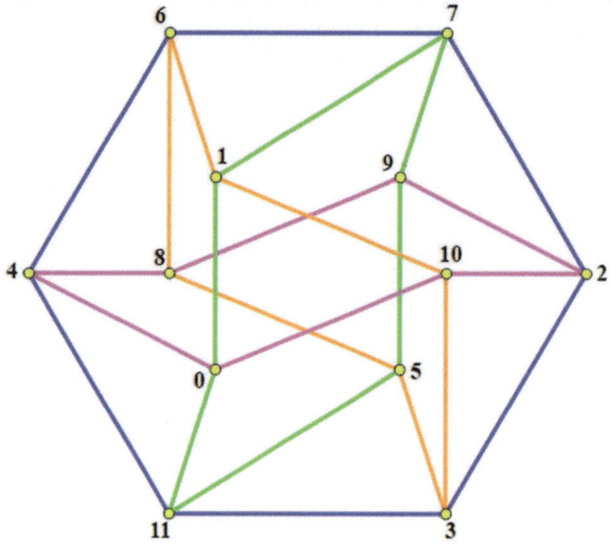

II

Figure 9.2

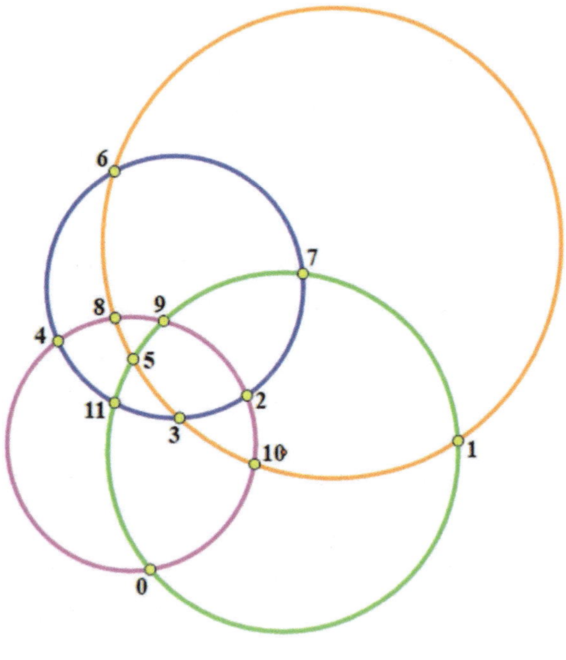

III

Figure 9.2

In Figure 9.3(I) a new kind of hexagon inversion is illustrated. In Figure 9.1(I) the three tiles labeled 0, 7, 11 surround a single vertex similar to the tile structure which occurs in the usual *plane* simple hexagon. As in the plane simple hexagon, the edges incident with the central vertex are broken and replaced by edges which join the alternative set of three boundary points. The result illustrated in Figure 9.3(I) is no longer a polyparallelogram since some of the faces are nonparallelogram quadrilaterals. Nevertheless, stream sums and thus magicality are preserved. All tile streams have the same length and intersect each other in exactly two points as in a cylinder. Due to these likenesses it seems important to include them in our consideration. In Figure 9.3(II) the corresponding graph contains an unusual 2-cycle. This is due to the fact that tiles 9 and 0 have two common sides.

The graph G(P) corresponding to a cylinder has the following three properties (i) it is regular of degree 4; (ii) it has a factorization into cycles each with the same even length, and (iii) any two of these cycles are either disjoint or have exactly two points in common. We will call a graph satisfying these three properties *cylindrical*, although, as we have seen in the example in Figure 9.3, no immediate relation to a cylinder need exist. It is an open question as to whether every cylindrical graph can be obtained from a cylinder by means of a sequence of hexagon inversions. It is also an open question as to whether every cylindrical graph can be represented by a system of circles as in Figures 9.1(III), 9.2(III) and 9.3(III).

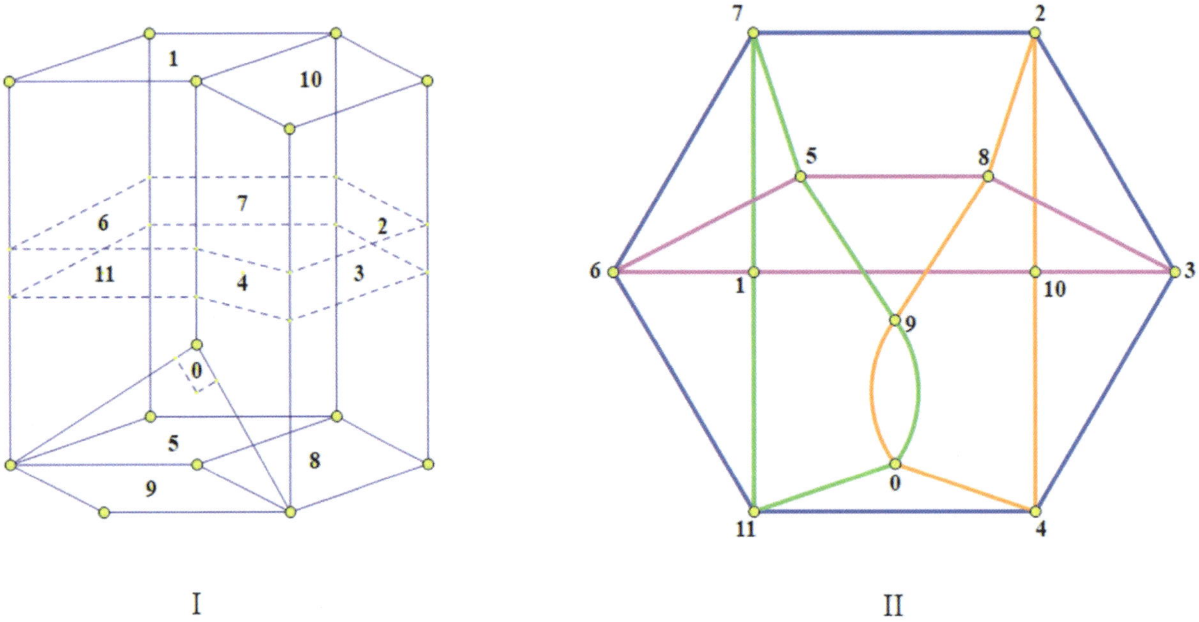

Figure 9.3

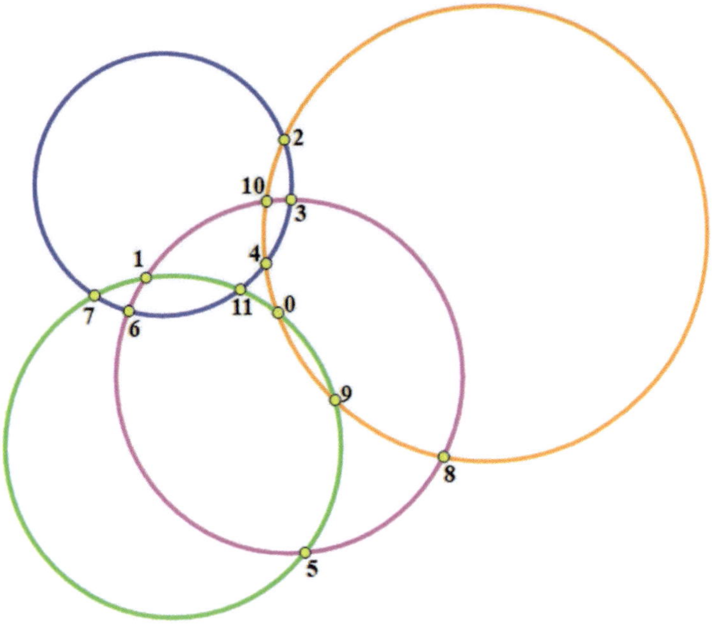

III

Figure 9.3

In Figure 9.4(I) a magic octagonal cylinder is illustrated with its corresponding graph in Figure 9.4(II), and corresponding *magic circle system* in Figure 9.4(III).

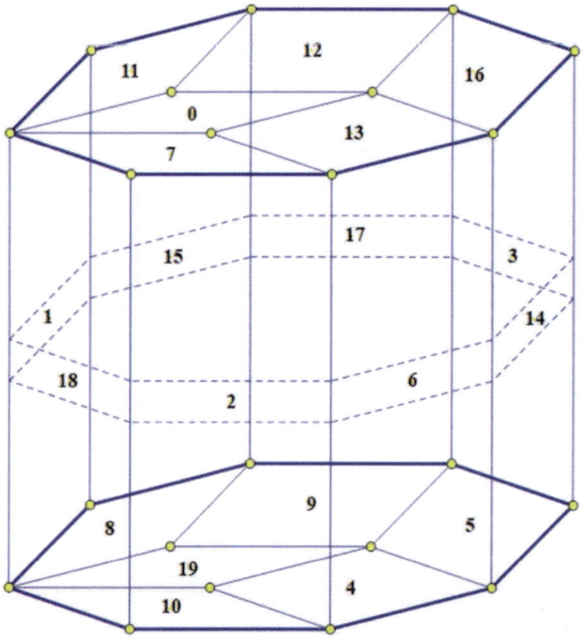

I

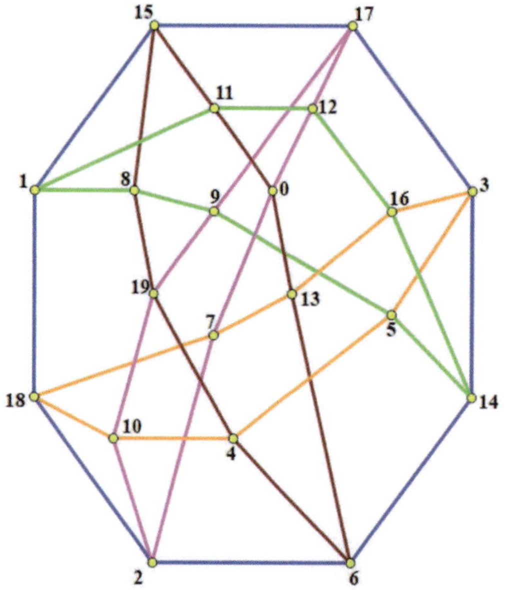

II

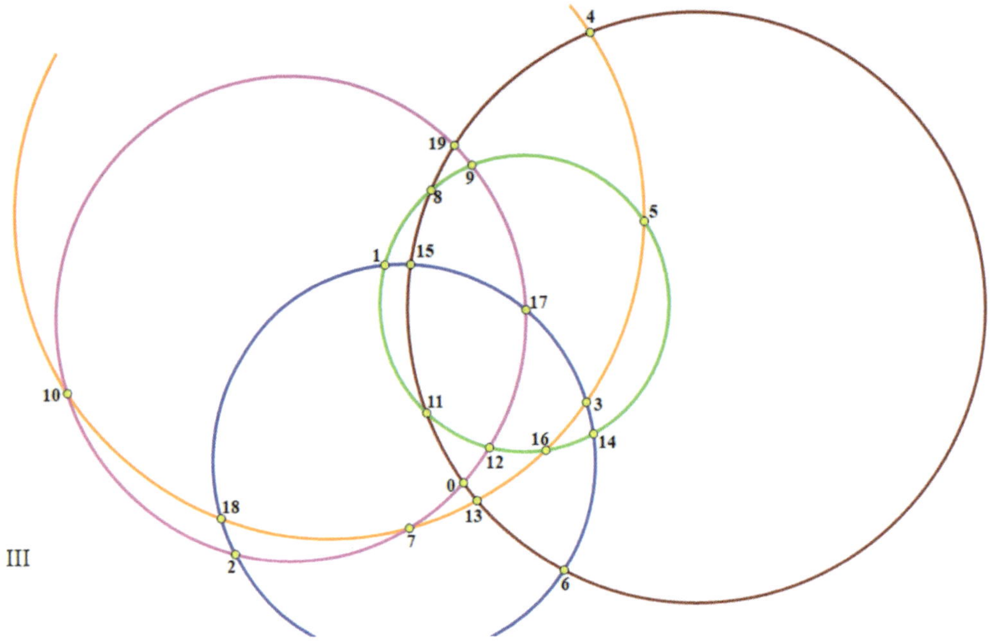

Figure 9.4

Based on these and other examples we make the following Conjecture which is analogous to Conjecture 2.3 regarding the linear representation of semiregular magic polygons. Let G be an abelian group. A *magic circle system over G* is a collection of circles each of whose contact points is labeled with a distinct element of G, and such that the following properties are satisfied: (i) the circles are either disjoint or have exactly two points of intersection; (ii) all circles have the same (even) number of contact points; (iii) no contact point is common to more than two circles; (iv) there is a constant value S in G which is the sum of the assigned values to each circle in the system. A system of circles which satisfies (i), (ii) and (iii) we shall call a *circle system of degree 2.*

Conjecture 9.1: *Every magic cylindrical graph can be represented by a magic circle system.*

A magic circle system has some properties in common with the *circle systems* described in [10]. The circle systems there have contact points which are also labeled with the elements of an abelian group G, and each circle is required to have the same sum S in G. But here the similarity ends; in the circle systems in [10] only four contact points are labeled; more than two circles usually pass through a given point, and every four element subset of G corresponds to four cocyclic points.

9.2 Complemented Magic Circle systems

Let G be an abelian group of even order n and let g be an element of G which does not lie on the main diagonal of a Cayley table for G, i.e., there is no element x in G such that x+x = g. The elements of G can then be partitioned into n/2 two element subsets each of whose pairs sum to g. If {a,b} is such a subset then we say that a, b form a *complementary pair with respect to g.* Now suppose Ω is a circle system of degree 2 with m circles and an even number n of contact points. The point set of Ω can be partitioned into n/2 two element subsets with the property that each such set consists of a pair of contact points of two circles in the system. Each complementary pair of G can then be assigned as labels to a unique pair of points in the partition of the point set of Ω. All circles in Ω have the same even number 2n/m of contact points, and when labeled as described will have common sum S = (n/m)g. We have thus obtained a magic assignment for Ω over G and proved the following Theorem.

Theorem 9.2: *Let G be an abelian group with even order n and let g be an element of G which does not lie on the main diagonal of a Cayley table for G. If Ω is a degree 2 circle system with n points and m circles, then there exists a magic assignment for Ω over G with sum S = (n/m)g.* □

We call the magic circle system Ω constructed in Theorem 9.2 *complemented with respect to g.* To obtain a complemented natural magic circle system over Z(n) we take g = n-1 and obtain the common sum S = (n/m)(n-1). In Figure 9.5 a complemented version of the magic circle system of Figure 9.3(III) is illustrated.

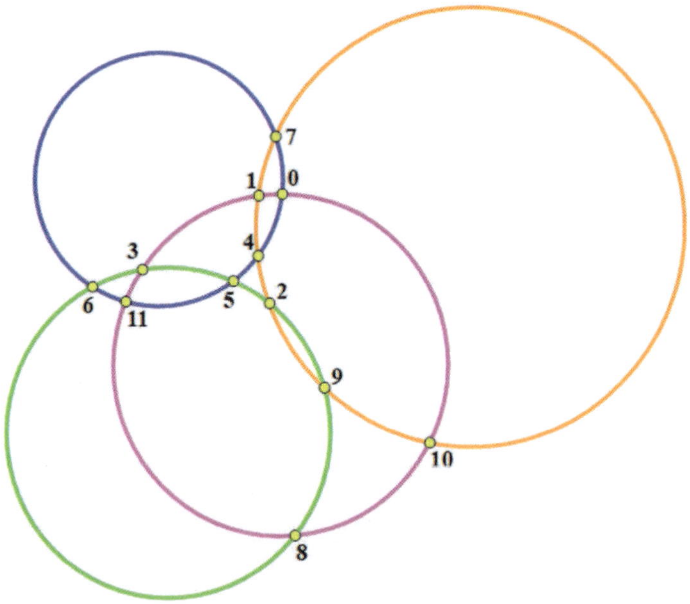

Figure 9.5

Many complemented versions of a magic circle system can be obtained since, firstly, values assigned to the pair of contact points between any two circles can be exchanged without affecting magicality and secondly, any correspondence between complementary pairs of elements in G and circle contact pairs can be chosen to effect a magic assignment. The procedure described in the first paragraph of this Section can equally well be used to construct complemented magic cylinders or magic cylindrical graphs. For magic cylinders the two tiles common to two intersecting streams are labeled with a complementary pair of elements from the base group G, and for magic cylindrical graphs, the two common points of intersecting cycles are assigned values similarly.

9.3 Matrix Representation

Let Ω be a natural magic circle system over Z(n) with m circles. We consider first the case in which Ω contains no pair of disjoint circles, i.e., each circle meets every other circle in two points. Let M(Ω) denote an m by m matrix whose entries *off the main diagonal* constitute the full set of elements from Z(n) . The circles of Ω can be given names, e.g., B for blue; P for pink, etc., and these names can be placed in the main diagonal of M(Ω). The values in the row and column containing a particular circle name constitute the elements from Z(n) assigned to that circle. The two values assigned to the contact points of two circles are placed in opposite positions with respect to the main diagonal. Since this can be done in two ways, the matrix representation of Ω is not unique. Since Ω is magic, each such row and column combination has the same sum. The same matrix can be used to represent the magic circle systems in Figure 9.1(III) and 9.2(III):

B	4	7	3
2	P	0	8
11	9	G	1
6	10	5	N

Here we use the name B for the blue circle, P for the pink circle, G for the green circle and N for the orange circle. The same matrix can also be used to represent the circle system in Figure 9.3(III) but with the positions of P and N interchanged on the diagonal. This matrix is a nice construct in and of itself without any reference to a magic circle system. It can be defined simply as an m by m matrix whose off diagonal entries constitute the full set of elements from Z(m(m-1)) and such that the sum of the elements in the combined i'th row and i'th column is the same for each i in {1, ..., m}. We call such a matrix a *circular magic square.* In case all pairs of entries opposite the main diagonal have the same sum we obtain a *complemented circular magic square.*

In case the magic circle system Ω contains circles which are disjoint then a corresponding matrix representation has a different appearance. A *parallel class* of circles in Ω is a subset of mutually disjoint circles. If we require that each parallel class contains the same number k of circles, then in M(Ω) these parallel classes give rise to a series of empty k by k blocks which cover the main diagonal. As an example, consider the magic *cubic* cylinder P in Figure 9.6(I) whose top and bottom polygons are 2 by 2 polysquares and the corresponding magic cylindric graph G(P) and circle system in Figures 9.6(II, III). In Figure 9.6(IV) a magic matrix is given for this system. The three 2 by 2 blocks along the main diagonal indicate the existence of three parallel classes each consisting of two circles; Only the color codes: R for red; D for dark blue; G for green; N for orange; B for light blue and P for pink occupy the appropriate squares in these central blocks, the remaining cells in these blocks are empty and are marked with an X.

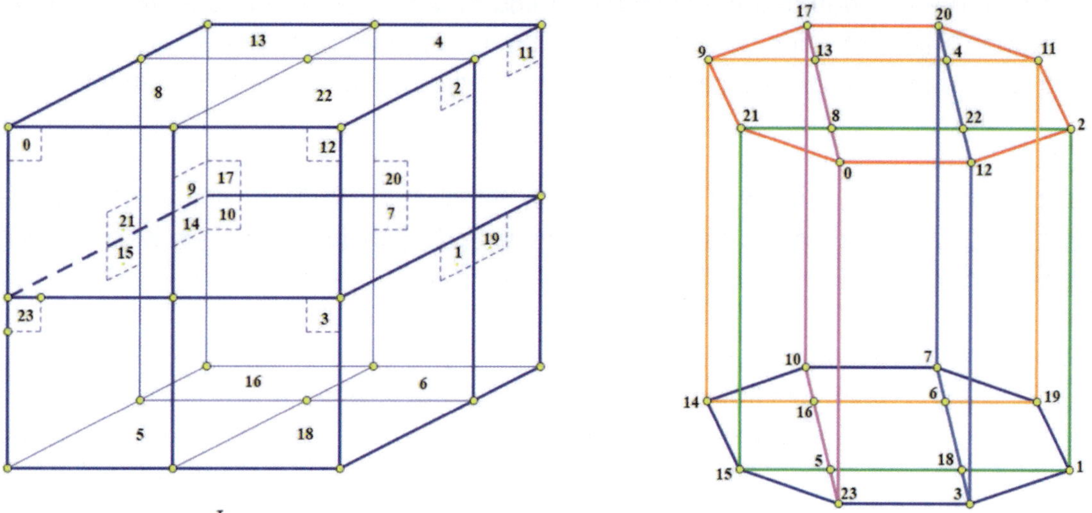

I

II

Figure 9.6

There are several simple transformations which carry one circular magic square M into another: (I) let a, a′ denote values which are opposite each other with respect to the main diagonal of M, and make the exchange a⟷a′; (II) Let a,b,c,d denote elements from the base group of M with a+b = c+d and let C_1,C_2, C_3, C_4 denote four distinct circles of M. If a,b ∈ C_1 ; c,d ∈ C_2 ; a,c ∈ C_3 and b,d ∈ C_4, make the exchanges a⟷c, b⟷d; (III) If a+a′ = b+b′, make the exchanges a⟷b, a′⟷b′. We are interested to know if it is possible by a series of transformations of Types I, II and III to convert any given circular magic square into a complemented version. Examples indicate that the following Conjecture is likely true.

Conjecture 9.3: *There exists a sequence of Type I, Type II and Type III transformations which carry any given circular magic square into a complemented circular magic square.*

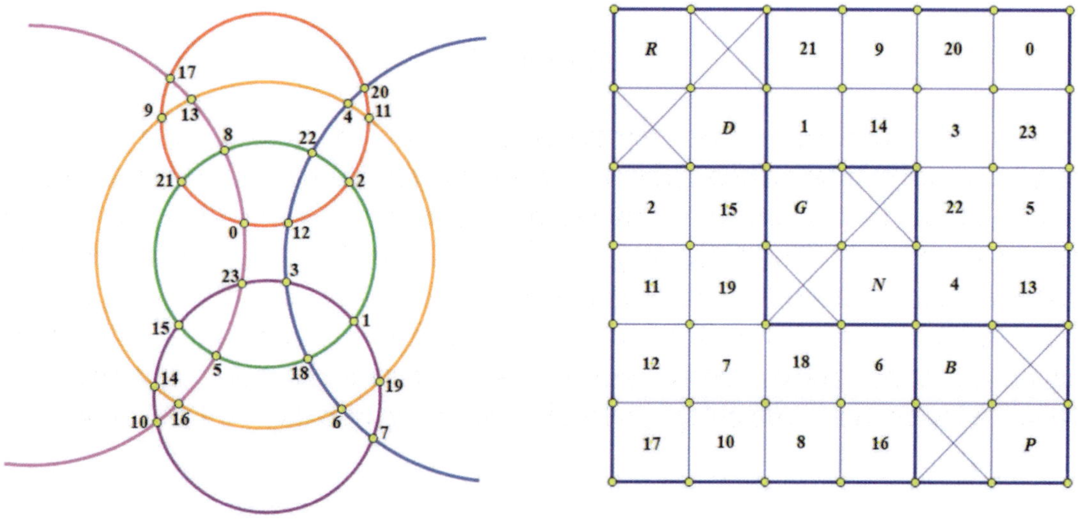

III

IV

Figure 9.6

As an example we start in Figure 9.7 with the magic circular matrix of Figure 9.6(IV) and indicate transformations by an arc joining the elements to be exchanged. The sequence of transformations is ordered left to right and top to bottom. The final complemented square is obtained by applying the indicated transformation to the last square in the sequence.

It is possible to construct a magic circle system in which the parallel classes *do not* all have the same number of circles. Such a system is illustrated in Figure 9.8(I) along with a complemented matrix representation in Figure 9.8(II).

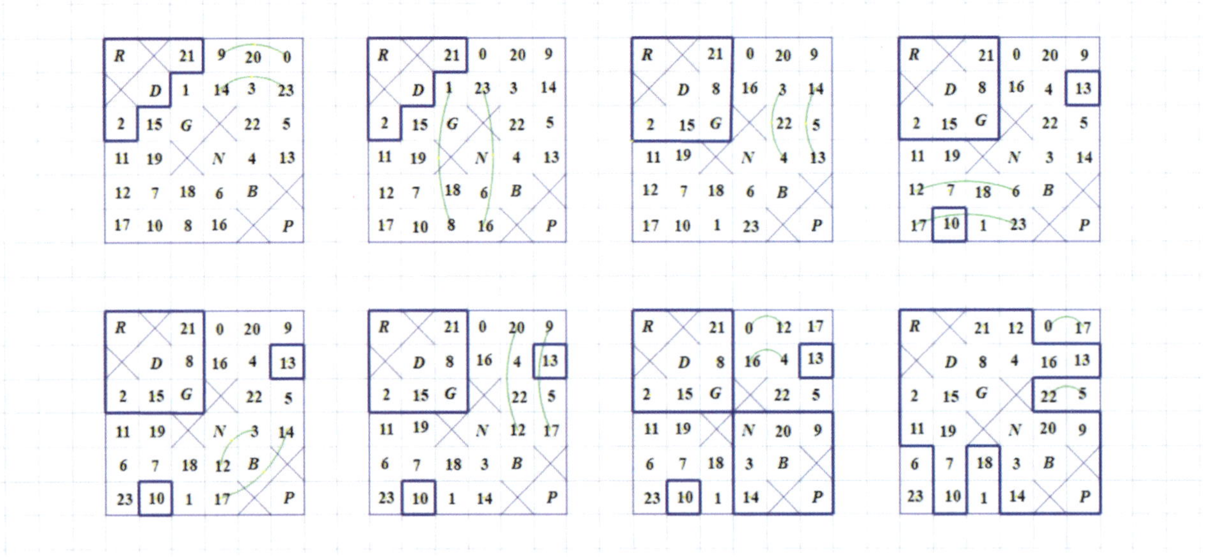

Figure 9.7

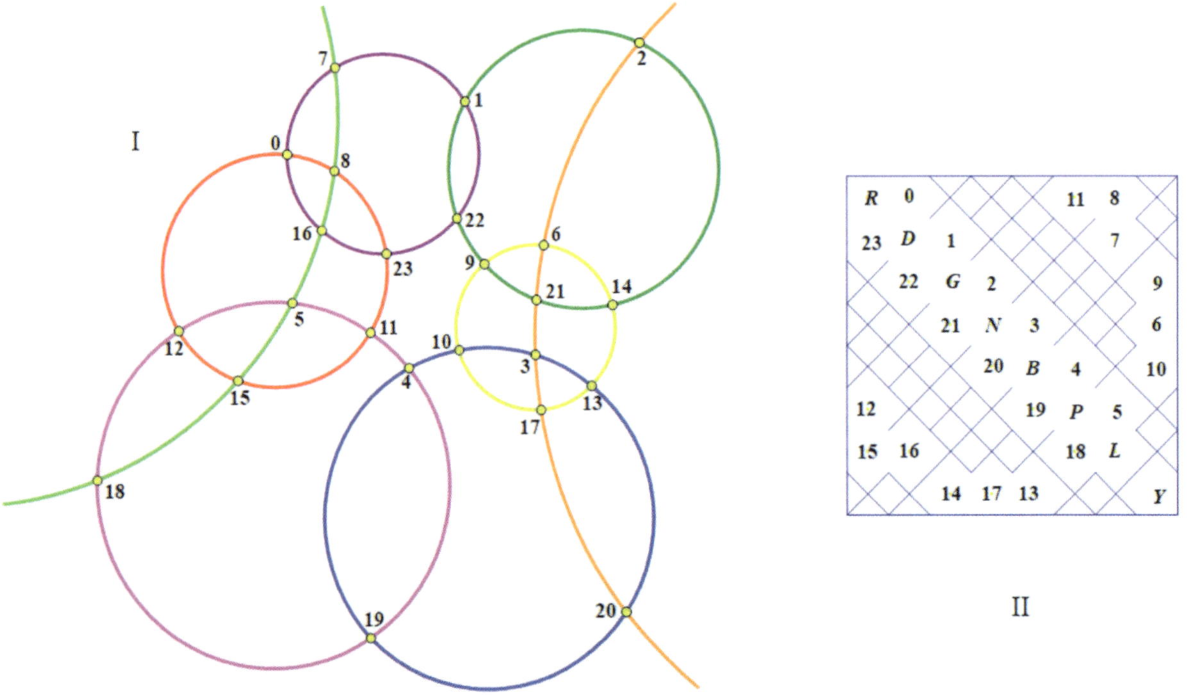

Figure 9.8

CHAPTER 10

LATIN TRIANGLES AND CIRCLE SYSTEMS

10.1 Circular Latin Squares

Let n = 2(m-1), and let Ω be a degree 2 circle system with m circles. We shall first consider the case in which each circle in Ω intersects every other circle. The contact points of these circles are labeled with the full set of values from Z(n) such that no circle contains two equal values, or equivalently, the contact points of each circle of Ω are labeled with precisely all the elements of Z(n). We call the resulting structure a *Latin circle system over Z(n)*, and will use an m by m square, a *circular Latin square,* for the purpose of representation. For each integer i ∈ {1,… m} circle values are represented by the combined off diagonal entries in row i and column i; the diagonal entries contain the circle names. The number m of circles must be even; the replication number is given by r = m/2 and the common circle sum is S = n(n-1)/2. In Figure 10.1(I, II) two circular Latin squares of order 4 are illustrated.

I

A	0	4	5
2	*B*	5	4
3	1	*C*	2
1	3	0	*D*

II

A	0	1	2
5	*B*	2	1
4	3	*C*	0
3	4	5	*D*

III

A	0	2	9	6	4
1	*B*	4	2	9	7
3	5	*C*	6	0	9
8	3	7	*D*	4	0
7	8	1	5	*E*	2
5	6	8	1	3	*F*

IV

A	0	1	2	3	4
9	*B*	4	3	1	2
8	5	*C*	0	2	3
7	6	9	*D*	4	1
6	8	7	5	*E*	0
5	7	6	8	9	*F*

V

A	0	1	2	3	4	5	6
13	*B*	2	1	4	5	6	3
12	11	*C*	0	5	6	3	4
11	12	13	*D*	6	3	4	5
10	9	8	7	*E*	0	1	2
9	8	7	10	13	*F*	2	1
8	7	10	9	12	11	*G*	0
7	10	9	8	11	12	13	*H*

Figure 10.1

In Figure 10.1(III,IV) two circular Latin squares of order 6 are given, and in Figure 10.1(V) a circular Latin square of order 8 is illustrated.

10.2 Type I Latin Triangles

Squares II, IV and V exhibit a special structure which we will call a *Δ-square*. These squares have the property that the upper triangular section contains all values in {0, …,m-2} and the lower triangular section contains the remaining values in {m-1, …,n-1}. Both of these sections are *Latin triangles,* the upper section will be denoted by Δ(0, …, m-2) and the lower section by Δ(m-1, …, n-1). If we include the circle names A, B, C, …,X then to each circle name a Latin triangle contains a corresponding linear or L-shaped polysquare. The names A, X belong to linear polysquares and the remaining names occur precisely at the bend of an L-shaped polysquare. These polysquares we shall refer to henceforth as *halfcircles.* All halfcircles meet each other in exactly one cell.

Once the upper Latin triangle Δ is constructed, the lower Latin triangle is easily obtained: if ϕ: {0, …, m-2} ⟶{m-1, …, n-1} is any bijection, then for each cell c in the upper triangle with value x insert the value ϕ(x) in the cell opposite c with respect to the main diagonal. If ϕ is chosen so that x + ϕ(x) = n-1 for each x in {0, …, m-2} then we denote the lower Latin triangle by Δ^C, and a *complemented circular Latin square* is formed by the union of Δ, Δ^C . Squares in Figure 10.1(II,IV,V} are each complemented.

The upper Latin triangle in Figure 10.1(V) exhibits an additional structural feature which is indicated by bold lines: the upper right quadrant is a magic square over {3,4,5,6} and adjacent to it are two Δ(0,1,2) Latin triangles which are reflections of each other across the diagonal which extends from the upper right corner. We shall refer to a Latin triangle with this composition as a *Type I Latin triangle.* The general structure of a Type I Latin triangle consists of a Latin square L((m/2) - 1, …, m-2) flanked by two Latin triangles Δ(0, …, (m/2)-2), Δ^R(0, …, (m/2)-2). Here the superscript R indicates that the Latin triangle Δ^R is a reflection of Δ. The largest value (m/2)-2 in Δ must be even and thus in a Type I Latin triangle m must be a multiple of 4. In Figure 10.2 a Latin triangle with m = 12 is illustrated. (For a Latin triangle the parameter m represents the number of *halfcircles, i.e.,* the number of constituent polysquares.) From this example it is clear that the existence of Δ(0, …, k) Latin triangle immediately implies the existence of a Δ(0, …, 2k+2) Latin triangle. We make the following Conjecture.

Conjecture 10.1: *A Type I Latin triangle can be constructed iff the number m of halfcircles is a multiple of 4.*

For certain values of m Conjecture 10.1 can be proved.

A	0	1	2	3	4	5	6	7	8	9	10
	B	2	3	4	1	6	7	8	9	10	5
		C	4	0	3	7	8	9	10	5	6
			D	1	0	8	9	10	5	6	7
				E	2	9	10	5	6	7	8
					F	10	5	6	7	8	9
						G	2	0	3	1	4
							H	1	0	4	3
								I	4	3	2
									J	2	1
										K	0
											L

Figure 10.2

Theorem 10.2: A Type I Latin triangle can be constructed when the number m of halfcircles satisfies $m = 2^n$, n = 2,3,4, ...

Proof. Consider: Δ(0,1,2), as constructed in the upper triangle of Figure 10.1(II), is a Type I Latin triangle with $m = 2^2$ halfcircles. Assume inductively that a Type I Latin triangle Δ(0, ..., m-2) exists with $m = 2^n$ halfcircles. Then, as per the remark immediately preceding Conjecture 10.1, the Type I Latin triangle Δ(0, ..., 2(m-2)+2) can be constructed. This Latin triangle has $2(m-2) +2+2 = 2^{n+1}$ halfcircles. The induction is complete and the Theorem is proved. □

The proof of Conjecture 10.1 depends on the construction of the set of Type I Latin triangles $\Sigma = \{\Delta(0, \ldots, m) : m \in \{2\} \cup \{4t : t = 1,2,3, \ldots\}$. Since m must be even and since the existence of a Δ(0, ..., m) Latin triangle implies the existence of a Type I Latin triangle Δ(0, ..., 2m+2), it can be shown that construction of the Latin triangles in Σ implies the existence of Type I Latin triangles with m halfcircles for every even integer m. A Σ Latin triangle with m = 2 is contained in the Latin circle system of Figure 10,1(V), and a Σ Latin triangle with m = 4 occurs in the Latin triangle in Figure 10.2. A Σ Latin triangle with m = 8 is illustrated in Figure 10.3.

A	0	1	2	3	4	5	6	7	8
	B	8	7	6	5	4	3	2	1
		C	0	2	7	3	4	5	6
			D	8	3	6	5	1	4
				E	1	7	0	4	5
					F	8	2	6	0
						G	1	0	2
							H	8	7
								I	3
									J

Figure 10.3

10.3 Type II Latin Triangles

In Figure 10.4(I,II) two *Type II Latin triangles* are illustrated. These are similar to Type I except that the upper right quadrant is a *deleted Latin square i.e.,* a Latin square except for the upper right to lower left diagonal. Another feature of a Type II Latin triangle Δ(0,..., m) is the alternating 0, m values which occur adjacent to the halfcircle names. As in a Type I Latin triangle the upper right quadrant is flanked by triangular structures which are reflections of each other across the diagonal which extends from the upper right corner. Type II Latin triangles are especially useful when the replication number is odd. Two more Type II Latin triangles are illustrated in Figures 10.5, 10.6. Due to space limitations the Latin triangle in Figure 10.6 is only partially indicated; the missing lower triangular flank is just the reflection of the upper one so the reader can easily fill it in. The Latin triangles in Figures 10.4, 10.5, 10.6 are all Σ triangles. It seems likely that the following Conjecture is true.

Conjecture 10.3: *Every Σ Latin triangle has a Type II Latin assignment.*

I

A	0	1	2	3	4	5	6	7	8
	B	8	3	1	5	6	7	2	4
		C	0	2	6	7	3	4	5
			D	8	7	1	4	5	6
				E	0	4	5	6	7
					F	8	2	1	3
						G	0	3	2
							H	8	1
								I	0
									J

II

A	0	1	2	3	4	5	6	7	8	9	10	11	12
	B	12	4	1	2	3	7	8	9	10	11	5	6
		C	0	5	3	2	8	9	10	11	4	6	7
			D	12	5	1	9	10	11	3	6	7	8
				E	0	4	10	11	2	6	7	8	9
					F	12	11	1	6	7	8	9	10
						G	0	6	7	8	9	10	11
							H	12	4	1	2	3	5
								I	0	5	3	2	4
									J	12	5	1	3
										K	0	4	2
											L	12	1
												M	0
													N

Figure 10.4

A	0	1	2	3	4	5	6	7	8	9	10	11	12	13	14	15	16
	B	16	1	6	5	4	3	2	9	10	11	12	13	14	15	7	8
		C	0	7	2	3	5	4	10	11	12	13	14	15	6	8	9
			D	16	6	7	4	3	11	12	13	14	15	5	8	9	10
				E	0	1	2	5	12	13	14	15	4	8	9	10	11
					F	16	7	1	13	14	15	3	8	9	10	11	12
						G	0	6	14	15	2	8	9	10	11	12	13
							H	16	15	1	8	9	10	11	12	13	14
								I	0	8	9	10	11	12	13	14	15
									J	16	6	1	5	3	4	2	7
										K	0	7	2	4	5	3	6
											L	16	1	7	3	4	5
												M	0	6	2	5	4
													N	16	7	6	3
														P	0	1	2
															Q	16	1
																R	0

Figure 10.5

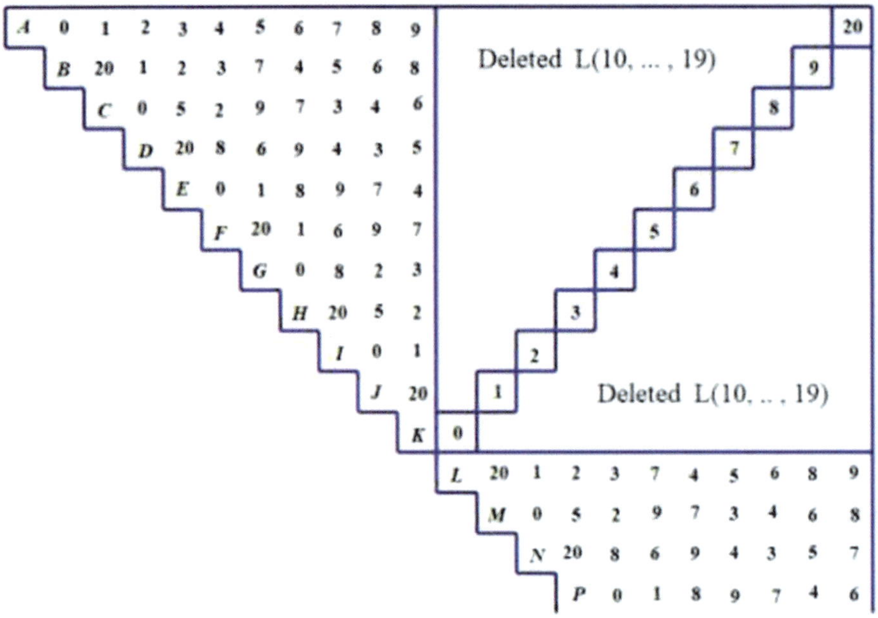

Figure 10.6

10.4 Parallel Halfcircle Classes

By a *parallel class of halfcircles* we mean a collection of mutually disjoint halfcircles. Let Δ denote a Latin triangle. In the following consideration of Latin triangles we will require that each parallel class contains the same number k of halfcircles with k ≥2. Let l denote the number of classes and thus m = lk denotes the number of halfcircles belonging to Δ. Here we consider only the group Z(n) as base where n = m-k. The number m of halfcircles must be even and the replication number is given by r = m/2. The notation Δ = Δ(l,k,S) will be used to represent a Latin triangle over the set S with the indicated parameters. Latin triangles with l = 2 are the simplest; they are essentially Latin squares as in the Δ(2,3,Z(3)) of Figure 10.7(I). In Figure 10.7(II) two copies of the Δ(2,3,Z(3)) are used to flank the Latin square L(3, …, 8) and thus create a Δ(4,3,Z(9)). It is clear that this flanking procedure can be applied to any given Latin triangle to produce a Latin triangle with twice the number of halfcircles.

Theorem 10.4: *Two copies of a Δ(l,k,Z(n)) can be used to flank an L(n, …, n+2l - 1) Latin square to create a Δ(2l,k, Z(n+2l-1)).* □

In light of Theorem 10.4 the problem of constructing Latin triangles devolves to creating those with an odd number l of halfcircle classes. In this case, since the number m = lk of halfcircles must be even, it must be that k is even.

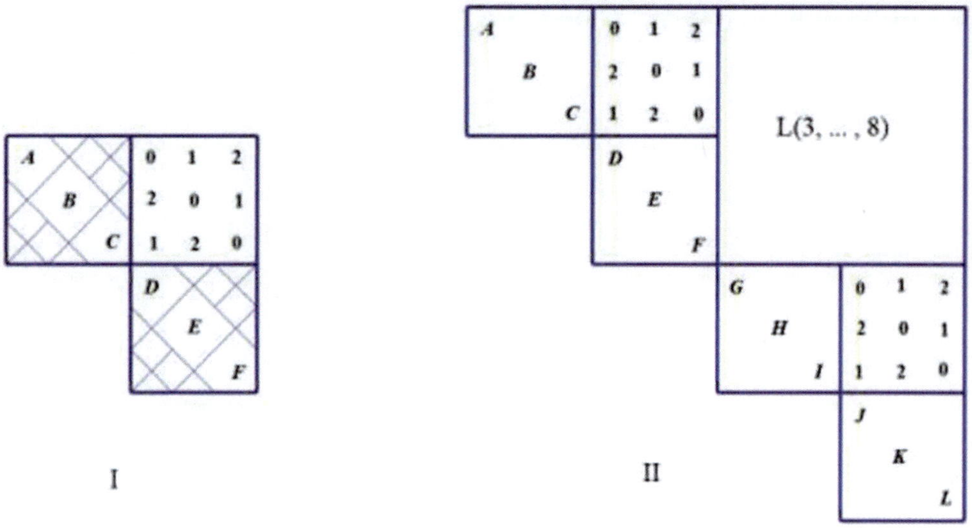

Figure 10.7

Four more examples of Latin triangles are given in Figure 10.8(I,II,III) and 10.9.

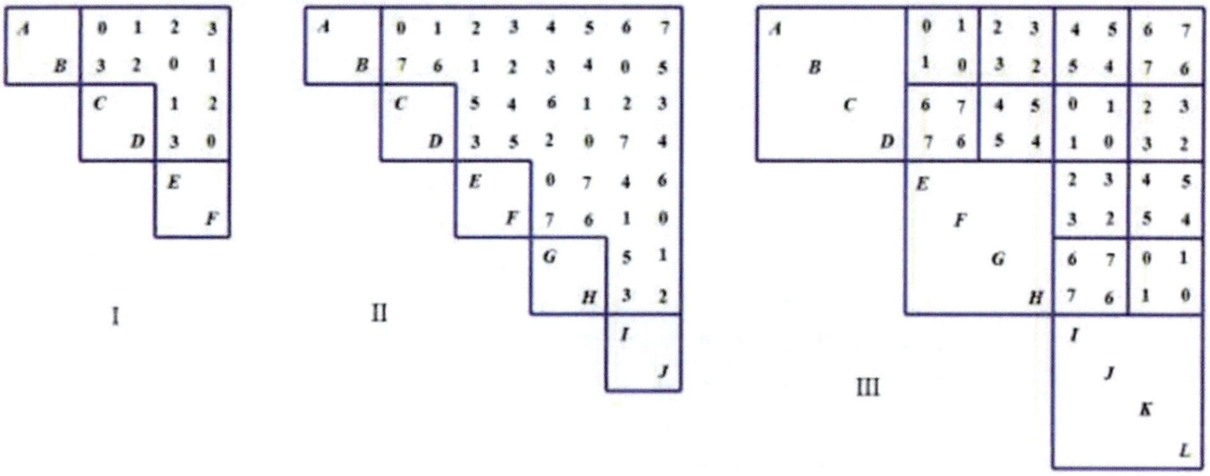

Figure 10.8

A		0	1	2	3	4	5	6	7	8	9	10	11
	B	10	11	0	1	2	3	4	5	6	7	8	9
		C		11	9	1	8	7	4	3	2	5	6
			D	4	0	6	7	8	10	2	3	9	5
				E		5	6	9	1	7	10	3	8
					F	11	10	2	8	5	4	6	7
						G		0	3	9	8	7	10
							H	11	9	0	1	2	4
								I		10	5	1	3
									J	11	6	0	2
										K		4	1
											L	11	0
												M	
													N

Figure 10.9

We shall refer to a Latin triangle with l = 2 parallel classes as *trivial.* Figure 10.7(I) is one example of a trivial Latin triangle. Trivial Latin triangles are just Latin squares in disguise, and since Latin squares exist of all orders, Latin triangles with l=2 exist over Z(n) for any integer n≥2. Besides the technique given in Theorem 10.4, the method of Latin substitution introduced in Section 8.2, can be used to create Latin triangles. If neither one of these methods can be used to create a given Latin triangle Δ then we call Δ *primary.* The Latin triangles 10.8(I, II)and 10.9 are primary and Latin triangle 10.8(III) was obtained by Latin substitution from triangle 10.8(I) using the replacements:

0 ⟶ L(0,1)   1⟶L(2,3)   2⟶L(4,5)   3⟶L(6,7).

A Latin triangle Δ is primary if l≥2 and the number m of halfcircles is twice an odd number, or if Δ is trivial with base Z(p) for some prime integer p. The Latin triangles 10.8(I,II) and 10.9 are the only nontrivial primary Latin triangles we have found.

10.5 Magic Triangles

Every Latin triangle is a magic triangle over the multiple group $(Z(n))^r$ where r is the replication number. Magic triangles over a single (nonmultiple) group are hard to find. In Figure 10.10 we illustrate the only three primary magic triangles we could construct.

A	0	1	5	
	B	2	4	
		C	3	
			D	

I

A	0	6	7	8	14	
	B	2	9	13	11	
		C	12	10	5	
			D	3	4	
				E	1	
					F	

II

A	1	5	9	7	20	18	
	B	2	8	14	19	16	
		C	12	17	11	13	
			D	15	6	10	
				E	4	3	
					F	0	
						G	

III

Figure 10.10

Further magic triangles can be constructed from the ones in Figure 10.10 by the method of *magic substitution,* see Section 3.4 .

10.6 Hybrid Latin Triangles

A *hybrid Latin triangle Λ consists of l≥2 circle classes.* Members of the same circle class have exactly two elements in common and members of different circle classes have exactly one element in common. Λ has a representation similar in structure to a regular Latin triangle except that there are no blank cells. In Figure 10.11(I,II,III) are given three examples of hybrid Latin triangles each with l=2 circle classes. Note that, as in a regular Latin circle system, the elements belonging to a circle consist of the values in the combined row and column through a particular circle name. The notation Λ(l,k,S) will be used to denote a hybrid Latin triangle over the set S with l circle classes and k circles per class.

A general construction can be given in case l = 2 and k≥4 is even . We will use the notation LC(S) for a Latin circle system over S. If two copies of LC(0, …, 2k-3) are used to flank the Latin square L(2k-2, …, 3k-3), then a Λ(2,k, Z(3k-2)) is obtained. This method was used to create the Λ(2,4, Z(10)) in Figure 10.12(I).

I

A	0	1	2
3	*B*	2	1
		C	0
		3	*D*

II

A	0	1	2	3	4
5	*B*	4	3	1	6
6	2	*C*	0	5	3
			D	4	1
			6	*E*	0
			5	2	*F*

III

A	0	1	2	3	4	5	6	7	8
9	*B*	8	7	6	5	4	3	2	1
10	11	*C*	0	2	3	12	5	6	4
11	12	9	*D*	5	10	6	4	1	3
12	10	7	8	*E*	9	11	1	4	0
					F	8	2	11	6
					1	*G*	9	10	2
					7	0	*H*	8	11
					0	3	12	*I*	5
					12	7	10	9	*J*

Figure 10.11

I

A	0	1	2	9	8	7	6
3	*B*	5	1	8	7	6	9
4	2	*C*	3	7	6	9	8
5	4	0	*D*	6	9	8	7
				E	0	1	2
				3	*F*	5	1
				4	2	*G*	3
				5	4	0	*H*

II

A	0	1	2	3	4
5	*B*	2	1	4	3
		C	4	0	5
		3	*D*	5	0
				E	2
				1	*F*

III

A	0	1	2	3	4	5	6	7	8	9	10
11	*B*	2	1	4	5	6	3	8	9	10	7
12	13	*C*	0	5	6	3	4	9	10	7	8
13	12	11	*D*	6	3	4	5	10	7	8	9
				E	8	9	10	11	12	13	0
				7	*F*	10	9	12	13	0	11
				2	1	*G*	8	13	0	11	12
				1	2	7	*H*	0	11	12	13
								I	3	4	5
								6	*J*	5	4
								2	1	*K*	3
								1	2	6	*L*

Figure 10.12

Figures 10.12(I,II) feature hybrid Latin triangles with l odd and k even. For this type of triangle we can present a general construction. Let S be a set with lk + k - 2 elements and partition S into l subsets $\{A_0, \ldots, A_{l-1}\}$ each with k elements and a set D with k-2 elements. For each $i \in \{0, \ldots, l-1\}$ let $B_i = D \cup A_i$ and let U(l) denote the upper triangular portion of the Cayley table for the group Z(l) along with the main diagonal. For each off diagonal element i of U(l) make the Latin substitution $i \longrightarrow L(A_i)$, and for each diagonal member i of U(l) make the *Latin circle substitution* $i \longrightarrow LC(B_i)$. The result is a Λ(l,k,S) hybrid Latin triangle. Note that this construction depends on the existence of circular Latin squares for which no general constructive procedure has been found. In Figure 10.1 circular Latin squares with 4,6,8 circles are illustrated. Also the Latin triangles in Figures 10.2, 10.3, 10.4, 10.5, 10.6 can all be extended to circular Latin squares by the technique described in the second paragraph of Section 10.2.

In Figure 10.13(I) an example of this type of construction is given with l=5, k=2, and S = Z(10). Split between Figures 10.13(II) and 10.14 another example is illustrated with l=5, k=4 and S = Z(22).

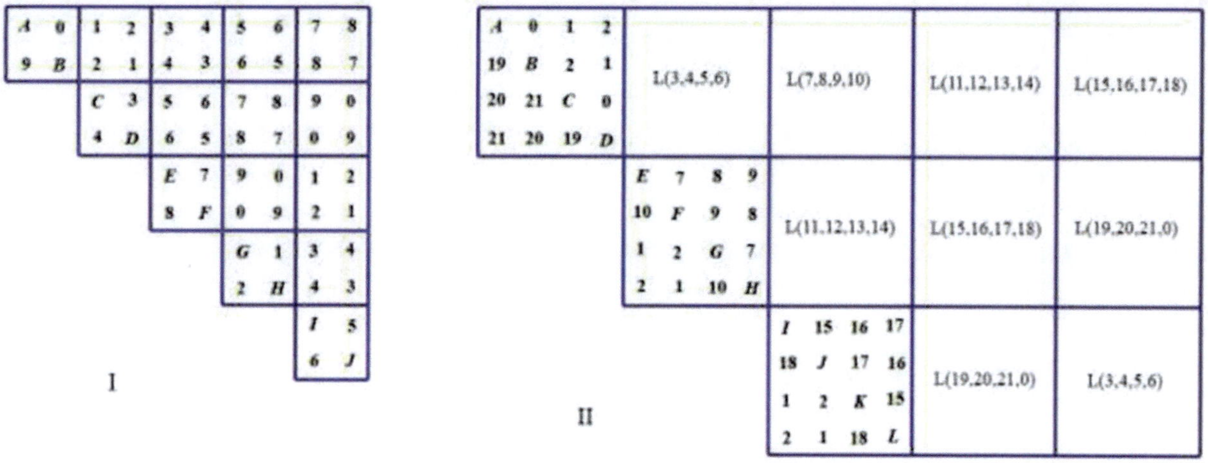

Figure 10.13

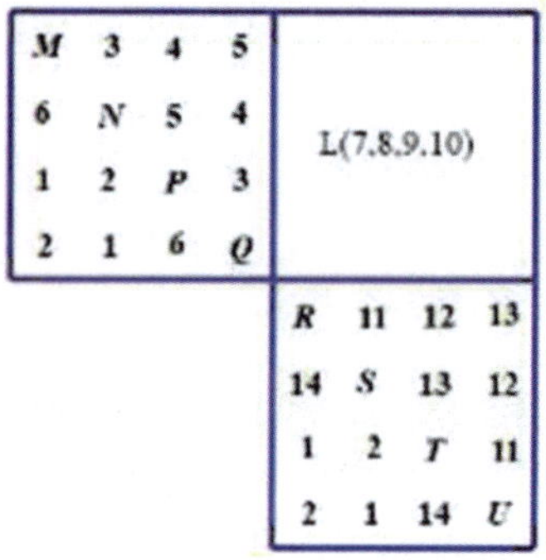

Figure 10.14

Besides the general construction of circular Latin squares, the general construction of hybrid Latin triangles with l=2 circle classes, as illustrated in Figure 10.11, remains an open problem.

10.7 Hybrid Magic Triangles

Hybrid magic triangles are hard to find. We have only the following example with l = 3 in Figure 10.15(I). There is also a general procedure for constructing hybrid magic triangles with l=2 and k≥3 typified by the example with l=2 and k = 3 in Figure 10.15(II). Let n = 2k(k-1) and partition Z(n) into two sets S_1, S_2 each consisting of k(k-1)/2 complemented pairs. Let MC(S_1) and MC(S_2) denote two magic complemented magic circle systems over S_1, S_2 respectively. Let L(Z(k^2) denote a magic square over Z(k^2) and then add n to each cell to obtain a L(n, …, n+k^2 - 1) magic square. Represent MC(S_1) and MC(S_2) by squares and place these adjacent to L(n, …, n+k^2 - 1) to obtain a hybid magic triangle over Z(n+k^2).

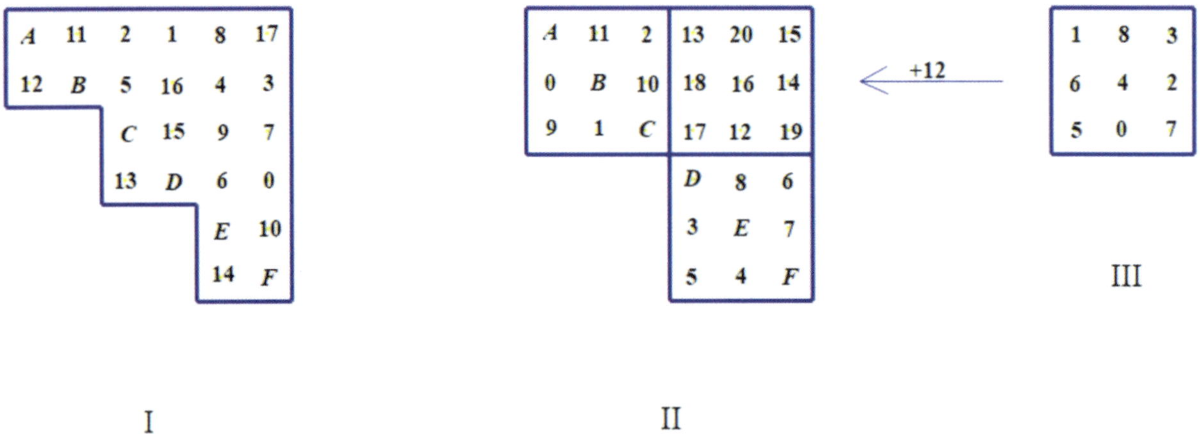

Figure 10.15

Appendix I

In the following series of transformations the chasms to be closed are bounded by the parallel broken line segments L_1 and L_2. In the following diagram the chasm has been closed and the emergent region tiled. The H(1,1,3) hexagon is indicated with a red boundary and the remaining regions all represent *0-dominoes*.

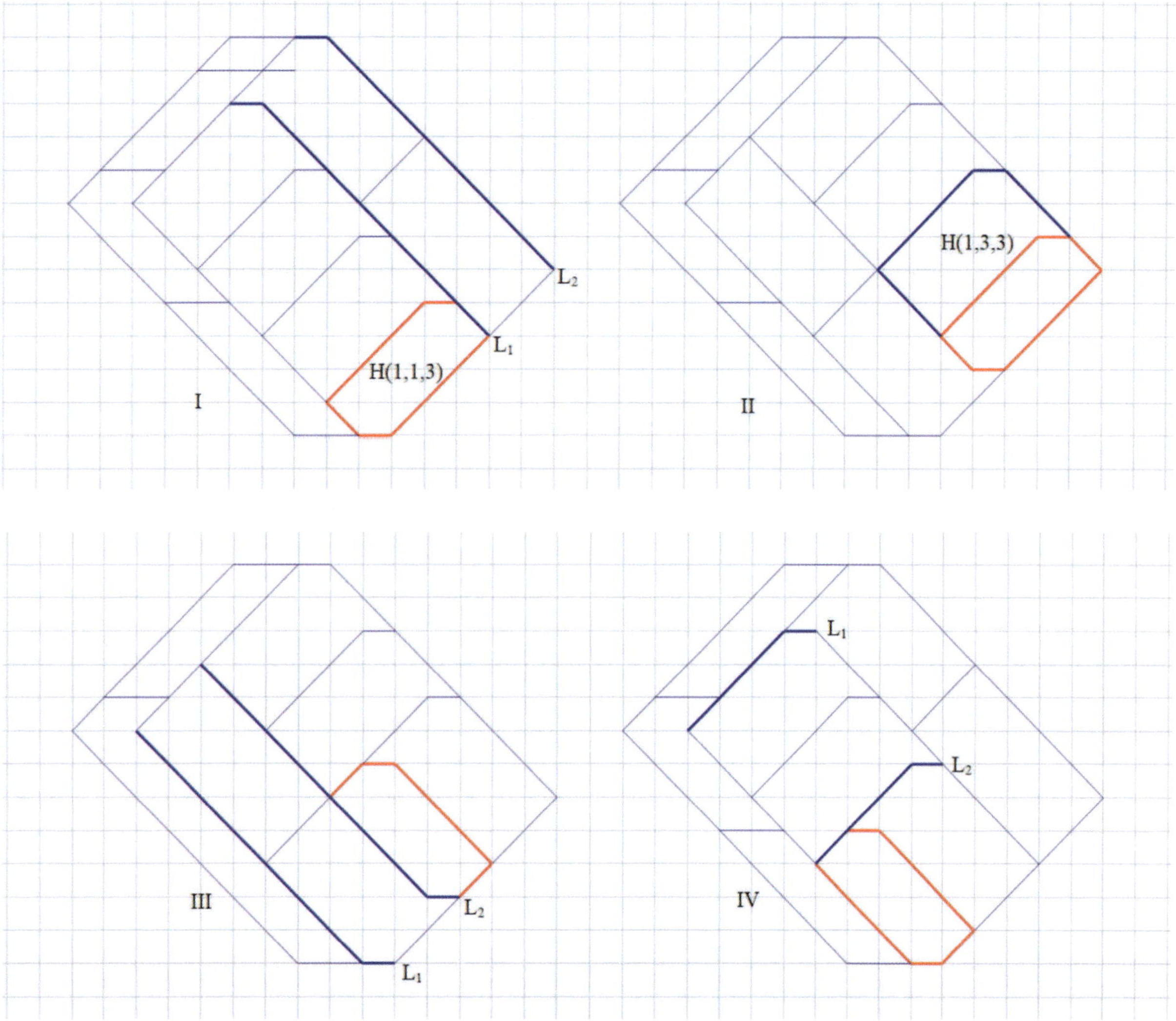

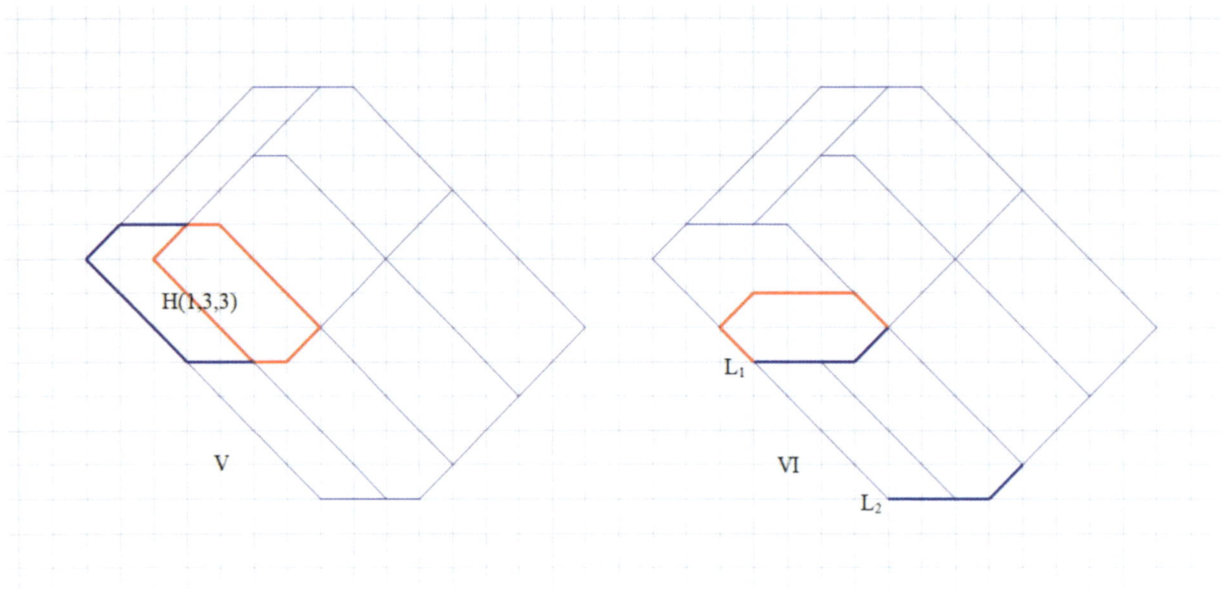

In moving from (II) to (III) and (V) to (VI) a small platform transformation has been applied. A 4 by 4 square can be tiled with two *0-dominoes* in two ways; sometimes the additional transformation which takes one such tiling to the other may be required.

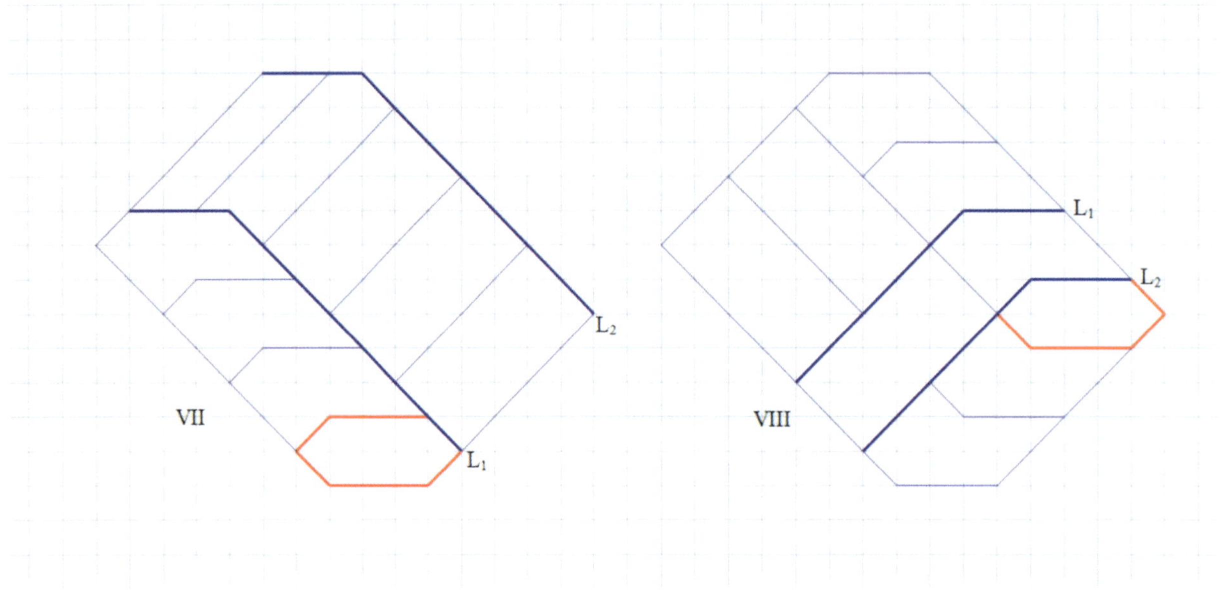

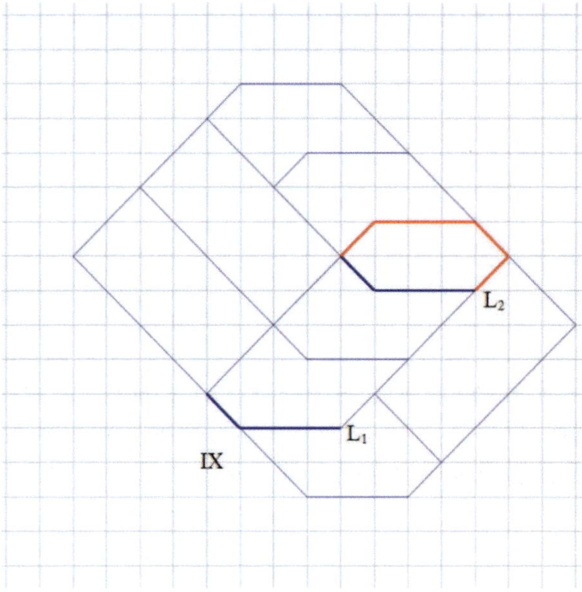
L2
L1
IX

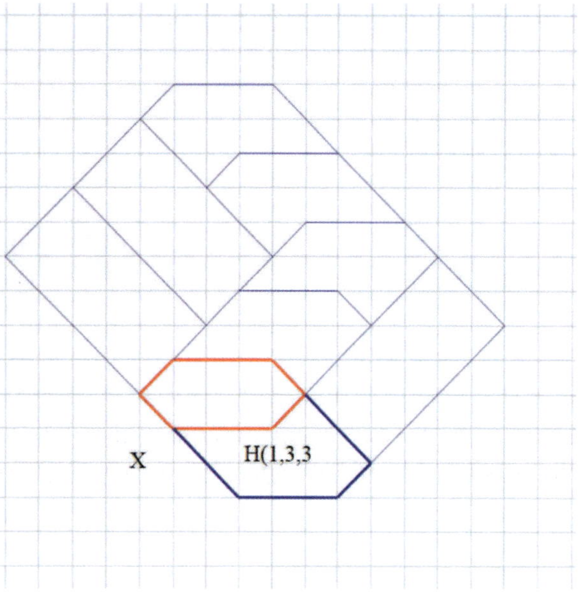
X
H(1,3,3

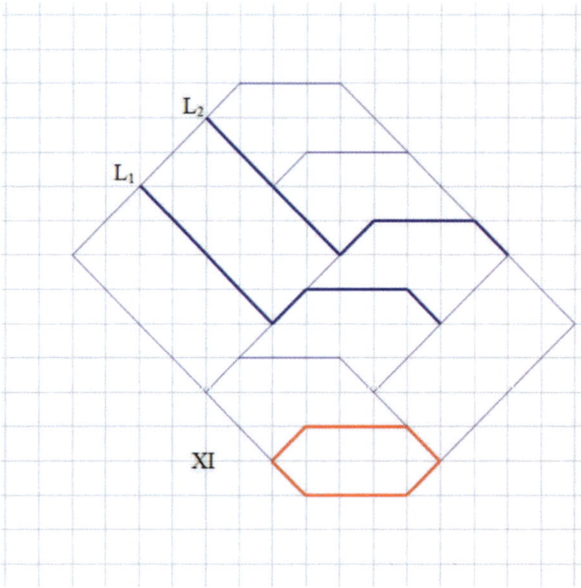
L2
L1
XI

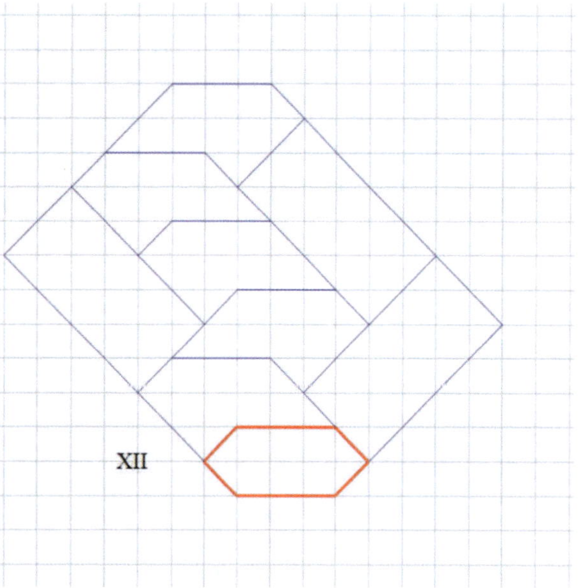
XII

Appendix II

Let P be a semiregular polygon. A *domino shell for P* is a sequence of *dominoes* which entirely surround P. Each *domino* in the sequence contains either 2 or 4 of the boundary edges of P and exactly two of the edges of each of its neighbors in the sequence. Some small examples of polygons with domino shells are given in Figure A1.

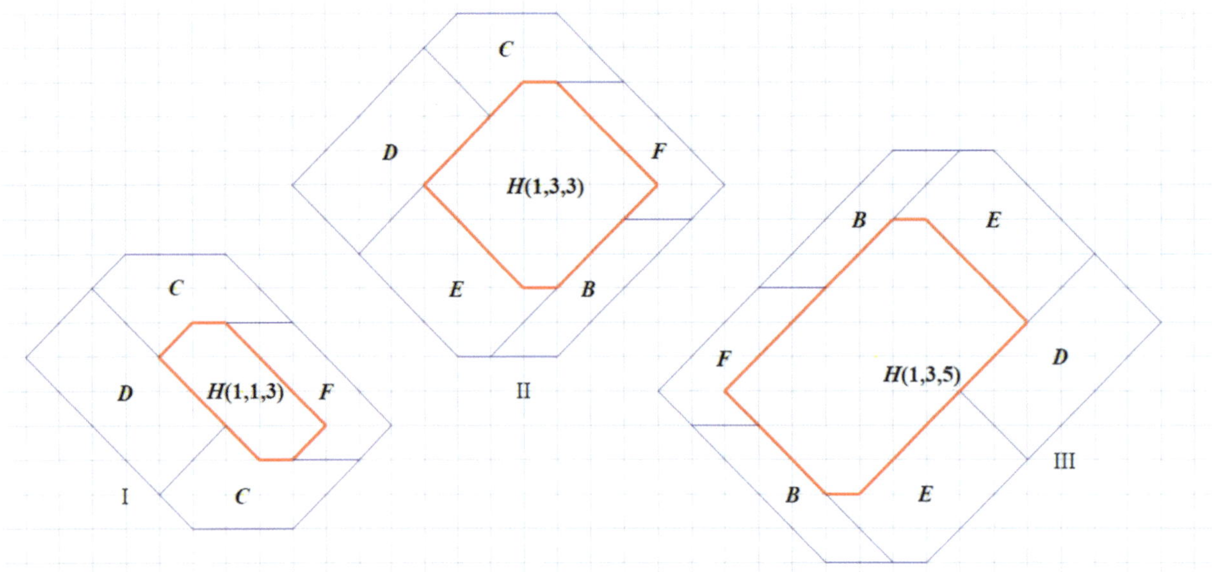

Figure A1

The shell construction is very useful for obtaining extensions of Class 1 hexagons as is demonstrated in the following Theorem. The different shape dominoes will be indicated by letters A, B, C, D, E, F as in Figure A1. All shells will contain the F domino, and the shell itself can then be represented by a counterclockwise sequence. Thus the shells in Figure A1 are denoted (FCDC), (FBEDC) and (FBEDEB).

Theorem A1 *Every Class 1 hexagon has a domino shell.*

Proof. Let H be a Class 1 hexagon .Thus each side of H has odd length. We will represent H, as in Figure A1, with two opposite 90° degree angles and remaining interior angles equaling 135°. The vertices of H will be labeled in counterclockwise order: p, q, r, s, t, u with vertex p at one of the 90° degree angles. The sides of H are thus pq, qr, rs, st, tu, up. There are four cases to consider depending on whether these sides have length of the form 4x+1 or 4x+3. We will just consider one Case; the shell constructions for the remaining Cases are similarly easy to determine.

CASE 1: $|pq| = |st| = 4x+1$; $|qr| = |tu| = 4y+1$; $|rs| = |up| = 4z+1$.

As in Figure A2 the F tile is placed at the 90° angle of P. There are two ways to place F at this point, both of which can be used to construct a shell. In Figure A2 it is placed so that it covers 1 unit of side up and 3 units of side pq. Once tile F is positioned, the entire shell sequence is determined. For this Case the shell sequence is given by ($FB^xA^yED^zD^xCA^yB^z$). For Figure A2 we chose x = 2, y = 3, z = 1. □

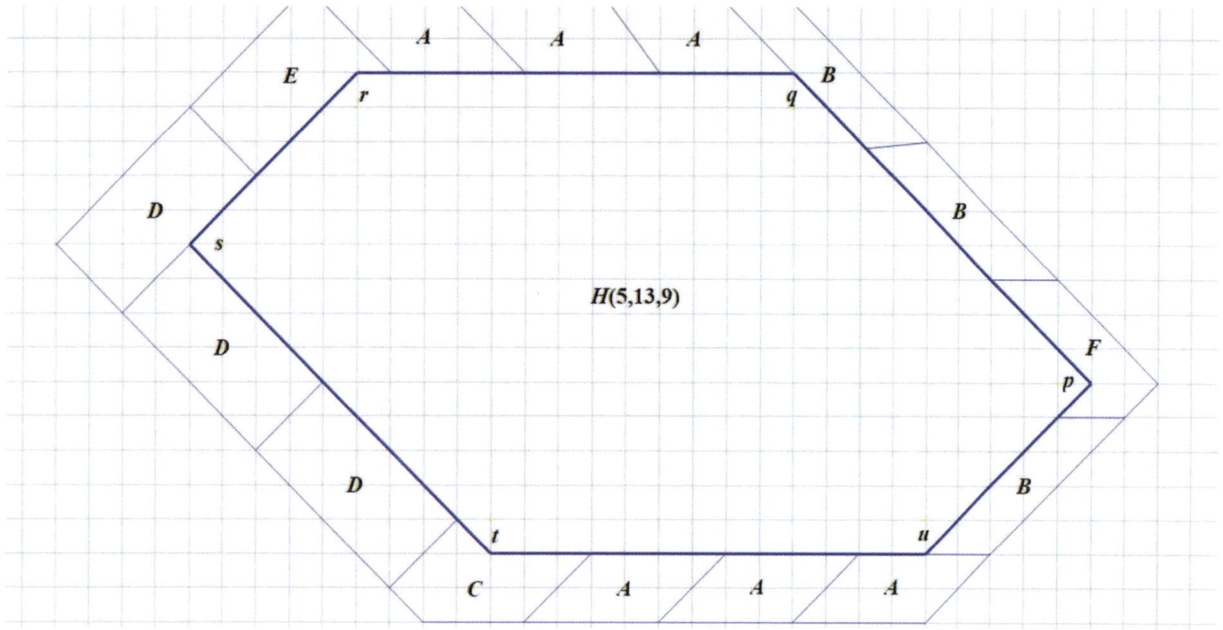

Figure A2

Note that Theorem A1 applies also in case H = H(e) is a regular hexagon with e odd. It even applies if e = 1, i.e., if H is a simple hexagon with one edge per side, or if H is any Class 1 hexagon with no integral solution, e.g., H = H(1,1,5). By a simple calculation no Class 2 hexagon has a domino shell.

Theorem A2 *No Class 2 hexagon has a domino shell.*

Proof. Since each domino has order 8, the order of a shell must be a multiple of 8. Let H = H(p,q,r) be a semiregular hexagon. The order of a shell for H is given by

$$(p+2)(q+2)+(p+2)(r+2) + (q+2)(r+2) - (pq + pr + qr) = 4(p+q+r +3).$$

This expression is a multiple of 8 iff p+q+r is odd. This is not the case for a Class 2 hexagon so the Theorem is proved. □

The expression p+q+r is odd in case p, q are even and r is odd. In this case H(p,q,r) is an even order semiregular hexagon which may not have an integral solution. In this case it appears that no domino shell can be constructed. Also in case p, q, r are all even no domino shell seems to exist. We leave these as open questions. Also it is an open question as to whether any semiregular polygons with more than 6 sides have domino shells.

As a consequence of Theorem A1 many nestings of Class 1 hexagons are possible. For example, in Figure A3 a nesting of regular hexagons using shell extensions is illustrated.

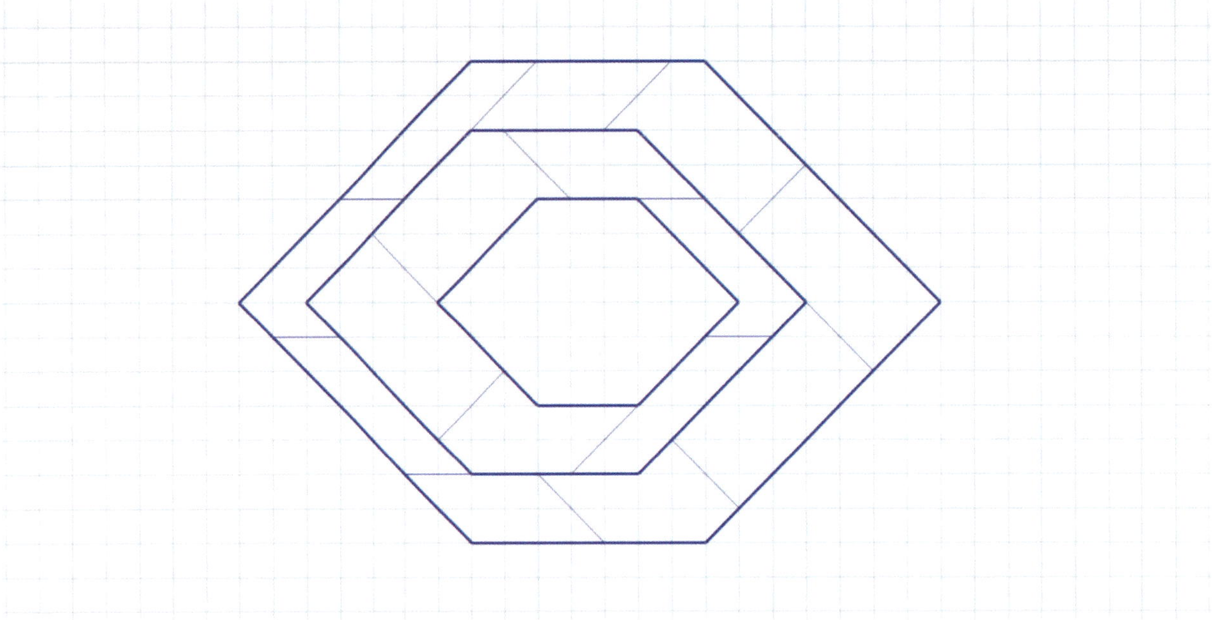

Figure A3

We may say that a shell extension leaves the kernel in a *central location.* It is interesting to ask if a domino extension is possible which leaves the kernel at a corner or touching a side of the extension. Consider the example in Figure A4. In this Figure note the chasm consisting of tiles F,G,B in A4(II) which is closed to obtain A4(I).

Every Class 1 hexagon has a *central unit* which lies at the midpoint of each pair of opposite sides. If H, K are Class 1, and H→K is a domino extension, the central unit indicates the precise location of K inside H. The collection of all possible central units for an H→K domino extension is the *domino extension locus (DEL) for H → K.* The DEL for H = H(1,5,5)→ K =H(3,7,7) is indicated by the red bars in Figure A4(I). Note that since K is symmetric with respect to a vertical and a horizontal axis, the DEL must have the same symmetry.

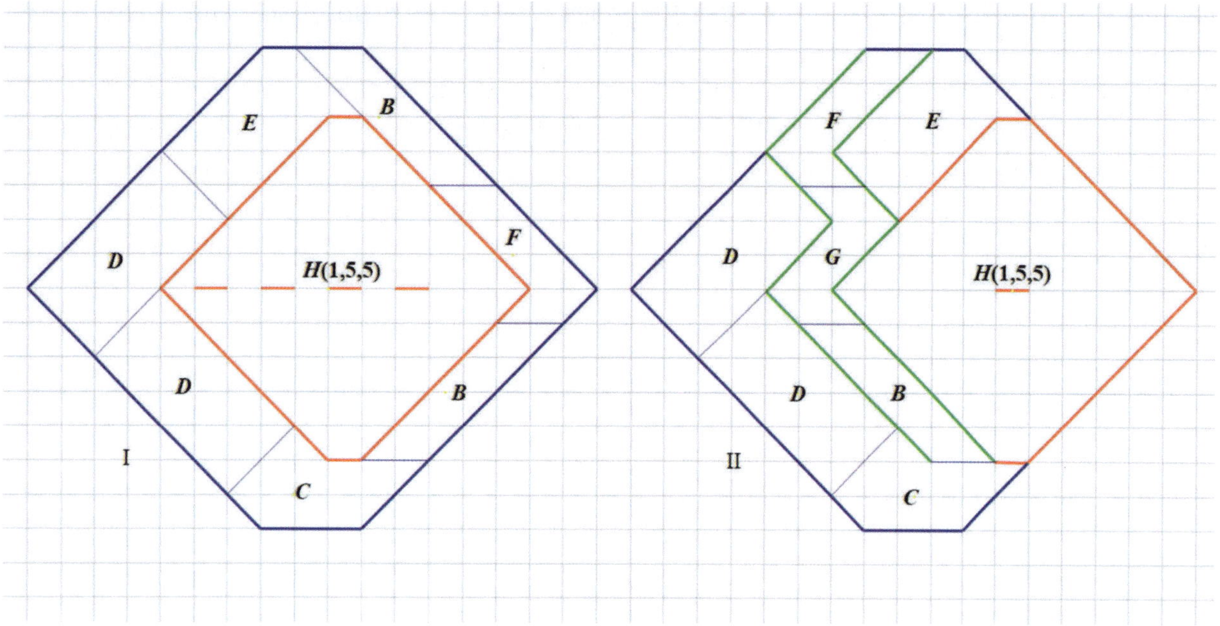

Figure A4

In Figure A5(I) the DEL for the domino extension H(1,3,3)→H(3,5,5) is illustrated, and in A5(II) the DEL for the domino extension H(1,3,5)→H(3,5,7) is given. We leave the constructions for the reader to find.

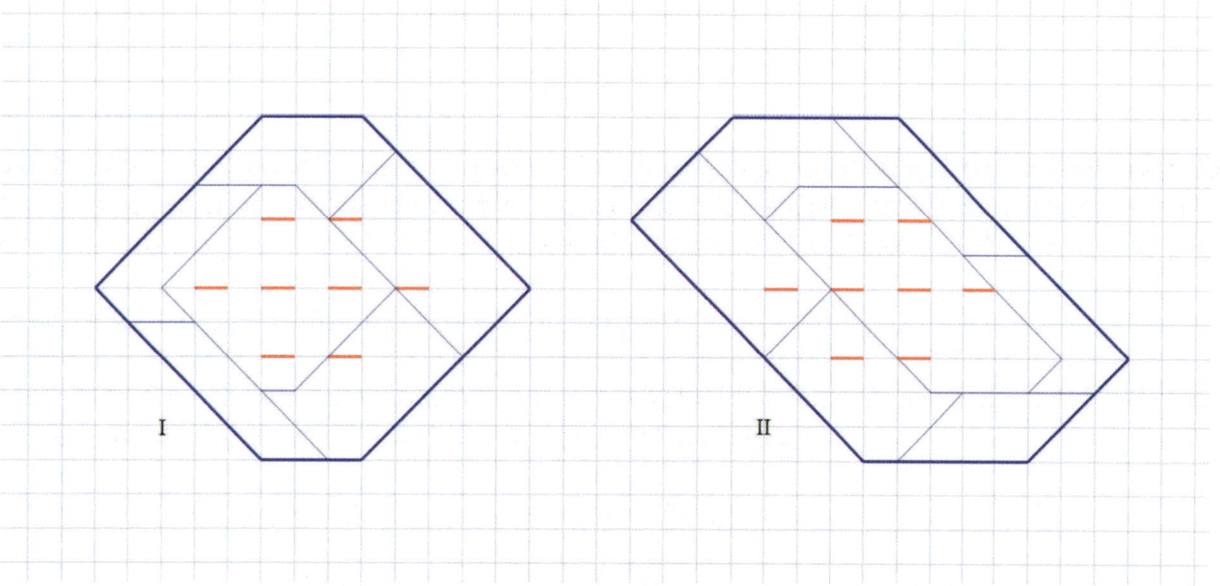

Figure A5

In Figure A6(I) the shell extension H(3)→H(5) is illustrated with red bars indicating its center as well as the entire DEL. The constructions in A6(II) and A6(III) along with their reflections and/or 180∘ rotations illustrate all possible H(3)→H(5) domino extensions.

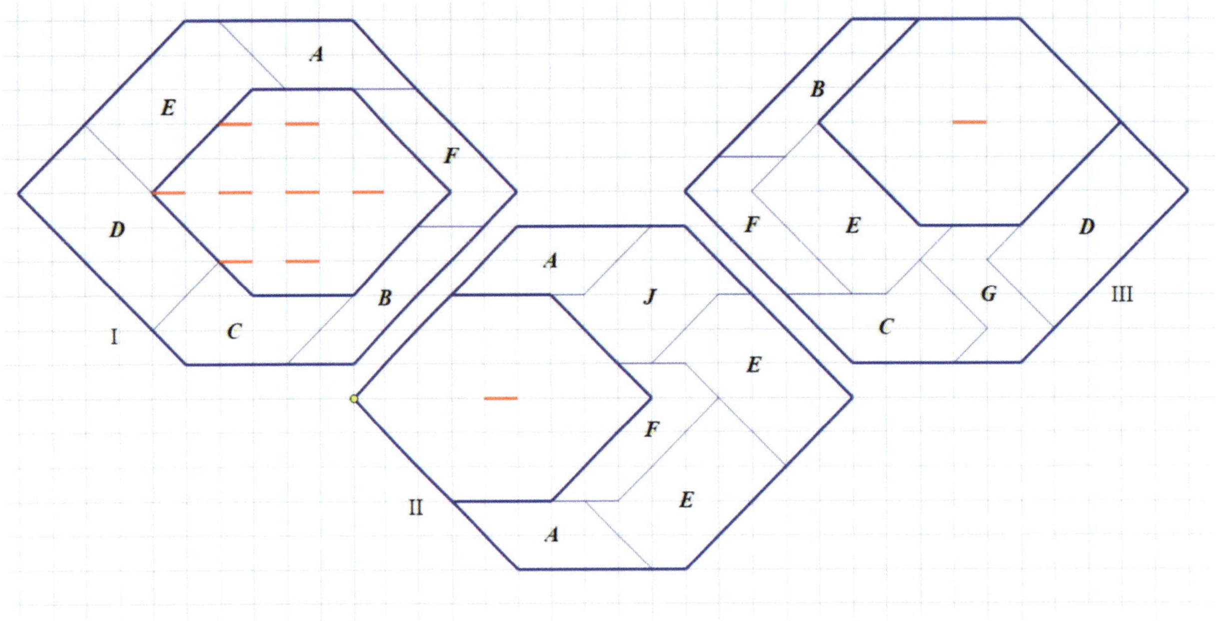

Figure A6

In Figures A7, A8, A9 the DEL for H(5)→H(7) is calculated. By combining the DELs for H(3)→H(5) and H(5)→H(7) we will obtain the DEL for H(3)→H(7).

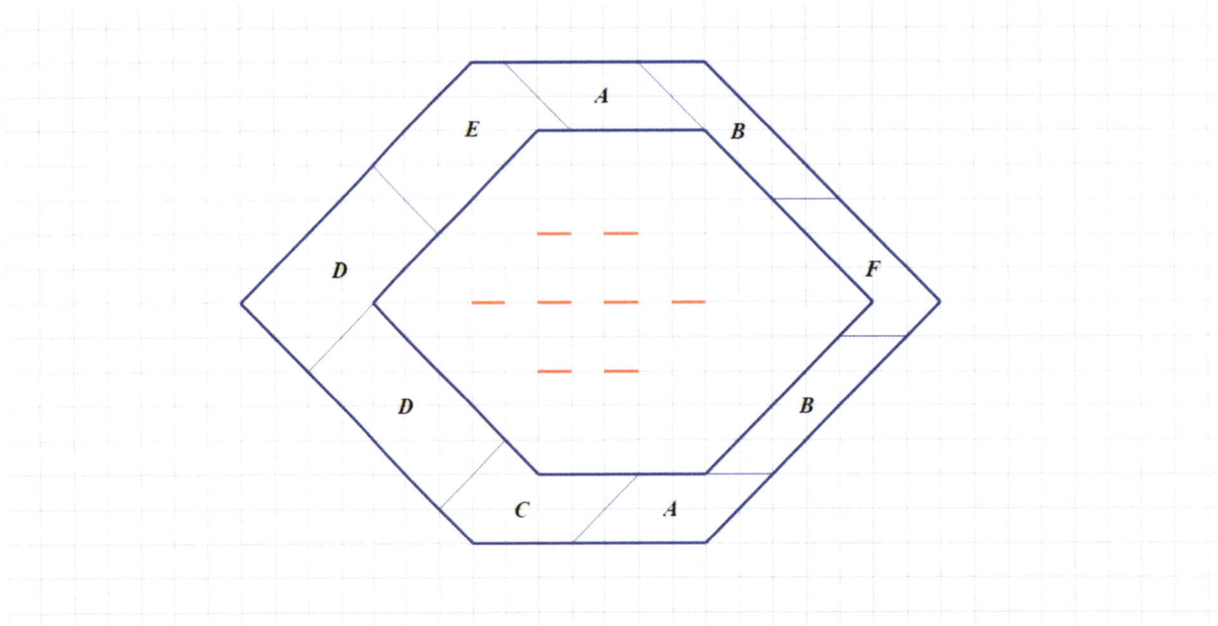

Figure A7

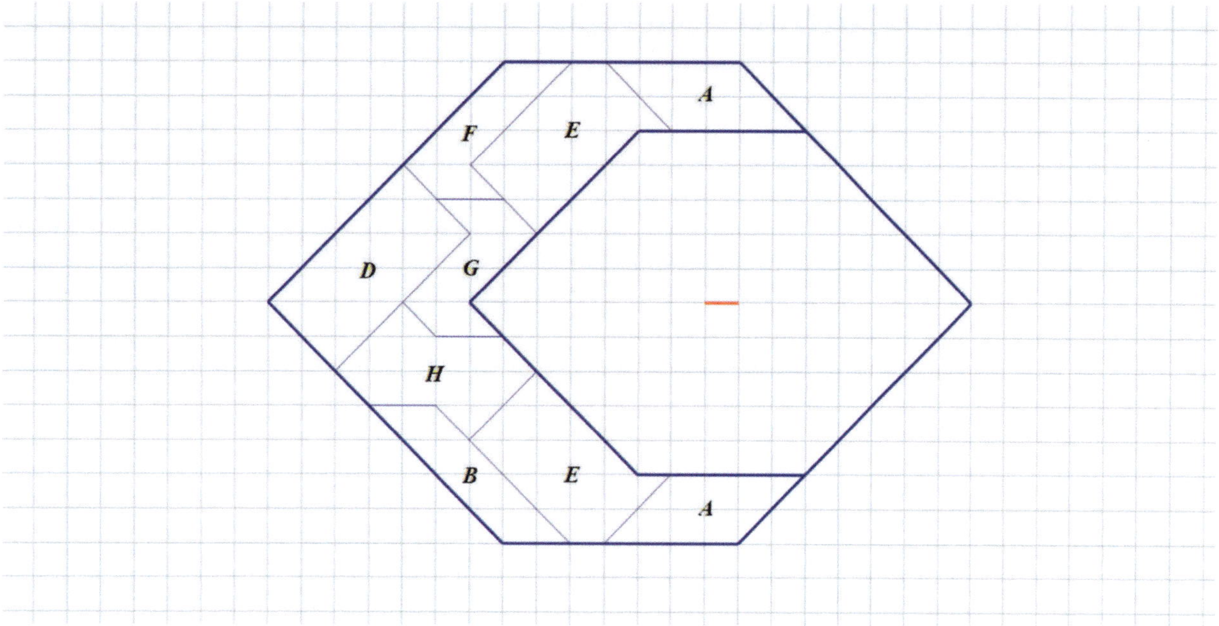

Figure A8

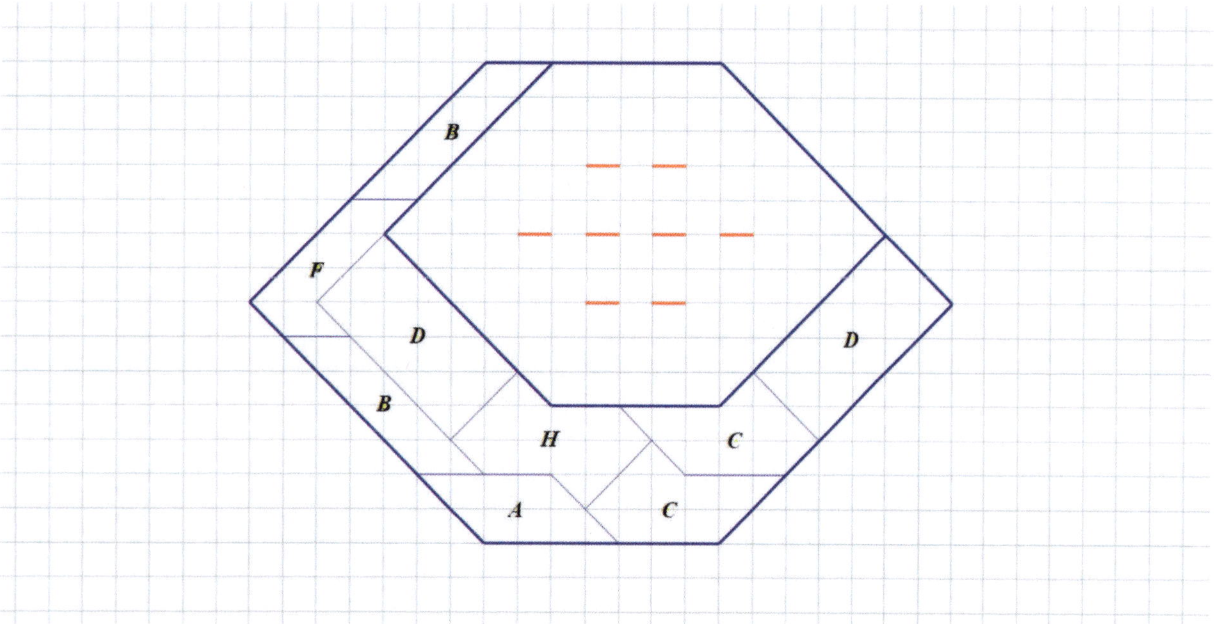

Figure A9

In Figure A9 the DEL for the embedded H(3) has been indicated by the red units. In Figure A10 these units are reflected across the vertical and horizontal axes of symmetry in order to obtain the central units for an H(3)→H(7) domino extension.

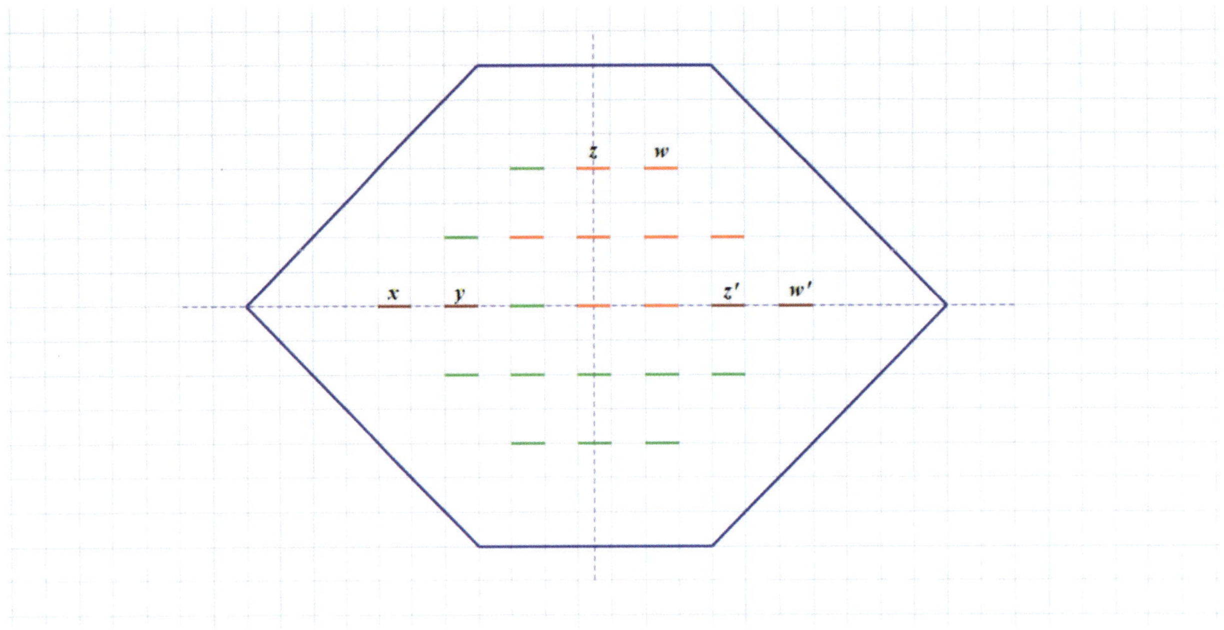

Figure A10

Before completing the determination of the DEL for H(3)→H(7), consider Figure A11 in which our usual rendition of a tiling of H(7) using square lattice parallelograms is placed alongside a rendition using an actual regular hexagon in a triangular lattice.

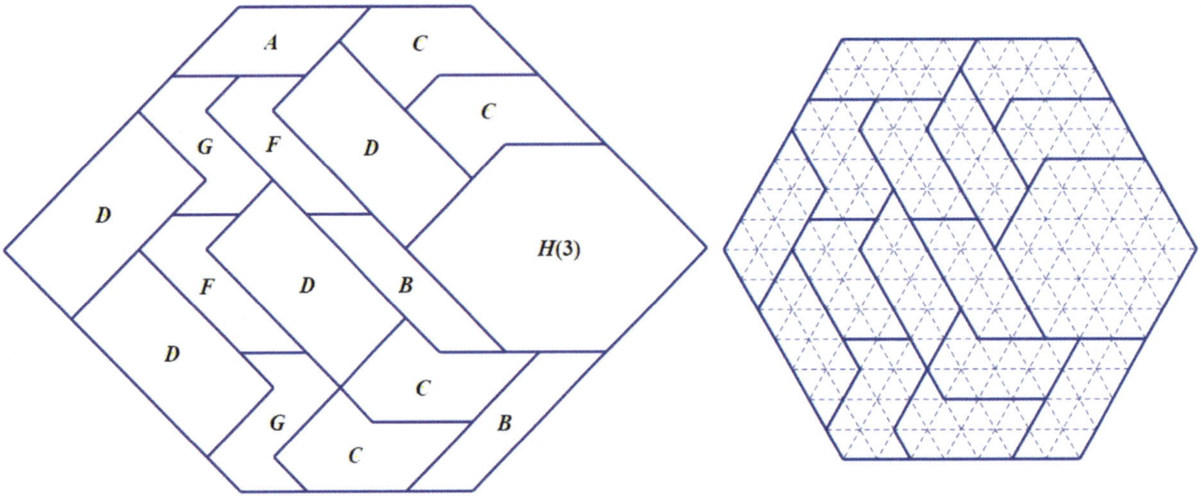

Figure A11

Although the square lattice version is easier to construct using Geometer's SketchPad, the more natural version based on a triangular lattice reduces considerably the number of different shapes for *dominoes*. The tiles D,B,A all correspond to the same shape in the triangular lattice, as do tiles E,F,C and J,H,G. One fact is made very clear by the triangular version: the tiling is preserved under a 60° rotation. This implies that there should be a corresponding square lattice tiling. In Figure A12 the tilings in A11 have each been rotated 60° clockwise.

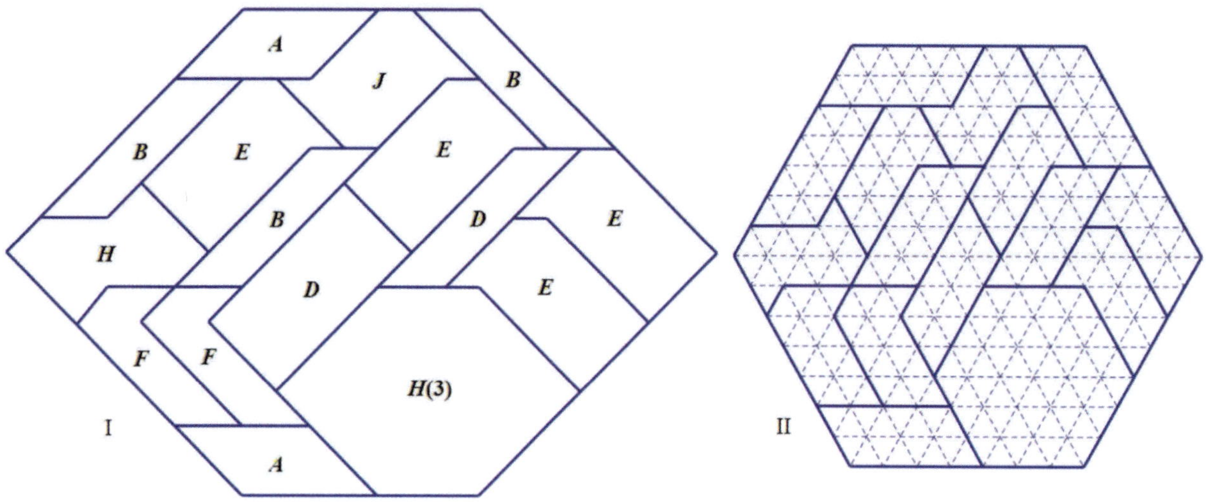

Figure A12

Now returning to Figure A10, a consequence of the above observations is that the H(3)→H(7) DEL, besides being closed under reflection through the vertical and horizontal axes of symmetry, must also be closed under 60∘ rotations. Thus the indicated central units z, w give rise to the central units z', w' respectively. The central units x, y are then obtained by reflection of z', w' across the vertical axxis of symmetry. The final result is that the H(3)→H(7) DEL consists of an hexagonal phalanx with 3 units per side.

Index

References

[1] S.A. Huggett, The Geometric Universe, Geometry and the work of Roger Penrose, Oxford University Press, (1998)

[2] Fletcher III, R.R., Polyparallelogram Puzzles and Tiling Problems, Xlibris (2020)

[3] Bose R.C., Shrikhande S.S., Parker E.T., Further results on the construction of mutually orthogonal Latin squares and the falsity of Euler's Conjecture, Canad. J.Math **12**, 189-203 (1960)

[4] Tarry G., Le probleme des 36 oficiers, Assoc, Franc. Paris **29** 170-203 (1900)

[5] Evans A.B., Orthogonal Latin Squares Based on Groups, Springer Int. Publishing (2018)

[6] Kelsey K., A Beginners Guide to Magic Squares and Cubes, Amazon

[7] Carus P., Browne C.A., Magic Squares and Cubes, Open Court Publishing Co. London (1908)

[8] Rees R., Wallis W.D., A class of resolvable pairwise balanced designs, Cong. Numer.**55** 211-220 (1986)

[9] Wanless I.M., Webb B.S., The existence of Latin squares without orthogonal mates, Des, Codes Cryptog. **40** 131-135 (2006)

[10] Fletcher III R.R. Circle Systems, Xlbris, (2020)

Printed in Great Britain
by Amazon

3119f272-08e0-411c-96f2-8294227fea11R01